Decision-Making GEOGRAPHY

Second Edition

Norman Law

Formerly Head of Geography at Waingel's Copse School, Woodley, Reading

David Smith

Head of Geography at Ranelagh School, Bracknell

Decision-making at a community meeting, or kgotla, *in Botswana.*

Stanley Thornes (Publishers) Ltd

Originally published in 1987 by Hutchinson Education (ISBN 0 7487 0238 5)
Reprinted 1988 (twice)

British Library Cataloguing-in-Publication Data

Law, Norman
Decision-making geography. – 2nd ed.
I. Title II. Smith, David
910.68

ISBN 0–7487–1111–2

Reprinted in 1990 by
Stanley Thornes (Publishers) Ltd
Old Station Drive
Leckhampton
CHELTENHAM GL53 0DN

Second edition 1991

Acknowledgements

Special thanks to the following for kind permission to reproduce figures and tables:
J Allan Cash Photo Library Ltd, Figure 1.46; Berkshire County Council, The National Trust, The Royal Borough of Windsor and Maidenhead, Figure 15; R W Bradnock, *Urbanisation in India* (John Murray) Tables 3.7, 3.8; BP Ltd, Figures 4.27, 4.28, 4.29, 4.30, 4.31, 4.32, 4.33, 4.34, 4.35, 4.36; T A Broadbent, The Town and Country Planning Association, Tables 6.8, 6.9; W Brogden, title page, Figure 3.29; *Bulletin of Indonesian Economic Studies*, Table 6.12; Jill Butler, Figures 1.34, 1.35, 1.36, Table 1.9; *The Ecologist*, Figures 2.18, 2.20, 2.24(b), (c), Tables 1.12, 2.4; Eurostat Publications, Tables 4.8, 4.10, 4.11, 4.12, T4.3, T4.4; Geofile MGP, Tables 3.5, 3.6; Gouvernement du Québec, Figures 1.38, 1.41, 1.42, 1.43, 1.44; Government of India, Tables 3.10, 3.11; Tony Gribben, Figures 4.24, 4.25, 4.26; *The Guardian*, Figure 4.8; Greenpeace, Figure 5.30; HMSO, *Strategic Plan for the South East*, reproduced by kind permission of the Controller, Figures 6.30, 6.31, 6.32, 6.33, 6.34, 6.35; R E Huke, *Journal of Geography*, National Council for Geographic Education, Figure 3.17, Tables 3.12, 3.13; R Leclerc and D Draffan, *Town Planning Review*, **55**(3), 1984, p 347, Tables 6.7; R R Montfield, *Geography*, Journal of the Geographical Association, Figure 5.27; J M Williams, Figure 2.8; G Neale, Figure 6.16; *New Internationalist*, Figures 2.21(b), 4.22, 4.23, 6.40; Dr S Nortcliff, Figures 2.19(b), (c), (d), 2.22(a), (b), (c), 2.23(a), (b), 2.24(a); *The Observer*, Figures 1.7, 1.13, 2.16; 2.19(a), 4.58; P E Ogden, *Geography*, Journal of the Geographical Association, Figures; 6.25, 6.26, 6.27; *Punch*, Figure 6.23; P J Riddell, Figures 6.2; Royal Commission for Jubail and Yanbu, Figures 6.8, 6.9; Sicillian Sun, A Member of the British Island Airways Group, Figure 4.52; D J Spooner, *Geography*, Journal of the Geographical Association, Figure 5.20; M S Travis and The High Commission of India, Table 4.2; Elisabeth and Jacques Trotignon, Figures 1.28, 1.30, 1.31; The Welsh Tourist Board, Figure 1; Tony Williams, Figure 1.30; M B Hewitt et al, Canadian High Commission, London; C N Devas.
And to Dr R S Bradley for his invaluable help in the preparation of this book.

Text set in 10/11pt Palatino Roman at Input Typesetting Ltd, on Linotron 202
Printed and bound in Great Britain at The Bath Press, Avon.

Contents

Introduction

Theme 1 Managing Physical Systems: The Example of the Drainage Basin

Theme 2 Managing Ecosystems

Theme 3 Agriculture and Rural Development

Theme 4 Some Important Issues in Industrial Development

Theme 5 Energy

Theme 6 Urban Issues

Introduction

We are making decisions all the time. Many of them are about very mundane or simple problems such as what to eat for breakfast. Others take longer and are more complex.

In deciding where to go on holiday, people make important geographical decisions. As well as distance and scenic beauty, they also consider climate, places of interest to visit and not least of all cost, which is often a function of all the other elements in the decision. The advertisement for holidays in Wales was obviously placed to satisfy two important considerations. The first is travelling time (and therefore cost), and the second is scenic beauty. If the advertisement were very successful, however, do you think that some holidaymakers would return the next year? Might they be disappointed to find crowds there? In this way we can see that the results of one decision might influence the environment, thus changing it before the next decision is made. Such decisions are the theme of this book.

Geographers study the locations and repercussions of things such as factories. They are therefore involved in analysing the patterns made as the result of decisions. Infrequently they are also involved in the actual decision-making process, yet they are in an ideal position to contribute to such a process. In this book we will look at the way decisions are made and encourage the participation of the geography student in them. Through an investigation of a variety of case studies we hope to foster the idea that as a worker, a manager or as a member of the public you have an important part to play in decision-making. Decisions will still be made without your participation, so it is better that you make an informed contribution to the geographical decisions that will undoubtedly affect your life.

Fig. 1 A Welsh Tourist Board advertisement

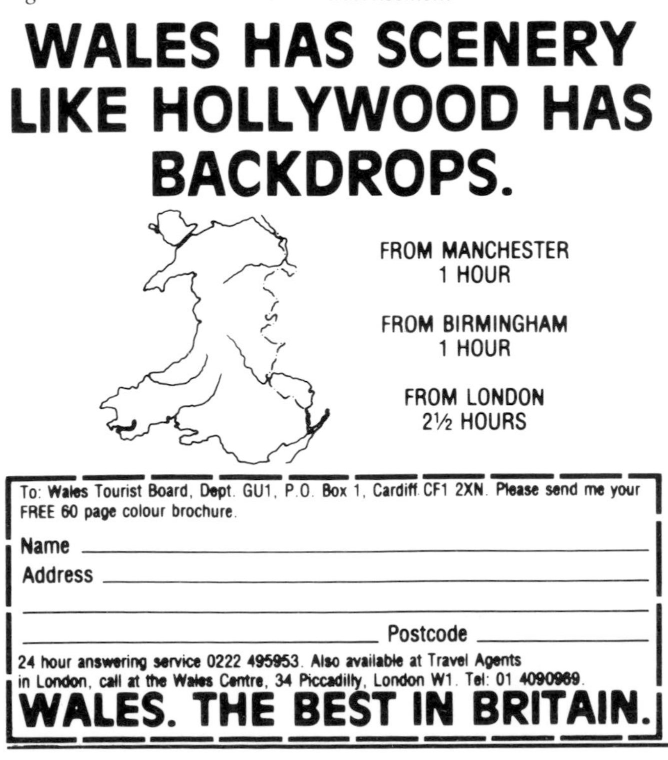

Using decision-making as a framework for this book allows the active participation of the student. A-level students will be able to see their studies in a context which makes them more meaningful. Physical features, for instance, are only discussed here when they have relevance to the human occupation of an area and where they became an important aspect of decision-making.

Decision-making involves moving through a series of steps to evaluate a number of possible alternatives to decide upon a particular course of action. Fig. 2 shows the steps or 'filters' through which information passes on the way to making a decision.

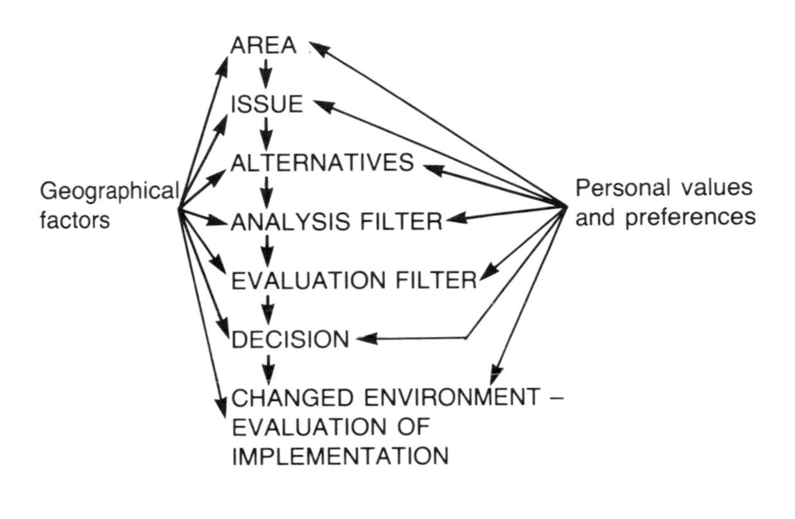

Fig. 2 A decision-making path

How to Use This Book

Each chapter (or Theme) of this book is set out in such a way that examples of geographical problems and decisions are analysed, followed by material challenging students to use the knowledge, skills and techniques gained through studying that material to solve another actual problem by making their own decisions. In each Theme there are pages marked with a triangle where there is discussion of important techniques or study skills which may be used by the student in conjunction with the material. In this way the student or teacher will be able to find such information quickly and it will not intrude unnecessarily into the text. The keynote of this book is involvement in challenging real-life examples. It is to be hoped that through this type of study the student will be challenged much more by the material, but will be able to see its relevance easily and will become as a result much more involved in decision-making in the real world both during and after such studies. Focusing in this way on the relevance of geography will, it is hoped, stimulate the A-level student to achieve well in the examinations.

The Process Leading to Decision-making

In order to illustrate the process leading to decision-making, let us investigate a particular example.

Case-study: a Farm in Central France

Monsieur Michel is a farmer who lives in the small town of St Gaultier in central France. He owns a small farm but lives in the town, 2 km away. Consequently, the old farmhouse on his land is at present unoccupied. To what use should he put this building? He considered several alternatives. Should he renovate the building and live there? Should he instead let the house as holiday accommodation? Should he demolish the building and increase the area under cultivation? In order to come to his decision, which was to renovate the building and let it as holiday accommodation, he considered the information outlined below. As you look through it, write a list of the geographical factors and the personal values and preferences that led him to his decision.

Monsieur Michel lives with his wife and two children in a modern detached house on the northern edge of the small market town of St Gaultier (population 2400). His wife is a teacher at the infant school just across the road and his two children, Jean-Charles and Marie-Pierre, attend school in Chateauroux, about 25 km to the north-east and connected by a main road and bus service. Each morning he leaves by car for his farm, called Le Breuil, which covers 84 ha. He bought the land quite cheaply in 1978 because it seems that fewer people in the area are going into farming. In fact, his neighbour, who owns a farm of just 40 ha, has said that when he retires Monsieur Michel may take over from him. In this way, he hopes to enlarge his holding gradually. He will not expand much more, however, without taking on workers and they are hard to come by in the area. He relies a lot on his machinery. He has two tractors, a combine harvester, a chemical fertilizer spreader, two seed drills, two ploughs, a harrow, a roller and a trailer for transporting the harvest.

Monsieur Michel commences work at 8 o'clock, depending on the season. In summer, during the harvest, he sometimes stays at the farm all day. It is during this time that Jean-Charles sometimes helps with the harvest, although he is much more interested in playing tennis!

In the woods surrounding the farm there is abundant wildlife, such as wild boar, pigeons, pheasants and deer. One of the major activities in France is 'la chasse', or hunting. The Michels themselves do not partake in this, partly because a farmer's free time is very limited. The family usually goes on holiday in December because work on the farm is least pressing at that time. The main tasks at various times of the year are shown opposite:

The climate is usually kind to the farmer for the low winter temperature kills off the soil pests, but still enables an autumn sowing of cereals to be made. The early spring and warm summer temperatures allow the harvest to begin in July and it is usually over by mid-August. The main problem in the years Monsieur Michel has owned the farm has been the variability of the rainfall. There have been times when he was unable to sow sunflowers because of the ground being too wet and there were other times when he had to irrigate in June.

Fig. 3. Le Breuil, showing cropping patterns for 1985

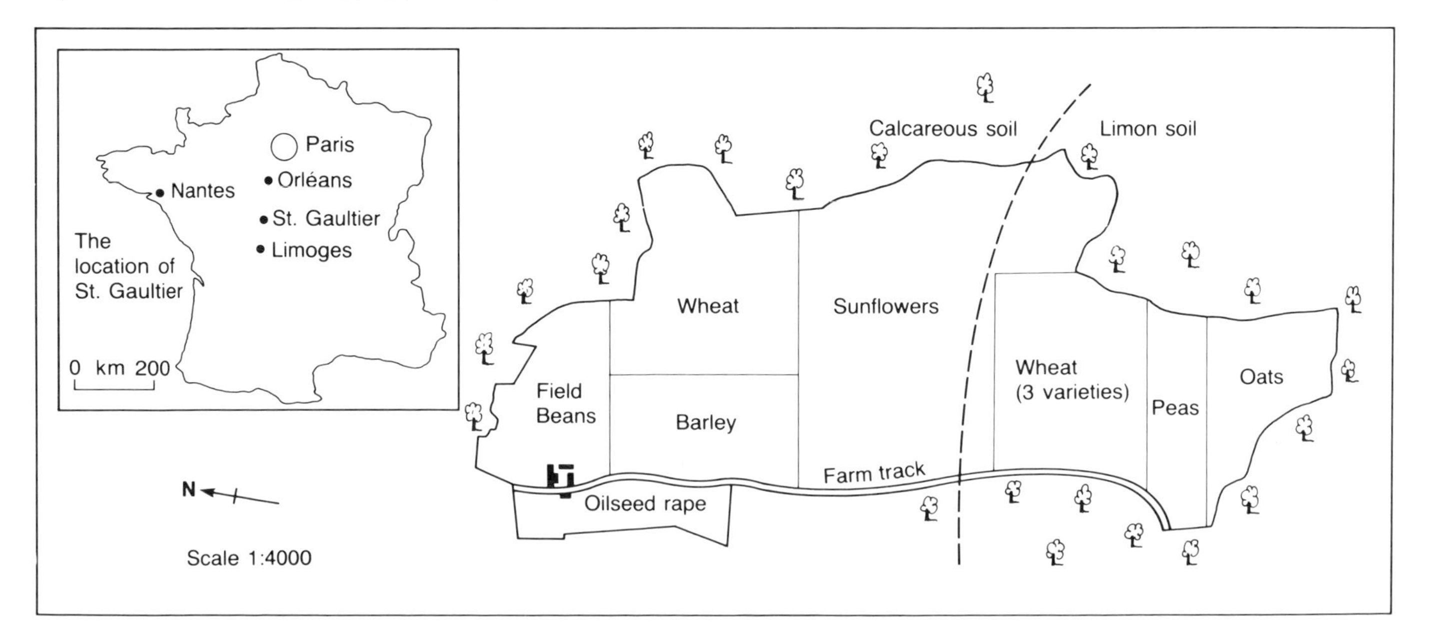

Fig. 4 Monsieur Michel and family

Fig. 5 Machinery at Le Breuil

January	Finish winter ploughing	Machinery maintenance
February	First application of chemical fertilizer	
March	Application of chemical weed-killer	
April	Second application of chemical fertilizer. Sowing oilseed rape	
May	Sowing of sunflowers	
June	Fungicide treatment of cereals (perhaps irrigation as well)	
July **August**	Cutting of oilseed rape, oats, barley and wheat. Harvesting of 'petits pois' and field-beans	
September	Ploughing in stubble	
October	Ploughing. Sowing oats, winter wheat, barley	
November	Sowing field-beans	
December	Preparing ground for sunflowers. Family vacation	

The co-operative, from which Monsieur Michel purchases most of his seeds and chemicals and to which he sells his harvest is in St Gaultier. For any machinery or major repairs he has to make to his present equipment, he visits Chateauroux. He is a forward-looking farmer, constantly trying out new methods of cultivation and at least one new seed variety per year. He keeps a scientific eye on the structure of his soil, which is principally calcareous (based as it is on oolitic limestone) near the farm building and sandy (limon) in the southern part. The soil is never more than 25 cm deep near the buildings, but is deeper than 40 cm on the limon. Peas and beans are best suited to the sandy land, whereas the heavier, moister, calcareous soils suit the cereals and oilseed rape the best.

Monsieur Michel likes to spread his risks. He has no field boundaries to worry about actually within the farm. The hedges were grubbed out by his predecessor. He therefore grows a wide mixture of crops, rotating them every year, but keeping an eye on prices in order to predict the most profitable mix for the next year. Prices can fluctuate a great deal. In 1984, for instance, his profits were down by 10% on the previous year, whereas his costs were up by the same amount. The Common Agricultural Policy of the European Economic Community guarantees him a minimum price for his products, but he would obviously like to get more than the 'intervention price', which is only paid when there is a glut of a particular product on the market. He therefore needs to predict which products will be in short supply the following year – not an easy task. Since his wife is earning money as a teacher, the family has more financial security than some farmers. The next-door neighbour, who is over seventy years of age, runs a farm on which methods are almost the same as those employed fifty years ago. The machinery is run down and ancient and the farm building is the one unit for cow stalls, living and eating quarters and storage for harvested products. The living here is at just a little more than subsistence level, but there can be no cover for any illness the farmer may have.

Monsieur Michel has two horses near his farm, from which he is hoping to breed racehorses. He has two in training in Normandy, and one has been a winner at trotting events. Any income from this interest is a welcome bonus, but it is done mainly for fun.

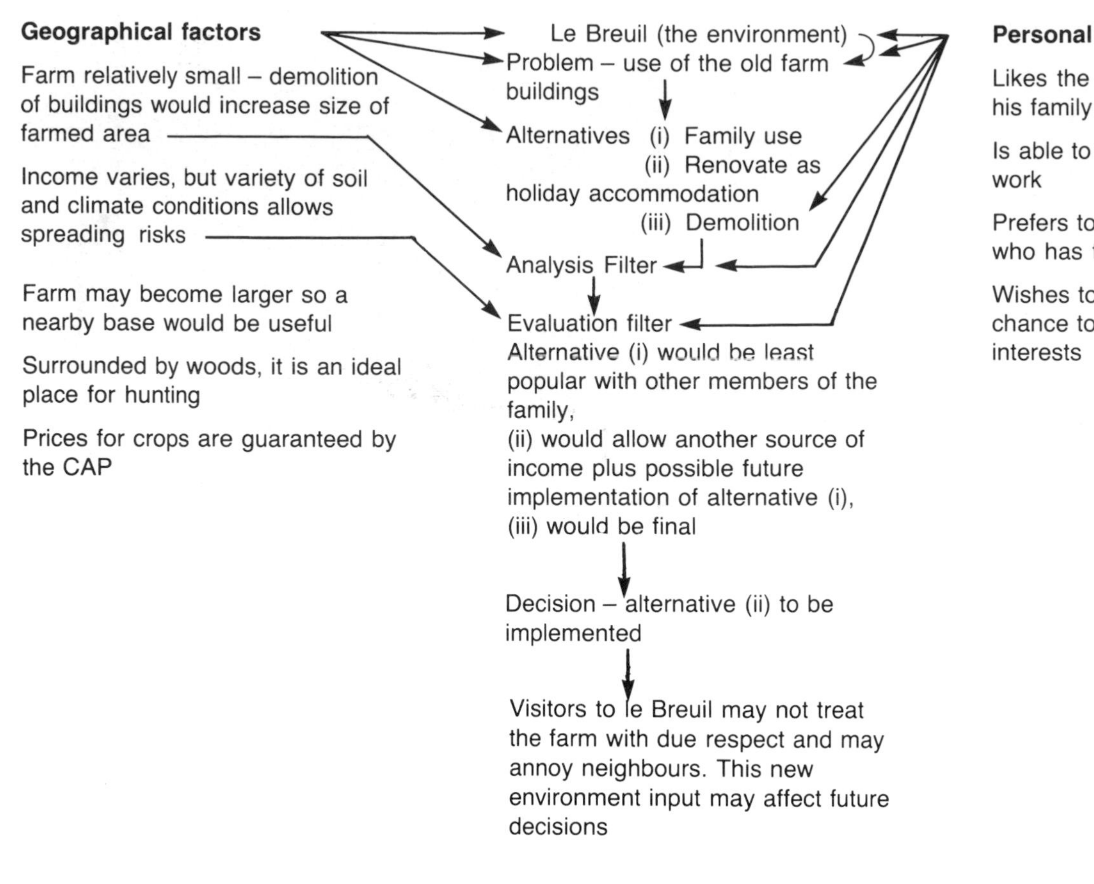

Fig. 6 The most important factors influencing Monsieur Michel's decision

Most of the farm buildings, apart from the machinery barn, are no longer in use. They are stone built and structurally sound. Sewerage and running water are installed but are, like the decor, in need of modernization. Apart from the neighbouring farm, which is about 200 m away and hidden from view, there is no other dwelling for about 2 km.

This thumbnail sketch of the background to an important decision should have revealed to you examples of geographical factors influencing the decision, but also the importance of personal values and preferences.

Some of the most important ones are outlined in Fig. 7.

Does your list of geographical factors, personal values and preferences concur with that shown in Fig. 7? If not, perhaps it is possible that different people view the same information in different ways. If they did not, after all, everyone would make the same decisions and geographical issues would not be half as interesting as they are!

Planning a New Road – an example of the decision-making process in action

Geographers study transport and communications because they link settlements and enable goods, for example, to be moved between places of production, storage and consumption. However, in the local area, the actual location of the line of the new transport link is significant. The role of the geographer is an advisory one, i.e. where to build the road to cause least disruption to the environment, being aware of the financial implications of any decision.

The problem defined

In East Berkshire the transport network is well connected with good quality A and M roads. However, the network has one weak link, the A423 linking the A404 with the A4 and the A423(M) as shown on the sketch map (Fig. 8). You can see that this road is the final link completing the network of the M4 with the M40 with good quality dual carriageway. Why is this fast link required? Look in your atlas to check on the importance of this link. Find Heathrow Airport, for example, and consider its location in relation to this link.

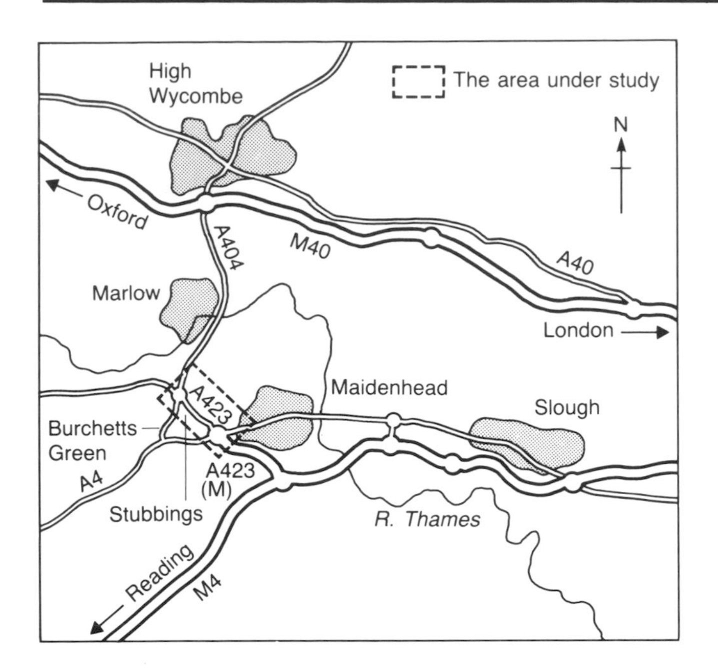

Fig. 7 Sketch map of the locality

Fig. 8 The over-used A423

The present situation is indicated in Fig. 9. The road is constantly in use with a high proportion of industrial vehicles. The noise from the road affects 39 houses at present, and 12 of these homes are only 50 m from the road. The road is difficult to cross for pedestrians and accidents are frequent. This has the effect of slowing the traffic up further and causing greater hold-ups. It is, then, a much overused road, which can be slow going for people driving along it and a hazard for the people of the area wishing to walk along it or to cross it.

What has been proposed
The County Council and the Department of Transport have devised the following alternatives shown in Fig. 10.

Table 1 (overleaf) gives the effects of each of the schemes using information prepared by the Department of Transport.

This information can be analysed to help in selecting the preferred route. Make a copy of Table 1, omitting the figures. Instead of the indicated savings, substitute figures from 1 to 5 to represent the rank order for each route; 1 should stand for the most beneficial route and 5 for the least beneficial. If, for instance, there is equal importance between the first four routes, add together the ranks they would occupy (i.e. 1, 2, 3 and 4) and divide by four, giving each a total of 2.5. For those items where qualitative rather than quantitative information is given, devise a way of showing the ranks so that each route can then be given a total figure.

Fig. 9 The alternative routes

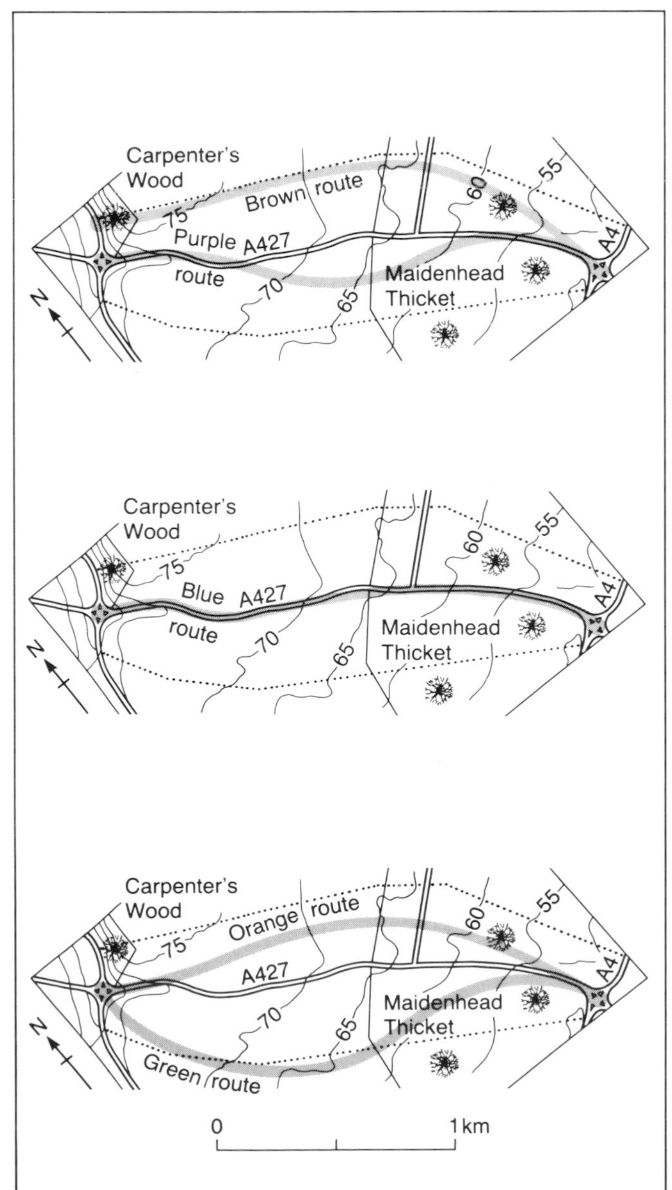

Table 1 The economic effects of the five schemes

Group of People	Effect	Purple	Blue	Green	Orange	Brown
Vehicle Travellers	Time savings (£m)*	4.92	5.84	5.52	4.84	5.80
	Vehicle operating costs (£m)*	–0.38	–0.77	–0.68	–0.60	–0.30
	Accident savings (£m)*	0.48	0.48	0.46	0.47	0.49
	Reduction in casualties: fatal	5	4	4	4	4
	serious	55	55	53	54	57
	slight	172	171	167	171	177
	Traffic delays during construction	slight	severe	slight	slight	slight
Local residents	Properties demolished	0	5	0	0	0
	Private gardens required (ha)	1.1	2.0	0	0.1	0
	Noise with new road 0.50 m from road	4	9	0	3	0
	(houses) 50–100 m	5	7	3	2	3
	100–200 m	9	2	9	8	10
	200–300 m	14	16	27	20	22
	Visual obstructions: severe	0	5	0	3	0
	(houses) significant	6	6	0	1	3
	slight	6	5	6	2	0
	Disruption during construction	slight	severe	slight	slight	slight
Farmers	Number of farms working land	3	3	3	3	3
	Area of land (Grade II) (ha)	2.5	0.4	6.2	4.6	5.0
National Trust	Loss of land (ha)	6.6	3.5	6.6	7.4	8.7
Other Public Buildings	Loss of land (ha)	0	0.2†	0	0	0
County Council	Developing hierarchy of loads in the county			Big improvement		
Department of Transport	Improve connection between motorways			Big improvement		
Walkers, riders	Visual intrusion, noise, loss of land	Reduction	Little change	Reduction	Reduction	Reduction
Congregation of St James's	Noise and access	Reduction	Increase	Reduction	Reduction	Reduction

* Cost calculated as savings (+) or costs (–) over a 30 year period.
† Including demolition of the School Hall.

1 Which route do you consider best by this method?

2 Since some factors are obviously much more important than others, could you devise a weighting scheme to enhance the value of obtaining a low rank order in a very important factor rather than in a relatively minor one? In this case, does your decision change?

3 Do you think that a firm decision could be made based on this technique, or do we have to consider other factors?

Refinements of the method

To refine the method, it is necessary to look at the actual landscape and consider the degree of interruption each scheme will make to the natural environment. A model which looks at the degree of interruption of the natural landscape is **Pickering's Principle**. This is illustrated by the **scattergraph** shown on Techniques p. 13 and means in simple terms that as human use of an area increases the importance of the purely physical environment decreases. At the high human impact end of the scale, an office block, there is little physical impact save the spiders and houseplants, while in the middle of virgin tropical rain forest human impact is very small. The values for the various environments can be calculated using a complex formula, but, as you can see, they add up closely to ten (Table 2).

Table 2 Value given to environment by Pickering Principal

Situation	Natural environment	Human impact
1. Nature reserve (closed)	9.90	0.00
2. Nature reserve (open)	9.80	0.51
3. Forestry (closed)	8.20	0.14
4. Forestry (open)	8.10	1.26
5. Agricultural land	7.56	0.90
6. Public open space (woodland)	6.70	4.35
7. Public urban park (50 ha)	6.70	4.40
8. Private camp site	6.65	4.90
9. Playing fields	6.40	3.60
10. Riding school	6.10	4.05
11. Gravel workings (active)	5.60	3.45
12. City open space (1 ha)	5.15	6.15
13. Rural road	4.70	5.50
14. Private garden	4.30	4.95
15. Rural footpath	3.90	5.62
16. Motorway	3.70	6.60
17. Low density houses	2.00	7.57
18. Urban footpath	1.95	8.32
19. High density houses	1.60	8.55
20. Lorry park	1.50	8.70
21. Public building	1.30	8.85

4 Make a list of the factors that you think make up Pickering's Principle.

5 Why is there not a perfect relationship on the scattergraph (graph (c), Techniques p. 13)?

6 Look at Fig 11(a) (b) and (c) and apply Pickering's Principle yourself. Use the values given in Table 2 to help you and also consider the human impact in your assessment.

7 Trace the environmental values from Fig. 12 and add your values.

8 Draw an **isoline map** from your completed version of Fig. 12 and shade it as shown on Techniques p. 14.

Your answer to Question 8 has given you a numerical pattern for the environment of the area under consideration. We can now see what happens when we consider the building of each of the proposed road schemes. Fig. 13 shows the effect of the construction of the Green Route. Using Pickering's Principle, a motorway standard road, such as the new A423 will be, has a value of 4. If we give a value of 4 along the entire length of the Green route and superimpose this on the tracing made for Question 8, we can now draw a new isoline map to show the effect the road would have.

Fig. 10 Look at these photographs of natural landscapes and apply Pickering's Principle

(a)

(b)

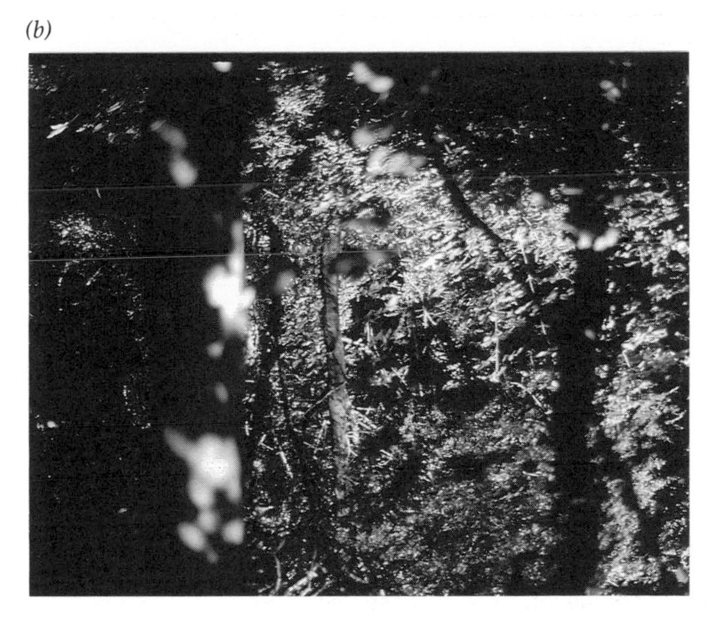

(c)

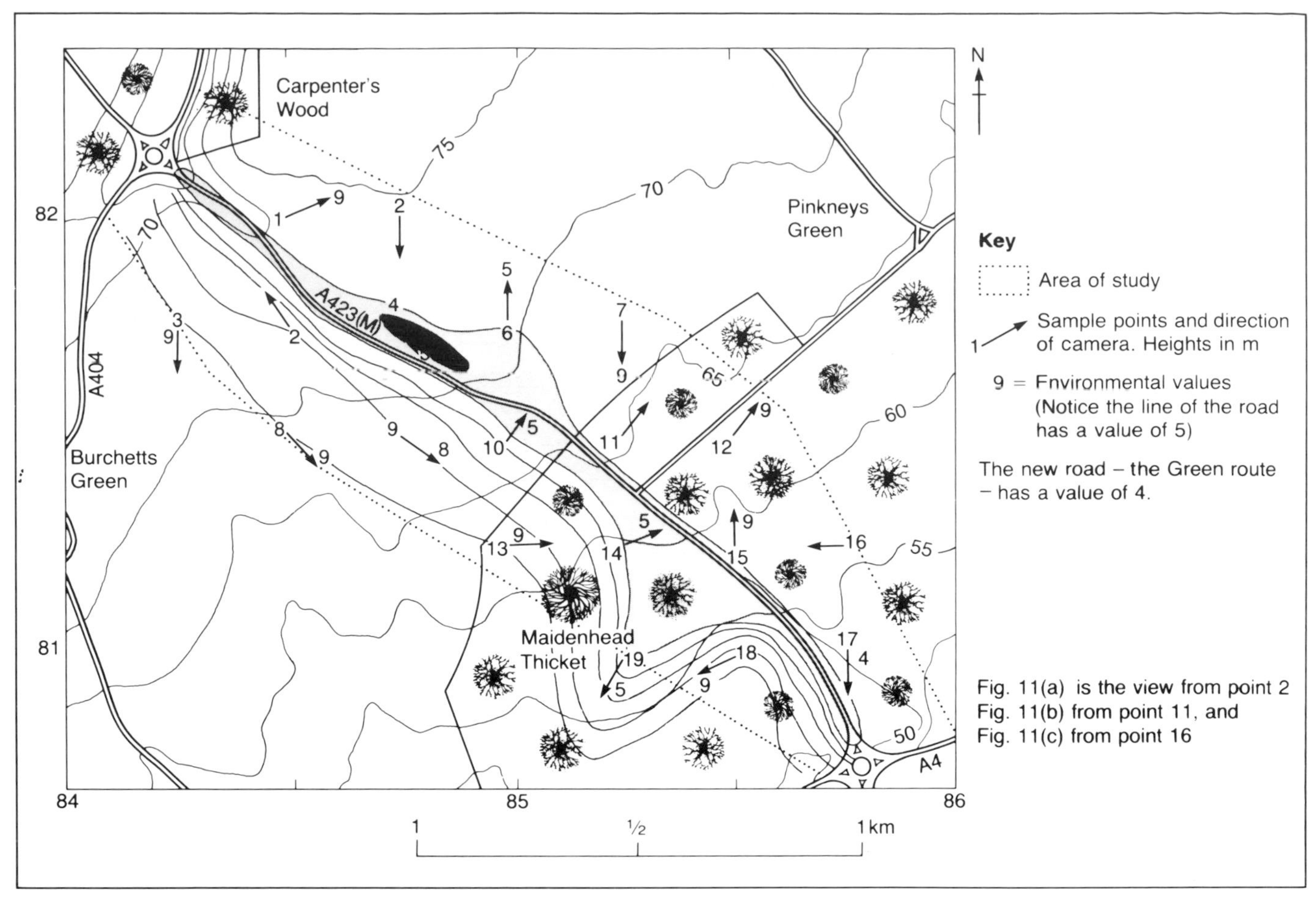

Fig. 11 The A423(M) road – Maidenhead Thicket to Burchetts Green

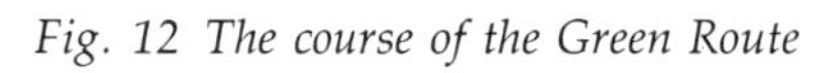

Fig. 12 The course of the Green Route

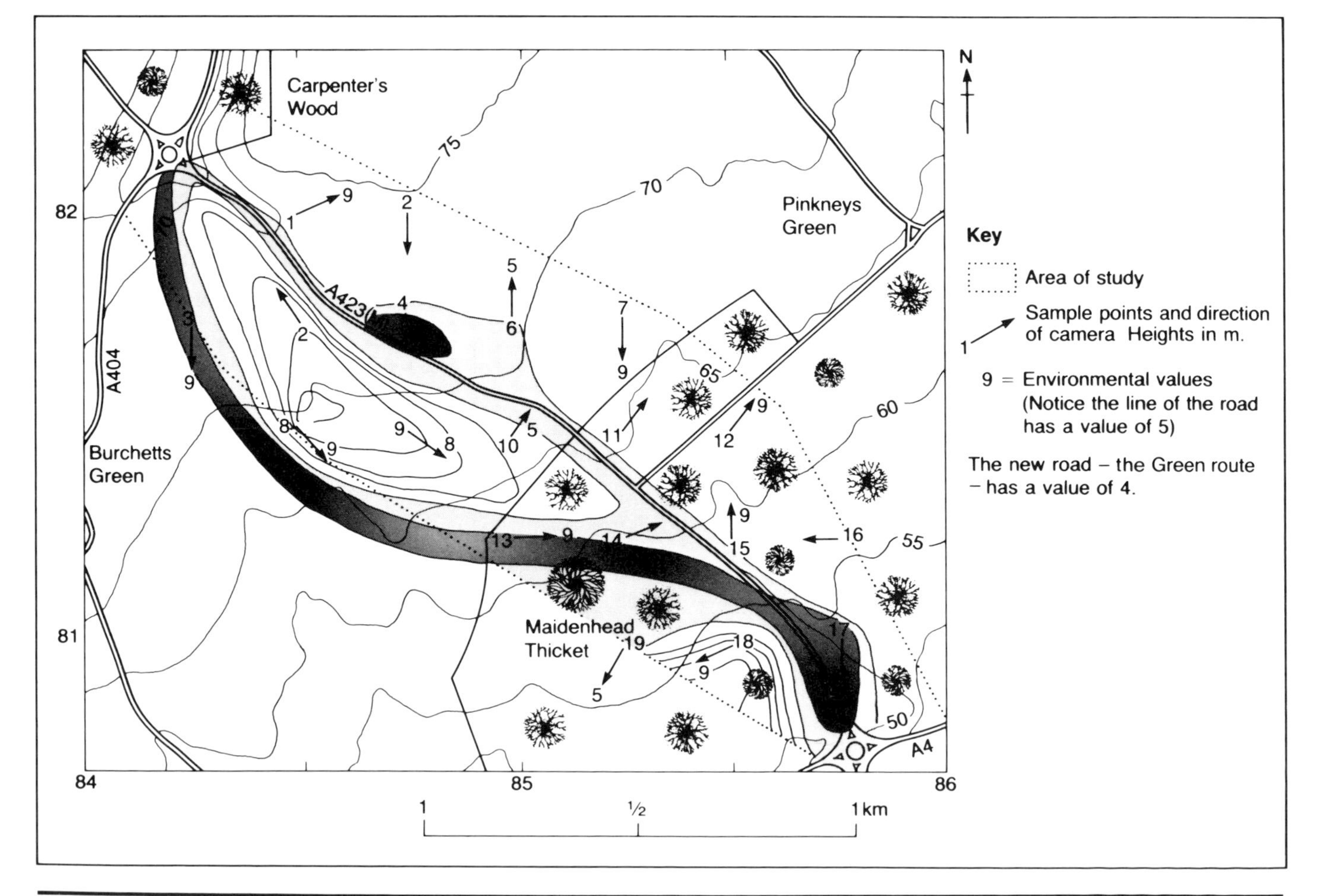

9 Repeat the exercise of drawing isoline maps for the Purple, Blue, Orange and Brown Routes. You may wish to share the work out among a group of you and compare the finished products.

10 Rank the routes on the degree of fragmentation to the physical areas, with 1 being the route which fragments the area least.

11 Write a short summary of this method and try to think of the reasons for trying to minimize fragmentation of the physical environment. You may wish to consider some of the ideas in Theme 2.

There are now two sets of information: the effects of the scheme from an economic point of view and its impact on the environment. What degree of agreement is there between the two sets of information? A third input into the decision comes from the attitudes and values of the people concerned in the decision.

12 Figs. 14 and 15 give information about the people concerned with the development and a selection of views given. Look at the information carefully and complete Table 3 to summarize their attitudes and values. Try to compose the likely attitudes of the local resident, the road user and the parish priest for yourself.

13 Place the routes into rank order according to the views of the people concerned.

14 Obviously, a factor not considered up to now is the cost of each proposed route. These costs are shown in Table 4. Convert these into rank order with 1 being the cheapest.

Fig. 13 The people concerned

Table 3 Analysis of the attitudes and opinions

People concerned	Summary of view point	What aspect concerns them most?	What evidence do they offer?	Which solution do they favour?
The National Trust				
Burchetts Green Village Association				
Berkshire CC				
Local councillor				
Local resident				
Road user				
Parish priest				

Table 4 Cost of the proposed routes

	Purple	Blue	Green	Orange	Brown
Economic analysis					
Environmental analysis					
Personal preferences					

Route	Cost (£ m.)	Rank order
Purple	6.156	
Blue	6.610	
Green	6.184	
Orange	6.321	
Brown	6.890	

Fig. 14 Views about the proposed routes

The fifth route, Blue, lies along the line of the existing Henley Road and is generally regarded to be the least acceptable.

FACTS

The Brown and Orange Routes:

1. Will entail the building of two roads, not one, across the open Common between Pinkneys Drive and Thicket Corner – a dual carriageway and a relief road running parallel to each other.
2. Will involve the building of a fly-over and banked approach to take Pinkneys Drive over the new dual carriageway constituting a major eyesore.
3. Will involve the loss of one of the few remaining safe open spaces available for recreation and leisure.
4. Will permanently damage the rural environment of Pinkneys Green and the quality of life enjoyed by so many.
5. Will introduce the hazards of noise and air pollution to a large residential area.
6. Will isolate many parishioners from their Parish Church at Stubbings.
7. Will cause greater inconvenience to the public at large due to a longer construction period.
8. Will cost considerably more money than either the Green or Purple Routes.
9. Will represent an unnecessary and avoidable hazard to children and animals playing upon and using the Common.
10. Will have a detrimental effect on a greater number of residential properties than either the Green or Purple Route.

The Department has indicated that local opinion will be a major factor influencing the final choice of route. The facts show that a decision to adopt either of the northern routes, i.e. those crossing the common land close to the Village, will result in permanent damage to the environment.

You can help to ensure the continued protection of that environment by stating your preference for the Green Route as a first choice and the Purple Route as a second choice.

ACT NOW BEFORE IT IS TOO LATE

Published by Robin Austin, Councillor, Royal Borough of Windsor and Maidenhead

The National Trust

Maidenhead Thicket is held inalienably under the National Trust Act 1907 and the Trust is, as a result, required to resist the use of the land for any purpose other than its present use. If and when a preferred route is published by the Department of Transport, the Trust will object formally.

If after a Public Enquiry the Trust still feels that the case for the new road has not been satisfactorily established, then it can resist compulsory purchase and invoke the Special Parliamentary Procedure. This means that the decision whether or not to use the Trust's land may ultimately be made by the Houses of Parliament.

Burchetts Green* Village Association

The Association makes the case that the preferred route for the new trunk road is that coloured Brown in the plans submitted for public inspection. We believe that the cost differences between the routes are not material and should not influence the choice of route. The Brown Route is the most effective for linking the M4 and the M40, since it provides a new access to the Marlow bypass. This leaves the maximum of existing roads for local traffic in Burchetts Green, Pinkneys Green, Stubbings and Hurley.

We believe that impacts on inhabitants in the Stubbings, Pinkneys Green and Burchetts Green area are minimised by the Brown or Orange Routes. Any routes will require some sort of screening from high-speed traffic noise. We believe the land configuration and location of houses makes this easiest, most effective and lowest cost in the case of the Brown Route.

We believe the other routes, that is Purple, Orange, Blue and Green, would result in a significant deterioration in the quality of the environment in the Burchetts Green and Stubbings area and would do little to improve that for people living in the nearby area of Pinkneys Green. Whilst the preferred route passes closer to Pinkneys Green, we consider that any adverse effects on that area can easily be overcome.

**Burchetts Green is the village to the West of Stubbings*

Berkshire County Council – Environment Committee Report

Assessment of the environmental impact of the alternative schemes, in terms of the issues relevant to the County Council as strategic planning authority, shows that two of the five alternatives – the Blue and Purple routes – are unacceptable on environmental grounds. The minor disadvantages attaching to the remaining three routes are not considered to be so severe as to justify an objection on environmental grounds to the selection of any one of them as the preferred route. None of these three stands out as a clear 'best route' from the point of view of the strategic planning authority.

It is therefore considered that the County Council should indicate that it has no objection on environmental grounds to the Green, Orange or Brown route being selected as the general line for the road improvement, and that on this aspect of the proposals no preference should be expressed out of the three. It is suggested, however, that the Council should request that, when the Department of Transport is considering the environmental aspects of the alternative road lines, they should pay particular heed to the views of the National Trust as the major landowner concerned and as the authority responsible for the management of the area's principal public recreation resource, Maidenhead Thicket.

On local traffic grounds, the Blue Route – which would require local traffic to share the entire length of the A423 with long distance traffic – is considered unacceptable. The Green Route stands out from the remaining alternatives as minimizing the increase in local journey lengths, and is therefore recommended as the preferred route on transport grounds – though if this route is eventually selected, some attention may need to be given to discouraging the continued use of the existing A423 alignment by long-distance traffic starting or finishing its journey in Maidenhead.

It is considered that the remaining three routes are not so unsatisfactory as to be objectionable on local traffic grounds. However, they are all estimated to be more expensive than the Green Route and do not appear to bring commensurate additional benefits to the County Council as local Highway Authority.

Fig. 14 Views about the proposed routes (contd.)

Once again, you may not consider that each of these inputs to the final decision is of equal weight. If that is the case, devise a weighting scheme which could be used to enhance the value of one factor. For instance, if you thought that the cost of the road was twice as important as the next most important factor, you should multiply the values of the other factors by two. The lowest figure is the most important. If you then sum your figures, the one achieving the lowest total would be the preferred route on all the considerations used.

The final decision for trunk road planning and building rests squarely with the Department of Transport. Which route do you think the Department will take? What other factors might the Department take into consideration? Consider, for example, the political force which could be exerted by such organizations as the National Farmers Union, by the Road Haulage Association and British Rail. Try to consider the likely reactions of these organizations and consider the effect they might have. Will the Department of Transport weight the factors differently? Which will benefit, the local environment or regional planning?

The wider context

Transport programmes have become very important. The building of the M25 and the over-use of that road have raised public awareness to the problems of congestion and to the difficulties of constructing such a long motorway link. The M3 extension near Winchester is another route which has been the subject of numerous public inquiries, not least because the preferred route passes through an area of high landscape value and the location of a valued archaeological site. These national concerns do show the importance of considering values when analysing developments. In this case the emergence of Green politics has made the Government think again before allowing some development to continue.

The controversy around the construction of the Channel Tunnel has been headline news. The actual route of the rail link, whether it is going to be in a tunnel or whether it will pass through land with high landscape value, has been hotly debated. Concern is also expressed as to the implications of such development on the immediate area of southern England and also further afield on the economic development of other parts of the UK. The impact of a relatively small development may spread far from the area in which the development is going to take place. Try to consider this when looking at the impact of any 'local' development.

15 Hold a public inquiry into the route of this road with people taking the following roles:
- Representative of the National Trust
- Berkshire County Council Planning Department
- Representative of Burchetts Green Parish Council
- Local residents and the local vicar
- Local farmer
- Representative of the Pinkneys Green Parish Council
- Representative of the Council for the Preservation of Rural England
- Representative of a local trucking company
- Representative of the Department of Transport.

You should refer to p. 188 for details on the working of a public inquiry.

16 Write a newspaper report summarizing the arguments expressed at the public inquiry. How would the national press report differ from the local press report?

17 Should the public have a say in something which is being planned 'in the national interest'? Refer to the information later in this book on Sizewell B (pp 184–186) and the Chico Dams (pp. 43–45). Do the locals get a fair hearing or is the public inquiry loaded against them?

18 Is the public inquiry the best way to listen to local opinion? If you consider the answer to be negative, can you suggest alternative approaches?

19 Investigate a local issue in your area using the decision-making path shown in Fig. 2. Make a presentation of your investigation to the rest of your class and argue the merits of your solution.

Drawing a scattergraph

The figures in Table 2, p. 7 are the scores derived from Pickering's Principle. What sort of relationship is there between the two columns? As the human impact figure increases, what happens as a general rule to the environmental figures?

A good way of looking for a relationship is to draw a **scattergraph**. You must make sure that you have thought about the reasons why a relationship may appear between these factors. The following guidelines will help you with the construction of the scattergraph.

1. You must decide which of the columns (or **variables**) is dependent upon the other. This will be the dependent variable, while the other is the independent variable. In this case, human impact is the independent variable, while the environmental value is the dependent variable.
2. Draw axes as shown on Fig. T1(a), with the independent variable along the x axis. You should always try to start from 0 on both axes.
3. Now start plotting the points as shown in Fig. T1(b).
4. When you have completed plotting the points, look for a general trend. It may help you to draw around the points to do this. Try then to draw a *best fit line* going through the main trend of the points. You should try to keep the distances from the line to the points equal above and below the line. Ignore points well away from the main trend. These are known as **residuals** and you may need to consider these oddities later (Fig. T1(c)).

This type of **correlation** is known as a **graphical correlation**. In this case the relationship is negative, i.e. human impact increases and the environmental value decreases. Is this a reasonable proposal? Can you try to work out why this should be? The relationship would have been positive if the best fit line had resembled Fig. T1(d). Think of two sets of variables which would produce a positive correlation if plotted on a scattergraph and a best fit line drawn.

Fig. T1 Scattergraphs

(a)

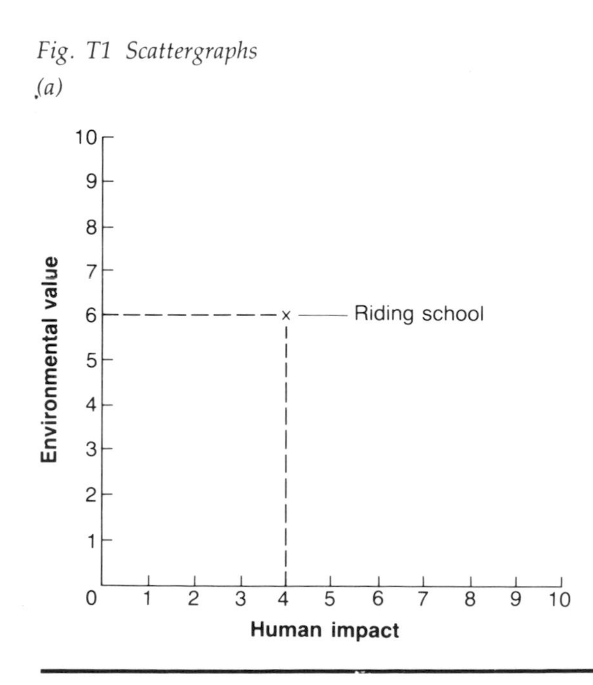

(b)

(c)

(d)

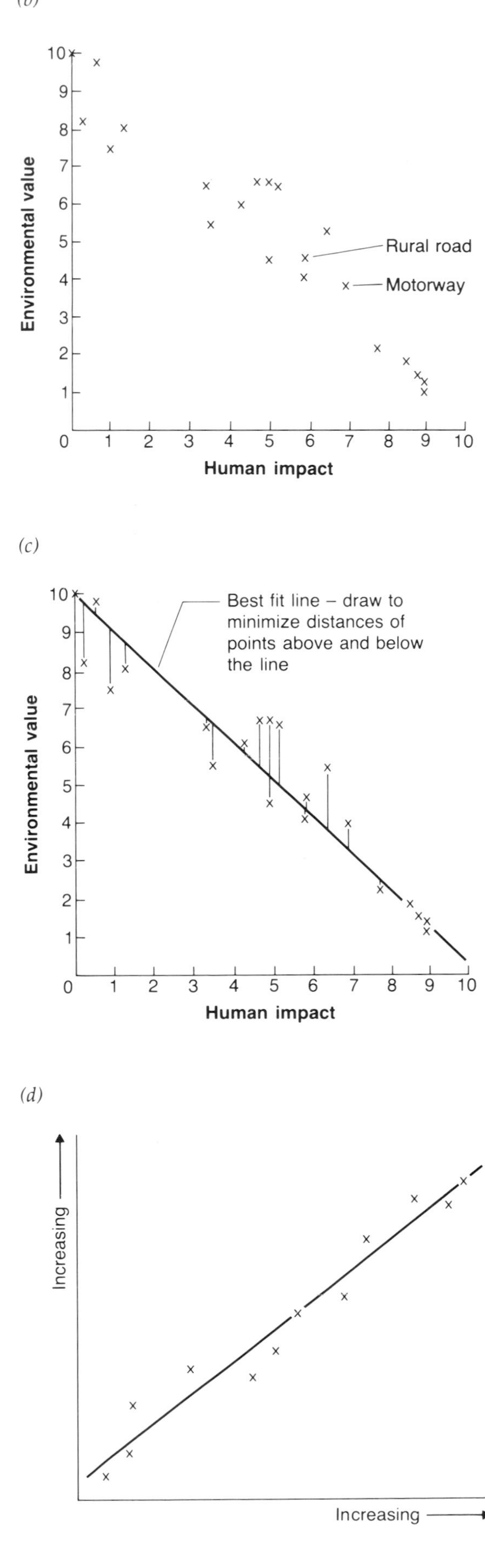

Drawing an isoline map

This is a useful technique for showing a pattern when you have information from a series of points or locations. It is the same technique used by cartographers in producing contour maps, and weather maps showing pressure and temperature.

A line joining places of equal height is known as a **contour**. What do we call lines joining places of equal temperature, equal pressure, and equal rainfall?

Let us assume that you start with a distribution of locations and values as shown on Fig. T2(a). The lowest value (3) is in the corner at the top of the map and the highest value (5.6) is in the bottom right corner. Let us draw the isolines at 0.5 intervals, starting with 3.

(i) Join the points value 3 as shown in Fig. T2(b). All the values within the 3 line should be higher than 3.

(ii) Now draw the 3.5 line. Note how you thread the line between 3.3 and 4.2, 3.1 and 4.9, and you go outside some of the high values to get to 3.5. You can then join it up with the beginning of the line (Fig. T2(c)).

(iii) Study Fig. T2(d) which has the rest of the line drawn in. Note carefully how the 5.5 line has been drawn to show isolated areas.

(iv) Now shade carefully between the lines; use light shading for the lower values and gradually get darker for the higher scores.

This technique is very useful for showing patterns from a meaningless jumble of figures!

Fig. T2 Drawing an isoline map

(a)

3	3.5	4	4.2	3
3.3	4.2	5	5.5	3.5
3.5	4.9	5.1	5.3	4.2
4.8	3.1	5.5	5.6	5.1

(b)

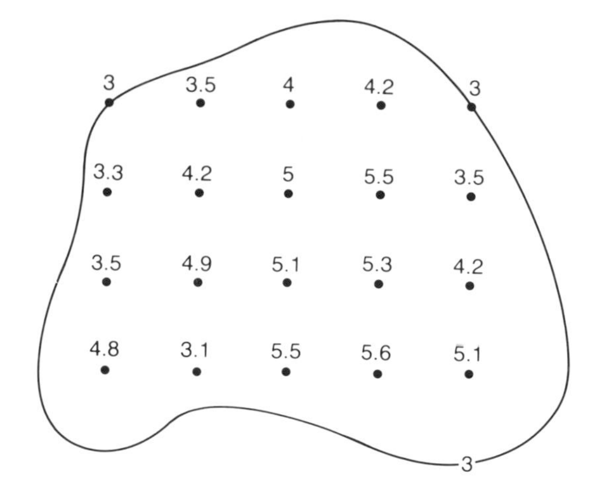

(c)

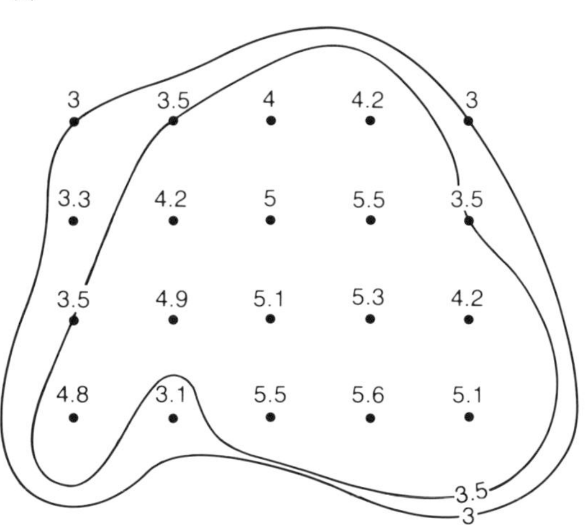

(d)

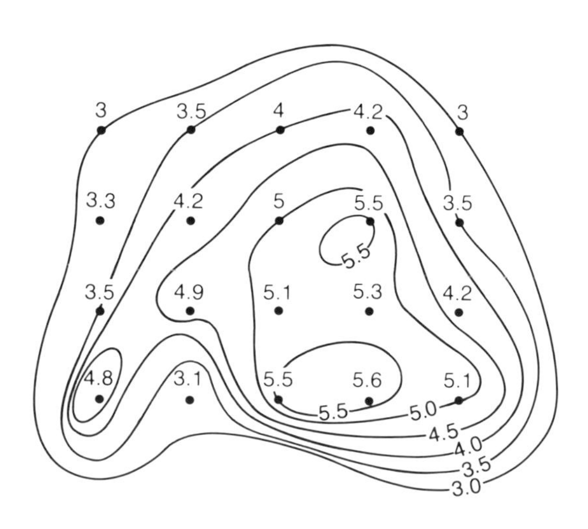

Theme 1 *Managing Physical Systems: The Example of the Drainage Basin*

Some Physical and Human Aspects of the Use of Drainage Basins

Inter-relationships between people and their environment are particularly well illustrated in the **drainage basin**. Streams and rivers, whose principal purpose is to remove water from the land, can be greatly influenced by human occupation. This influence may affect the channels themselves but may also extend to the sediment they are required to carry, for human occupation of an area may well lead to an increase in the amount of material being washed from slopes by rainwater. In this chapter we will investigate the processes at work in a drainage basin, the way they may be influenced by human activity and what can be done to ensure that land-use changes and urban development do not lead to unexpected and undesirable changes in both processes and landforms.

Some of the inter-relationships between physical processes and human occupation of the drainage basin are shown in Fig. 1.1.

1 (a) The sketch (Fig. 1.1) shows some beneficial and some problem-creating land-uses. Make two lists, one of the benefits and one of the problems for the catchment of different land-use practices.
(b) For any *two* from each side, explain why they bring problems or benefits.
(c) Are there any items which are difficult to place in one list or the other? If so, explain why it is the case.

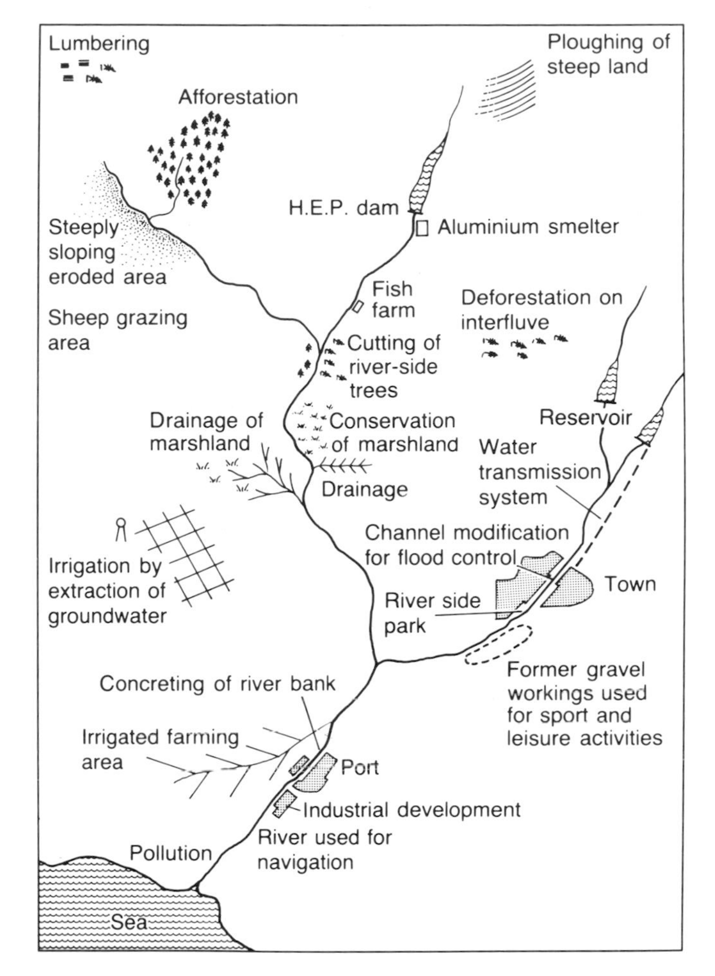

Fig. 1.1 Some possible uses of a drainage basin

Fig. 1.2 An area in northern England

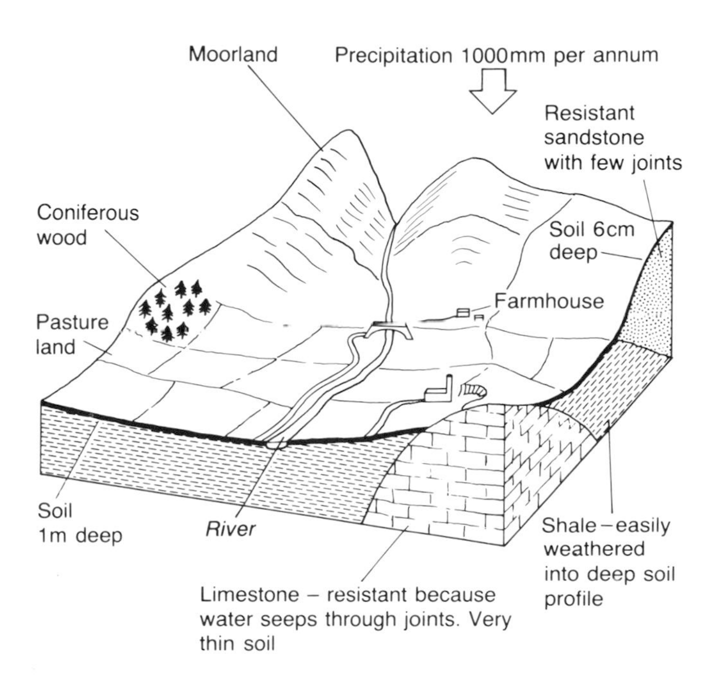

A preliminary investigation into some of the physical aspects of the basin will enable us to evaluate more precisely the effects that different land uses have on catchment processes.

The circulation of water in the environment is known as the **water cycle** (or **hydrological cycle**). Such a cycle can be illustrated by using a **systems diagram**. Since this is so useful in all studies emphasizing inter-relationships, a detailed explanation of systems diagrams is given on page 47.

Fig. 1.2 shows a composite picture of part of northern England, similar to the features found within a few square kilometres of Hope in the Derbyshire Peak District. The hydrological cycle for this area is shown in the systems diagram (Fig. 1.3). This highlights the importance of rock type both for its permeability and for its influence on soil type and vegetation. Vegetation can intercept precipitation, storing it for a while on its leaves and stems. The water may later be **evaporated** from these surfaces.

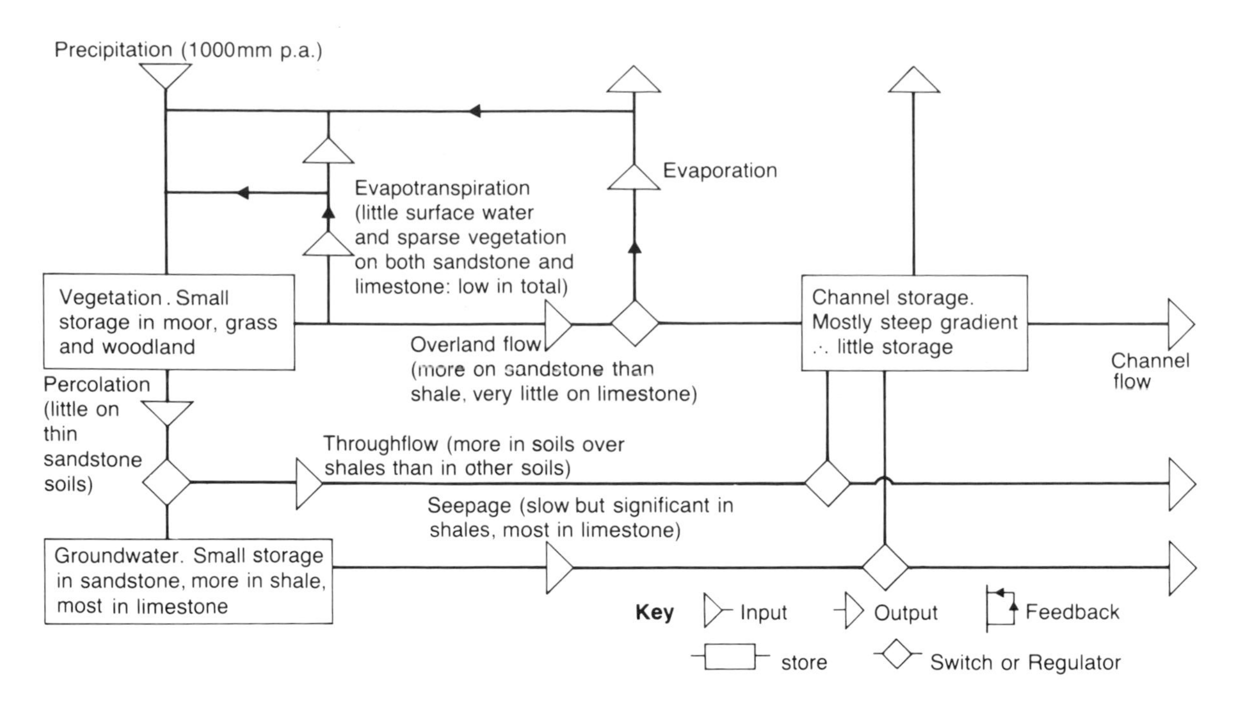

Fig. 1.3 A systems diagram to show the hydrological cycle for Fig. 1.1

At the same time the plants take in water through the roots. This passes up the plant and is then **transpired** (breathed out) through small cells on the leaves called stomata. The two processes together are known as **evapotranspiration** and this can be a major part of the circulation of water in an area. The most important way in terms of quantity that water is transferred down a slope in an undisturbed area is through the soil, a process known as **throughflow**. All these relationships are shown in the systems diagram (Fig. 1.3).

Obviously, Fig. 1.3 shows an area in which people are managing the land basically for agriculture. In Fig. 1.4, however, the management of the land has changed considerably. For instance, a dam has been constructed. This impounds a small reservoir. At the foot of the dam, a pumping house sends water to the nearby town. The steep slopes surrounding the reservoir have been afforested to stop the reservoir being choked with soil slipping down these slopes. A cement works has been built. This makes cement by baking shale and limestone together in large kilns which drive the water hitherto contained in the rocks back to the atmosphere in the form of water vapour. Surface water drains have been constructed alongside the new road which allow water to reach the river channel much quicker. Again all these developments can be seen within a few kilometres of Hope.

The relationship between elements in the hydrological cycle for an area is known as the **water balance**. This can be expressed as:

Q	=	P	−	E	±	S
(Flow in stream or discharge)		(Precipitation)		(Losses by evapotranspiration)		(Changes in storage)

It is clear from the previous example that this balance can be altered by development, but it also varies because of natural differences. In order to investigate differences in the water balance in two small areas, a Sixth Form geography party conducted a piece of fieldwork.

2 Construct a new systems diagram to describe the hydrological cycle for the area shown in Fig. 1.4. Your diagram should show new detail to include one more store and two more inputs.

3 Construct another systems diagram like Fig. 1.3 to show the physical *processes* at work in the area shown in Fig. 1.1. When you have drawn this, add the *human* aspects affecting the cycle in another colour to complete your diagram of the inter-relationships within the drainage basin.

Fig. 1.4 Developments in the area shown in Fig. 1.2

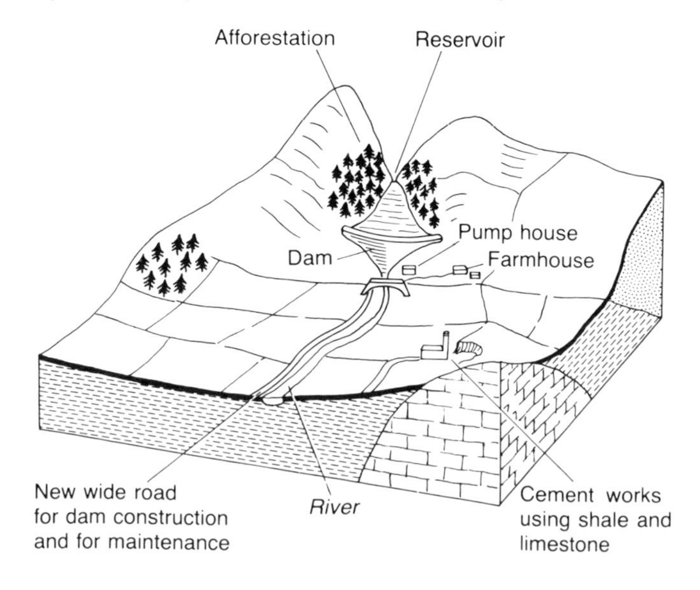

Case-study: an area in North Wales

Two small catchment areas were selected from Ordnance Survey 1:25 000 maps of an area of North Wales. They were within easy reach of the base for the work, which was a Youth Hostel in Dolwyddelan, North Wales. Each area was visited on the same day so that external variables could be kept to a minimum. Points were selected at the end of the main streams before they joined a more major stream. These points were both accessible and possible to wade across (as shown in Figs. 1.5(a) and 1.5(b)). The party also noted the vegetation and land use in the catchment areas of the two streams above the measuring point. Other statistics were revealed by study of the 1:25 000 maps. Armed with this information the two catchment areas could be compared (Table 1.1). The fieldwork techniques used to measure the channel cross section, velocity of flow and discharge are explained on page 48. It should be noted that the measurements were completed in the middle of a long dry spell and the discharges of the two streams are often much greater. In this particular case, however, since the results were being used to compare the two basins, this is of no consequence.

4 Complete Table 1.1 by reference to the sketch maps of the two catchment areas (Fig. 1.6).

5 What is surprising about the discharges of the two streams when compared to the area of the basins?

6 With reference to your copy of Table 1.1, attempt to explain why the water balance is so different in the two catchment areas.

7 Read Fig. 1.7, an article from the *Observer*. Imagine that you are Nalni Jayal. Write an illustrated submission for the International Union for the Conservation of Nature and Natural Resources setting out the reasons for water problems in India and measures that could be taken to overcome them.

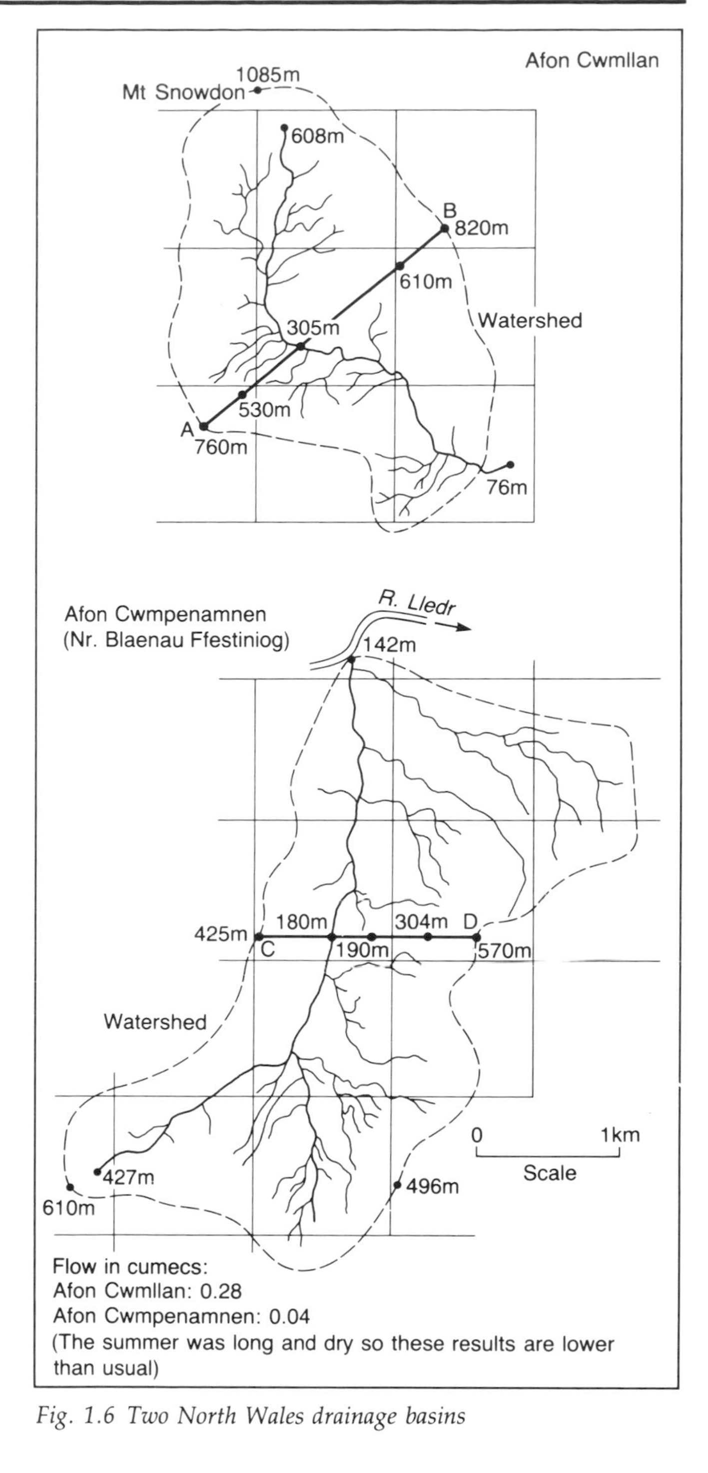

Fig. 1.6 Two North Wales drainage basins

Fig 1.5 (a) Afon Cwmpenamnen

Fig. 1.5 (b) Afon Cwmllan

Table 1.1 Summary of factors affecting the water balance in two North Wales drainage basins

	Afon Cwmllan	**Afon Cwmpenamnen**
Discharge (cumecs)		
Area of catchment (km²)		
Average gradient of long section		
Cross section	(draw on graph paper)	
Total stream length		
∴ Drainage density = Area of catchment / Total stream length		
Vegetation (see Figs. 1.5 a and b)	Largely moorland and bog. Some woodland in lower part of basin	Largely pasture and moorland. Much coniferous woodland in lower parts of basin
Settlement	None	13 buildings (mainly in lower part of basin)
Rock type	Mainly massive (poorly jointed) volcanics	Well jointed volcanics
Height of watershed		
Precipitation	High totals. In Snowdonia area therefore much relief rain	In rain shadow to east of Snowdonia
Soil depth	Mainly thin on steep slopes. Some bogs in valley floor	Thin on upper slopes but some deep glacial deposits on valley floor
Land-use	Mainly upland grazing	Upland grazing. Much afforestation. Good pasture in lower valley floor

The Development of Drainage Basins

The two small basins in North Wales have not been affected to any great extent by building. The run-off is related largely to natural factors. In many other cases, however, basins have been developed for settlement. Not surprisingly, the water balance has changed dramatically as a result.

Fig. 1.8 is a systems diagram of a basin which has been developed for housing and industry. It shows the main changes that have occurred from the situation shown in Fig. 1.3.

The flow of water in a channel can be described by use of a diagram known as a **hydrograph**. This shows the changes in flow over time and is normally related in some way to rainfall. Usually, after rainfall occurs, it takes some time for the water to reach the channel, passing as it does along the many paths indicated in Fig. 1.3. In an urbanized basin, however, the time between when the rain falls and the maximum flow in the channel can be drastically reduced (Fig. 1.9). This lag time can be very significant since the amount of water flowing in the channel for the peak flow period is also increased. Flooding and erosion may result unless sufficient planning has gone on.

In planning for urbanization, the resultant size and shape of the channel should be taken into account. Gregory and Park (1976) found that the channel capacities were up to 150% larger in urban than in rural sites in North West Yorkshire. Other studies by Hammer (1972) in a large number of small basins near Philadelphia, USA, suggested that channel area makes a gradual adjustment through the urbanization process (Table. 1.2).

India faces famine on Ethiopia scale

by GEOFFREY LEAN, Environment Correspondent

INDIA is facing an ecological crisis which will cause Ethiopia-style famine within the next decade.

Water sources are drying up, according to evidence gathered by Indian officials and presented at a meeting of world experts on environment and development in Madrid.

This seriously threatens food supplies and health in India, which is regarded as a third world success story for its achievements in increasing food production.

Nalni Jayal, adviser to India's Planning Commission, said at the general assembly of the International Union for the Conservation of Nature and Natural Resources: 'We are on the verge of an enormous ecological disaster. What is happening in Africa is going to happen in India within ten years.'

The first signs of crisis emerged a year ago when representatives of the state governments met in Delhi to discuss their progress in introducing measures under the UN Clean Water Decade.

India has set a target of providing clean water to every home by 1990. But state after state reported that although the engineering work of laying pipes and digging wells was going satisfactorily, springs and ground water sources were drying up, leaving more people without water than when the decade began.

Nalni said the Chief Minister of Maharashtra reported that his state had a target of completing water supplies to 15,000 villages: it found that 22,000 villages had *lost* water supplies.

Nalni investigated throughout India, and found similar crises everywhere, even in Kerala in the extreme south which has seven months of rains. In the northern state of Uttar Pradesh, 2,300 out of 2,700 water supplies provided by the Government had dried up.

Worse floods

Within 10 years there will be much too little water to satisfy demand in almost every state in the country, he said.

Nalni believes the reason for the crisis is the felling of forests. Trees help retain water in the soil, and in a country where the rains are seasonal people depend on such stored water for most of the year.

'Wherever the forests have been left intact there is water,' he told the conference: 'Wherever they have been cut down there is a crisis.'

When the forests disappear, the rains run straight off the bare hillside causing worse floods than before. 'Once it took a month of the monsoons to produce floods,' said Nalni. 'Now vast areas are flooded after the first rains.'

Another cause of trouble is limestone mining. The Doon valley in the foothills of the Western Himalayas gets heavy rains which normally percolate down into a belt of limestone. But the rock is being mined for industrial raw material. As a result, most of the springs have dried up and agricultural productivity has declined by half.

Big dams, often built to provide irrigation water for cash crops for export, contribute to the crisis. Forests have to be cleared upstream to provide land for people driven from the area of the reservoir.

As a result there is less rainfall and soil from the bare slopes silts up the dams often reducing their working life.

Fig. 1.7 From the Observer

Fig. 1.8 A systems diagram for an urbanized basin

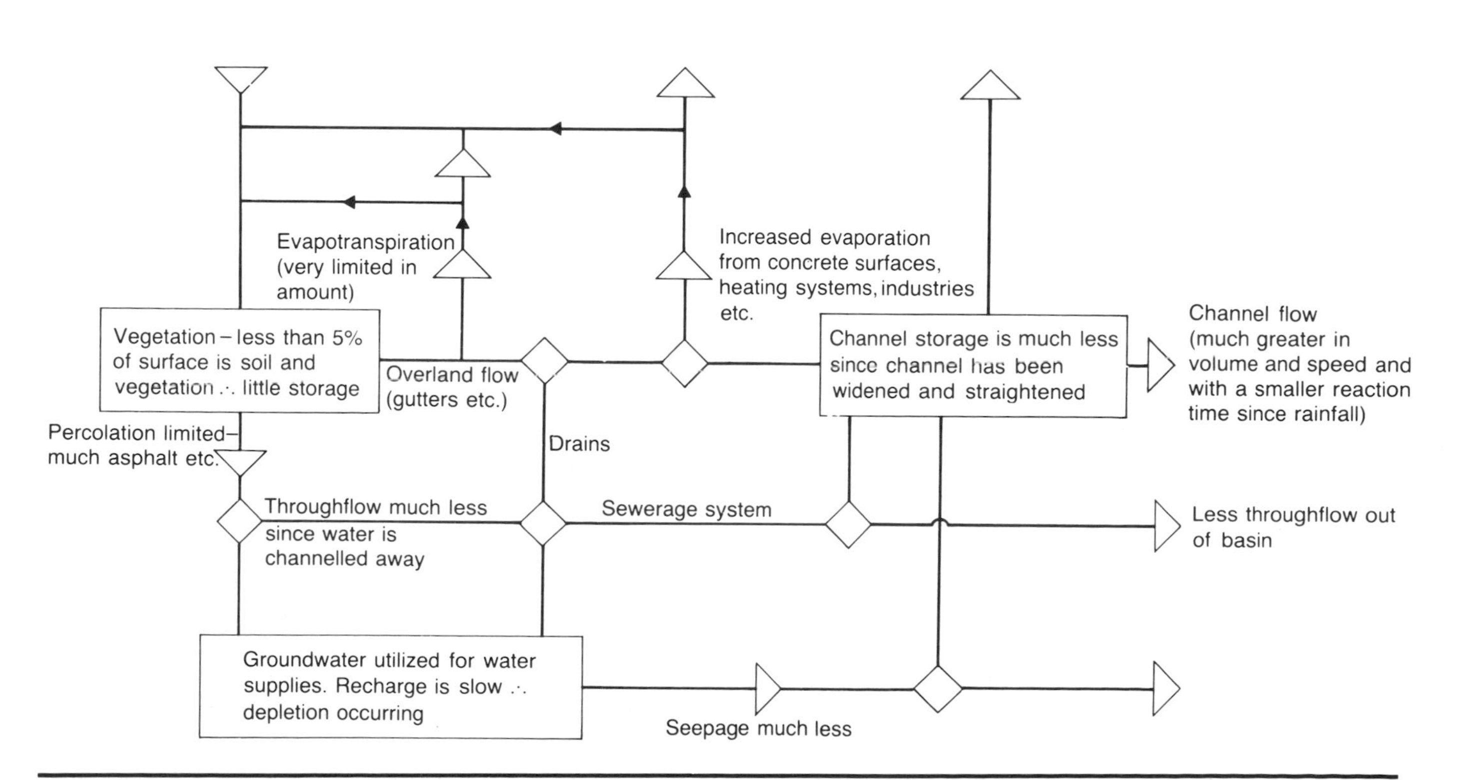

Table 1.2 Channel area change during urbanization in an area near Philadelphia

Land-use	Enlarged channel area/ Natural channel area
Impervious area less than 4 years old; unsewered streets and houses	1.08
Houses more than 4 years old, fronted by sewered streets	2.19
Sewered streets more than 4 years old	5.29
Impervious area more than 4 years old, fully sewered	6.79

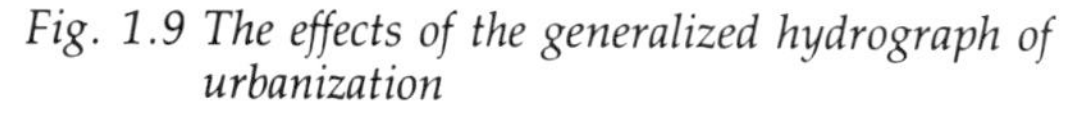
Fig. 1.9 The effects of the generalized hydrograph of urbanization

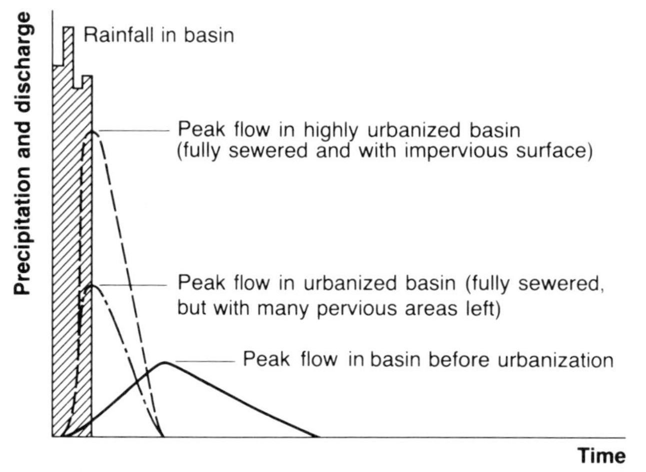

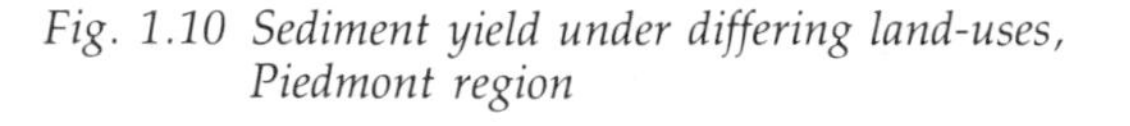
Fig. 1.10 Sediment yield under differing land-uses, Piedmont region

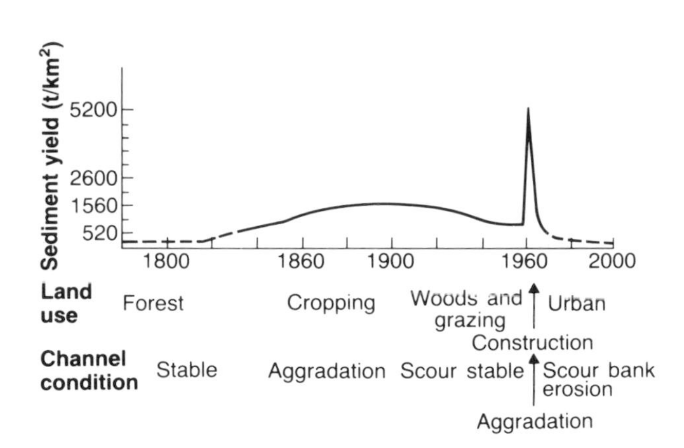

Fig. 1.11 Some aspects of the detention basin

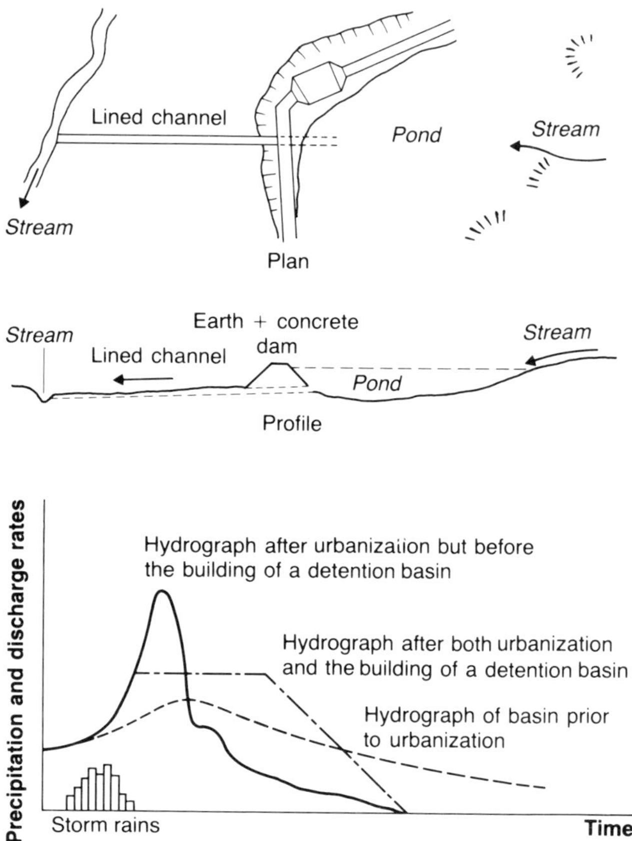

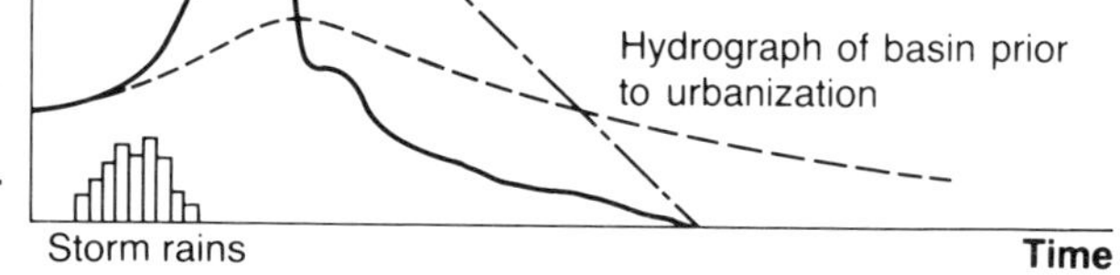

As far as sedimentation is concerned, urbanization generally leads to an initial huge increase in yield as newly disturbed surfaces are washed by run-off. As the surfaces settle down and are often made impervious, yield falls off dramatically so that later on in the urbanization process less sediment is carried in the stream than was the case in the original rural basin. This lack of sediment, together with the high peak discharges, is probably the reason for channel area changes, since the surplus energy of the stream can be used to scour the banks, unless they are protected (Fig. 1.10).

Urbanization brings with it an increased use of water. Often this comes from storage reservoirs in distant places. In this way, more water enters the catchment area and may add to the threat of flooding. Other possible changes are shown in Table 1.3 (opposite).

The traditional ways of dealing with a channel that is subject to flooding is to widen and deepen it, usually by dredging, and sometimes by shoring up the banks with metal or concrete piles. This sort of action is often taken in urbanized catchments. In the USA, use has been made recently of the **detention basin**. This is a wide, shallow basin hollowed out in the course of a stream. It is fronted by an earth embankment and forms a floodable area since no building is allowed in it. It may, however, form a useful resource, for when it is not flooded it can be used for sports pitches and in the winter it can be deliberately flooded to form a skating rink. One such basin in Chicago covers 11.5 acres. It is 21 feet deep at the centre and provides 7 million cubic feet of floodwater storage from a four-square mile residential area. It cost $1 million to construct but, apart from safeguarding against floods, it also provides valuable open space and recreational facility in the urban area. Some aspects of the detention basin are shown in Fig. 1.11. This is only one method of protecting against flooding. A number of other possible measures are shown in Table 1.4. Usually the flood plain itself is avoided for building

Table 1.3 Some changes in the catchment area due to urbanization.

Stage of urbanization	Process	Result
1. Early stage-separate 'pioneer' houses	Well construction. Construction of septic tanks and sanitary drains	Some local lowering of water table. Some local increase in soil moisture and water table. Perhaps some pollution
2. Transition from early urban to middle urban stage	Bulldozing of land for estate building. Some topsoil removal. Farm ponds filled in	Accelerated erosion and sedimentation in channels. Increased peak discharges. Decreases of lag time. Smallest streams eliminated
	Estate building, paving of streets and building of culverts	Decreased infiltration leading to greater peak discharges and lowering of water table. Occasional flooding in remaining channels where they are constricted (e.g. bridges, culverts). Occasional flooding or bank erosion of artificial channels on small streams
	Discontinued use and abandonment of shallow wells	Local rise in water table
	Diversion of nearby streams for public water supply	Decrease in run-off in affected area
	Disposal of untreated or poorly treated sewage into streams or ground-water	Pollution of streams and wildlife. Decrease in quality of water for domestic use and recreation downstream
3. Transition from middle to late urban stage	Urbanization completed: more houses plus public, commercial and industrial buildings	Creation of process of pollution and loss of amenity value of water
	Abandonment of remaining shallow wells through pollution	Rise in water table
	Increased importation of water from distant sources to cope with population	Increase in flow of local streams
	Creation of artificial channels and culverts. Construction of sewerage system	Changes in channel shape through erosion and silting. Removal of water hitherto used as recharge for aquifer
	Improvement of storm-drainage system.	Increased peak flow and decreased lag time
	Drilling of deeper, larger industrial wells	Decreased pressure in aquifer. Perhaps local subsidence. Possible inflow of salt water into aquifer if near coast
	Increased use of water in central heating and air-conditioning	Possible overloading of sewerage system. Possible local recharge to water table in the event of leakage
	Drilling of recharge wells	Raising of water pressure surface
	Wastewater reclamation and utilization	Recharge to ground-water aquifers. More efficient use of water resources

Table 1.4 Possible responses to flooding

1 ADJUSTMENT
- a) Do nothing
- b) Action on floodplain
- c) Emergency action
- d) Floodproofing of buildings
- e) Land Use Regulation
- f) Financial – Insurance

2 ABATEMENT
- g) Action in catchment
- h) Afforestation
- i) Other vegetational changes
- j) Changing agricultural practices
- k) Control in urban areas
- l) Management of snow-melt areas

3 PROTECTION
- m) Action along channel, e.g. walls and embankments, detention basins, channel improvements, diversion schemes, reservoirs, barrages and flood barriers

purposes. Trees may be felled, for they allow sediment to build up around their roots, thus altering the cross section of the channel. This sort of treatment, the traditional approach to flooding problems, can easily be improved upon from an ecological point of view, but often it might be found that remedial action is also more economic in the long term. Two possible approaches to planning the same channel are shown in Fig. 1.12.

Another important kind of development in catchment areas is the creation of reservoirs. There are many reservoirs throughout the world developed for industrial or urban purposes, but as well as being a vital part of the economic life of an area they sometimes pose a threat, however remote, to the very people they were built to serve. For instance, before 1940, many reservoirs were built in narrow river valleys which cut across the San Andreas Fault above Los Angeles with their dams actually on the fault line. In Switzerland, some of the highest hydro-electric power dams in Europe stand at points which were beneath glaciers at the time of the 'Little Ice Age' (about 1600 to 1850). Unfortunately, several dams in several areas have already given way due to natural disasters, such as landslides and earthquakes, or for human induced reasons such as ineffective monitoring.

On 19 July 1985, a dam in the Dolomite Mountains in Italy burst. It released a torrent of mud nearly 20 m high and 200 m across which surged down the narrow valley of the River Stava, destroying the village of Stava itself. It rushed on, brushing aside forest trees as if they did not exist and narrowly missing the busy little town of Tesero. After a journey of some 13 km the slurry came to a halt, but not before it had killed over 200 people (Fig. 1.13).

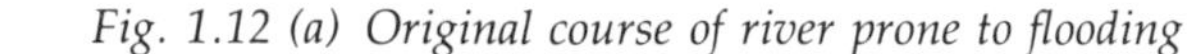

Fig. 1.12 (a) Original course of river prone to flooding

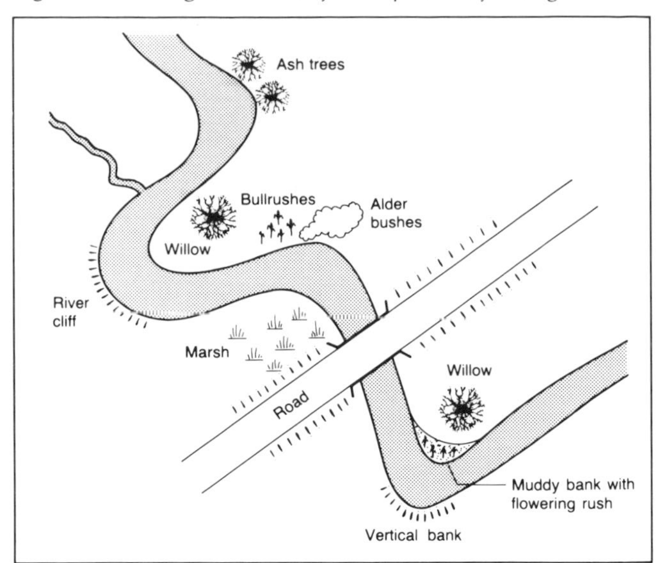

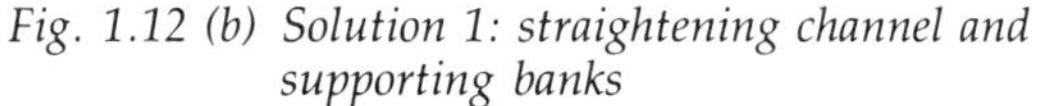

Fig. 1.12 (b) Solution 1: straightening channel and supporting banks

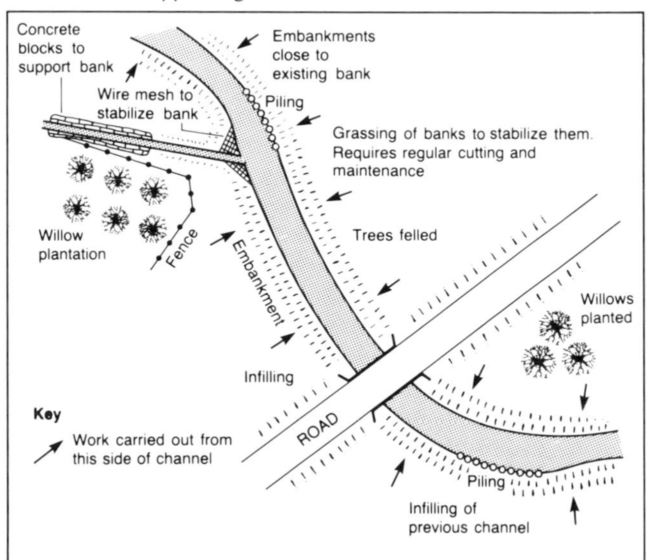

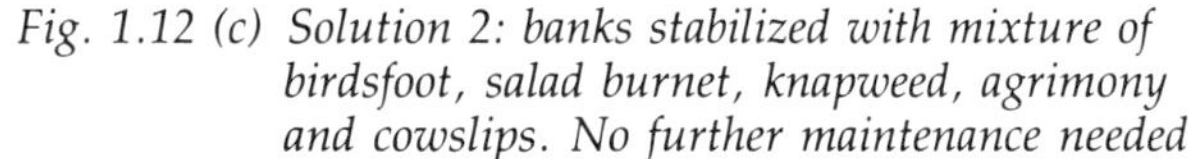

Fig. 1.12 (c) Solution 2: banks stabilized with mixture of birdsfoot, salad burnet, knapweed, agrimony and cowslips. No further maintenance needed

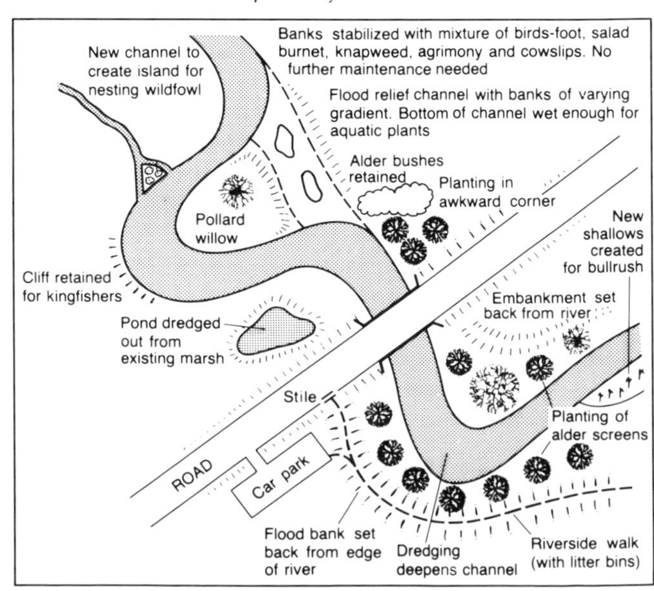

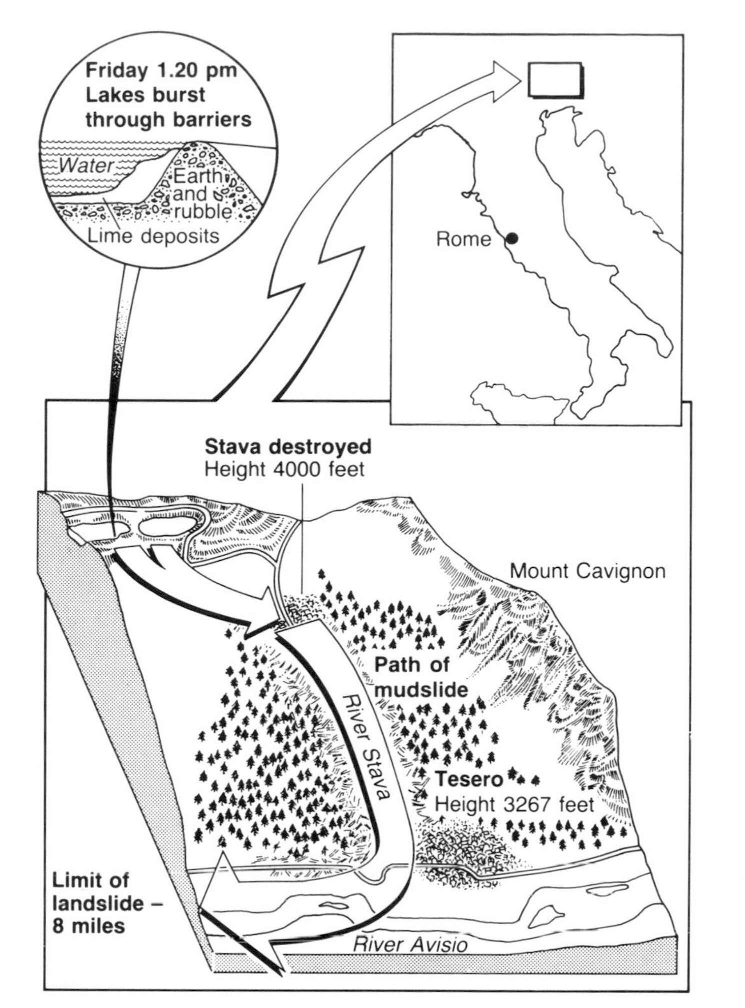

Fig. 1.13 The collapse of the Stava tailings dam

Table 1.5 Ownership and age of some UK reservoirs

Owner	No. of reservoirs	No. over 50 years old (where known)
Anglian W.A.	9	–
British Steel	(Too small for inspection)	
C.E.G.B	3	–
Newcastle & Gateshead Water Co.	11	–
North West W.A.	225	200
Northumbrian W.A.	16	–
Severn Trent W.A.	48	Several
Southern W.A.	14	–
Thames W.A.	60	–
Welsh W.A.	95	–
British Waterways Board	90	–
Yorkshire W.A.	146	123
National Trust	100	–
Scottish Hydro-Electric Board	80	8

The damage was very localized, in the floor of the valley, but since the main lines of communication also ran along the valley floor, it took several days to reach all the affected areas and complete the search for the dead and injured. The principal farming lands had also been on the alluvium of the valley floor, so they were virtually totally destroyed by the flood. Water supplies were badly affected by the damage done in the villages, which together with the problem of dead bodies not being buried led to fears of the outbreak of disease, although this thankfully came to nought.

The dam had been built of earth and rubble as part of a mining development. Fluorite mining had led to the build-up of residual deposits such as lime within the lake impounded by the dam. The weight of these deposits pressed down on the dam which was not capable of withstanding the pressure. Fifty-five million gallons of sludge roared off down the valley when the failure occurred. The twenty-year-old dam had not been upgraded as mining developments had occurred and the lack of regular inspections of the dam failed to bring the danger signs to the notice of the civil authorities. This was only a small dam in a relatively lightly populated area of the world, but great loss of life ensued.

There have unfortunately been far worse instances of dam collapse than this one. In the USA in 1928, the St Francis Dam, a 60 m high concrete construction, suddenly failed leading to 400 deaths and the equivalent of £4m of damage. In France, the Fréjus dam, some 66.5 m high, failed in 1959 releasing 25 million m^3 of water which swept down the valley of the River Reyran killing 421 and partly destroying the town of Fréjus. In Great Britain there are many dams over 50 years old, but the last major collapse was of the Dolgarrog dam in North Wales, which gave way in November 1925, killing sixteen people.

This fine record, compared with many other countries, is probably because the standards set down, partly as a result of the Dolgarrog collapse, have been strict and well adhered to. The 1930 Reservoirs (Safety Provisions) Act has been subsequently up-dated and modified by another Act in 1975, which should be fully enforced by 1988. This requires the County, Metropolitan District and Scottish Regional Councils to act as overseers and enforcement authorities and for there to be a supervising engineer appointed for every reservoir. However, the British Act only covers water impounding dams. It would not cover tailings dams such as the one at Stava.

New developments in monitoring techniques include the use of **piezometers**, which measure pressures and can be placed in the face of dams to give adequate warning. Floods can be planned for, but it is impossible to cope with the sudden and devastating effect that a dam burst can bring.

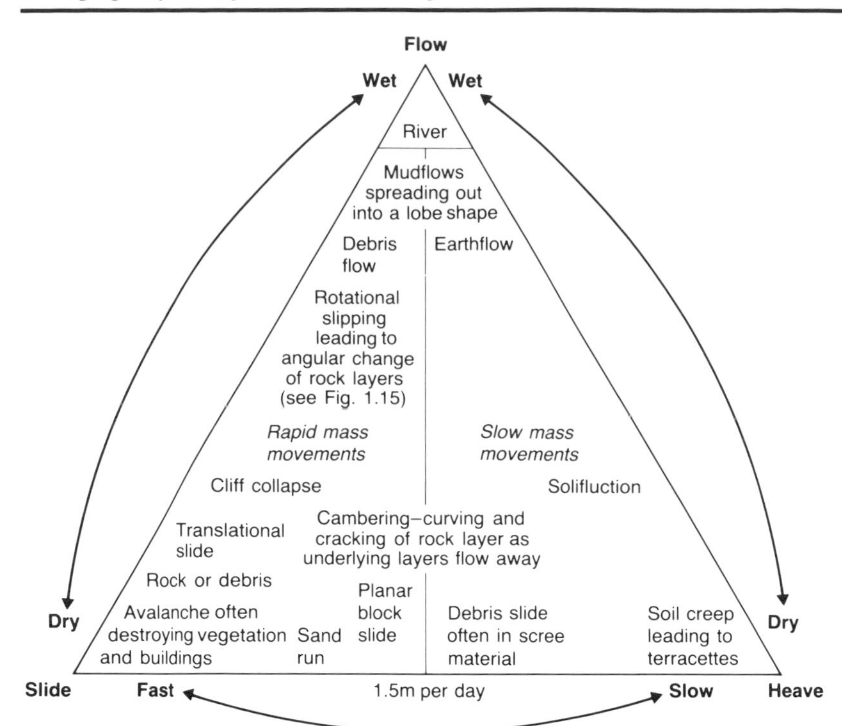

Fig. 1.14 A classification of mass movement

Fig. 1.15 Rotational slipping at Criccieth

Fig. 1.16 Terracettes

Surface Processes on the Slopes of the Drainage Basin

When it rains, the majority of the water transferred down the slopes of the drainage basin is transferred by seepage through soil, the process known as throughflow. A combination of water acting as a lubricant to soil particles and the force of gravity leads to the downslope movement of these particles. Such activity is known as **mass movement**.

Various attempts have been made to classify mass movement processes, but the majority of them depend on the speed of the movement and the amount of water involved. One possible classification is shown in Fig. 1.14. The sketch of a most unstable (and unlikely) slope shown in Fig. 1.17 illustrates the result of many of the processes of mass movement.

> **8** Using the classification of mass movement phenomena and the sketch of the slope, suggest what processes have been responsible for each of the features lettered A to G.

Building in areas which are subjected to rapid mass movement processes can be both hazardous and costly. In 1963, a major landslide sent 240 million m^3 of rock into the Vaiont lake impounded behind a major concrete dam. The dam did not break, but a huge wave over 70 m high swept over the dam devastating the valley and killing over 3000 people. The worst damage in the Guatemalan earthquake of 1978 was experienced in the shanty towns which were built in the most unfavourable locations, especially those where steep slopes added to the speed of the mass movement which followed the earthquake. A less spectacular but no less costly

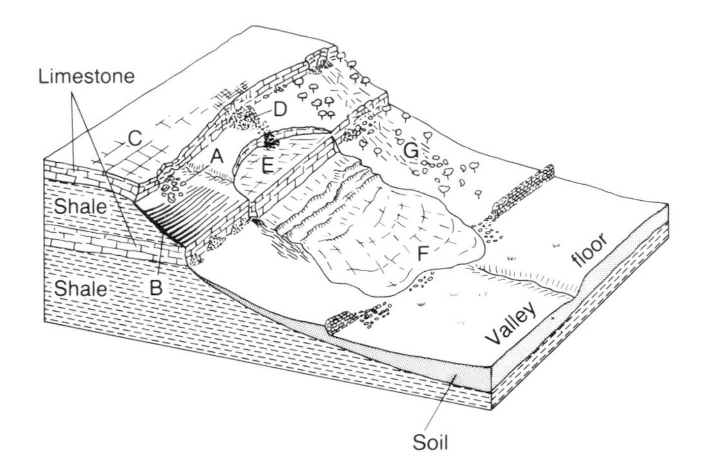

Fig. 1.17 Mass movement phenomena on a hillside in a humid temperate area

example of the effects of mass movement comes from the urban areas of Wellington, New Zealand. The slopes on the outskirts of Wellington are largely developed on greywacke, a fine grained sedimentary rock. Sometimes this is well weathered and sometimes it is overlain with material left at the time the area was recovering from glaciation, material known as periglacial deposits. These are again fine grained. In urban areas, there are also a good many slopes which have been created by the construction process. All of these materials may be subject to slipping (Fig. 1.18).

In a survey of the slips which occurred over a period in Wellington, R. J. Eyres, M. J. Crozier and R. H. Wheeler found that certain factors were important in the development and extent of landslips. One of these factors was, as has been suggested above, water. The amount of precipitation into the storm immediately preceding the failure was not however, the critical factor as is revealed in the figures in Table 1.6.

Fig. 1.18 A slope in Wellington, New Zealand

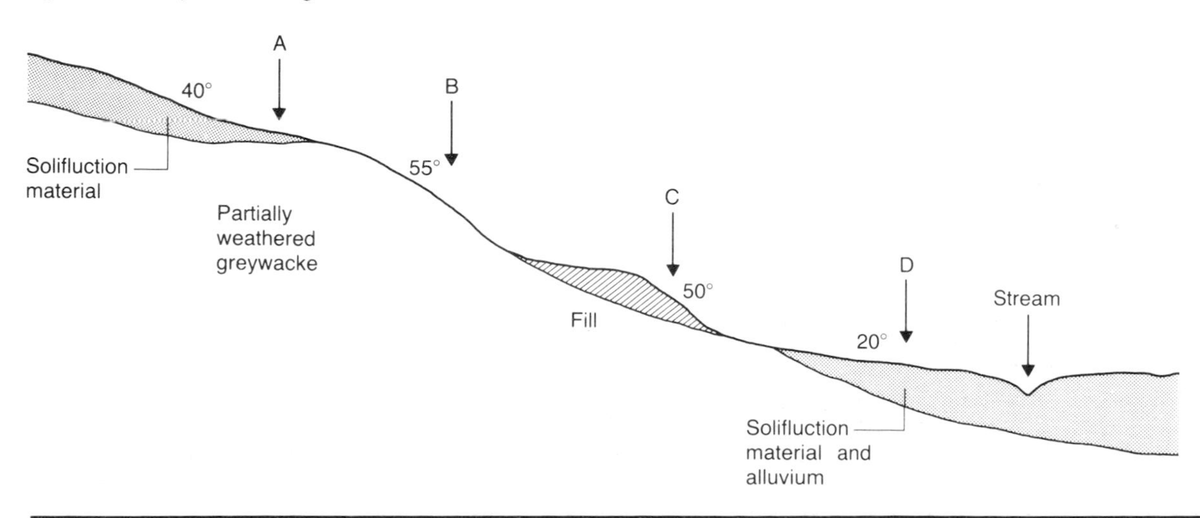

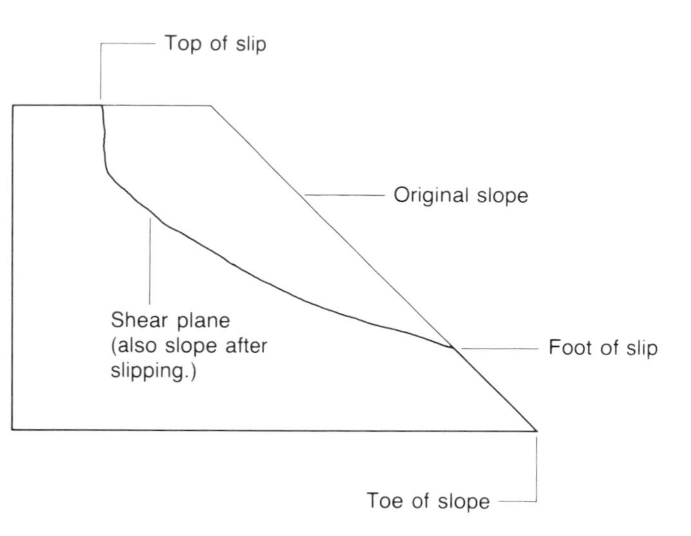

Fig. 1.19 Morphological features of a landslip

Table 1.6 Rainfall preceding landslips

Date of storm	Max Rainfall mm 24 hours	Max Rainfall mm 72 hours	Rain in preceding 4 months (mm)	Slips No.	Slips Volume (m^3)
7–10 Oct 1974	61	109	656	46	161
15–17 July 1976	60	115	413	9	24

The amount of rainfall in the period leading up to the storm in which most slope failures occurred is seen to be important. The greater the wetting of the material comprising the slope, the more likely it was to fail in subsequent storms.

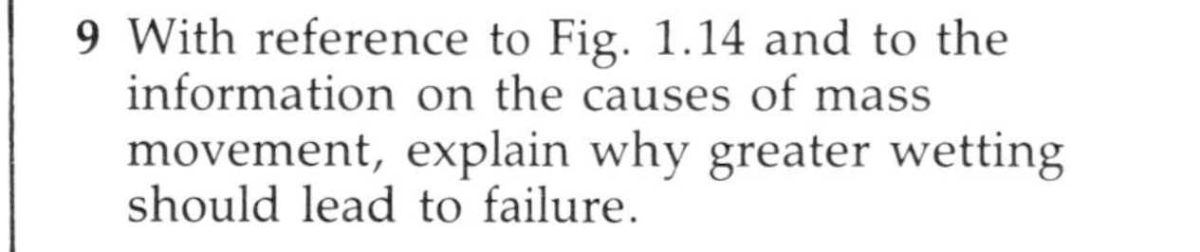

9 With reference to Fig. 1.14 and to the information on the causes of mass movement, explain why greater wetting should lead to failure.

The morphological features of a landslip are shown in Fig. 1.19. A great deal of information on landslips was gathered by the three workers, since they were keen to be able to warn developers which slopes were most subject to failure and under what conditions. A summary of this information is given in Table 1.7.

Table 1.7 Characteristics of slips and adjacent undisturbed surfaces, Wellington, New Zealand

Landslide number	Material	Volume of slip in m^3	Weathering grade	Median angle of shear plane	Median angle of original slope	Vegetation cover %	Vegetation type	Aspect °
1	G	51.8	4	53	58	100	T	71
2	G	10.0	4	59	50	25	Sh	27
3	G	27.4	3.5	34	67	20	Sh	46
4	G	18.7	3.5	53	75	100	Gr	208
5	G	9.2	3.5	68	87	–	Sh	111
6	G	13.5	3	57	60	100	T	136
7	S	8.9		54	67	100	T	31
8	S	8.5		59	65	5	Sh	7
9	G	65.1	3	46	66	0	–	279
10	S	5.1		49	61	100	T	93
11	S	18.8		59	60	80	T	131
12	S	7.5		62	57	100	Gr	180
13	S	80.0		47	43	95	Sh	345
14	F	5.5		34	36	0	–	78
15	F	16.0		43	43	0	–	76
16	S	31.2		35	57	100	Gr	26
17	G	54.7	4	64	54	60	Sh	266
18	S	19.3		43	62	20	Sh	79
19	S	45.9		38	68	80	Sh	47
20	G	18.6	4	50	71	70	Sh	256
21	G	6.4	4	60	69	0	–	8
22	G	15.6	5	50	64	80	T	48
23	S	6.7		44	64	30	Sh	11
24	S	122.4		31	52	60	Sh	7
25	S	0.75		44	71	90	Sh	67
26	S	15.4		47	55	95	Sh	236
27	G	26.3	3	56	72	80	Sh	337
28	S	17.6		49	60	70	Sh	3
29	G	3.7	2	54	69	0	–	277
30	S	95.5		27	22	100	Gr	360

Key

Material

G = greywacke
S = solifluction deposits
F = fill

Weathering grade

1 = only surface weathering
5 = deep, almost total weathering

Vegetation type

T = trees (greater than 2.5 m high)
Sh = shrubs – no trees
G = grass and/or herbaceous plants; no woody species.

Decision-making Exercise A: Buildings and Mass Movement

You are to write a pamphlet for builders in Wellington telling them what sort of slopes to avoid building on and explaining why. In order to produce your report you should complete the following exercises, before putting the information they reveal to good use in your pamphlet.

Illustrate your pamphlet wherever possible.

1. Produce a polar graph plotting median angle of slope failure against aspect (the direction the slope faces). (See Techniques, p. 51.)

Your plotting of the failed slopes in Wellington could be done in three colours, one for each of the parent materials, to see whether there is any significance in the type of underlying material as well as aspect or median angle.

Do you find any preferred aspect for slope failures in Wellington? Does a particular sort of material tend to fail more in certain aspects?

The people who provided the data on the Wellington slopes state that differential rates of weathering seem to be a factor in the development of landslips. The faster that drying takes place, the greater is the range of expansion and contraction of bedrock. Where expansion and contraction are greatest, keying or interlocking of the joint blocks becomes reduced and the slope eventually fails. Does this help to explain your results? (Remember which hemisphere Wellington is in.)

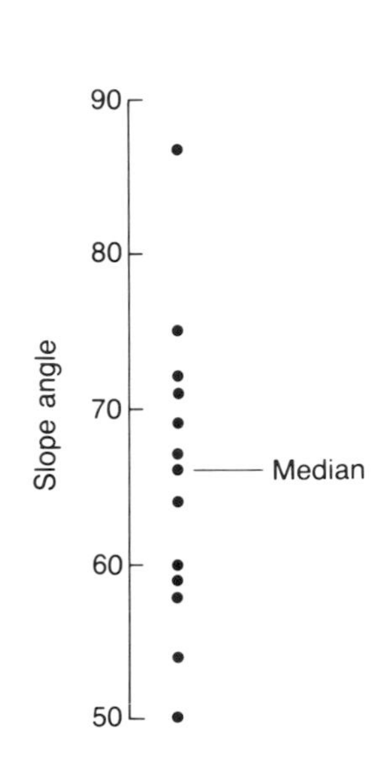

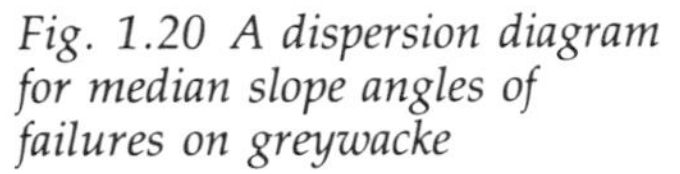
Fig. 1.20 A dispersion diagram for median slope angles of failures on greywacke

2. Vegetation has been shown in other areas to have a bearing on the susceptibility of a slope to mass movement. If a slope is stripped of its vegetation cover, soil creep and surface slope wash may become exaggerated. If large trees grow on the slope, their weight may decrease the stability of the weathered material unless they are anchored through to the bedrock. On the other hand, a lot of vegetation may help to dry out the slope, thus rendering it less liable to mass movement. Devise methods to test the following statements, which may or may not be true:

(a) Slips are less frequent on well vegetated slopes

(b) Slopes on unconsolidated material (fill, well-weathered greywacke and solifluction material) may fail at lower angles than those on more compact bedrock

(c) This is especially the case if large trees are growing on the unconsolidated material

3 Plot a dispersion diagram of median angles of slips for solifluction material to go with the one already drawn for greywacke (Fig. 1.20). What is the minimum slope angle that has failed on each parent material? Does the likelihood of failure increase as median angle increases? From the limited information available, how do slope failures in construction fill compare with those in solifluction material or greywacke?

The lubrication of the surface material has been shown to be very important for the failure of slopes. If houses are built with inadequate drainage systems, the water simply runs off roofs, down drainpipes and into the soil. If surface vegetation is altered or removed, the amount of water in the surface material may be increased dramatically. Your report should include reference to the construction and maintenance of drainage systems and the care and conservation of vegetation cover.

Hazards and Their Management

In 1985 there was a major landslip on the island of Mindanao in the Philippines. It killed a large but unknown number of people who were engaged in mining for gold through very unstable weathered material overlying the steep slopes of an extinct volcano. Before the year was out, people were mining through the slip material. Obviously the perception of the hazard by these people was not as strong as their need to earn money in an island which offers only hard won gains to the peasants. Could such occurrences be predicted and planned for?

Hazards are usually regarded as events which greatly exceed our expectations in terms of frequency or magnitude. To cope with them we can adapt or adjust. **Adaptation** entails the arrangement of land use to take account of extreme events while adjustment includes the measures which could be taken to cope with the events. **Adjustment** includes attempts to modify natural hazards, attempts to alter people's vulnerability to them and attempts to share or distribute losses. Some examples relating to flood events are quoted earlier in this chapter. Table 1.8 shows the degree of accommodation we have made to a number of hazards. It shows that mass movement is so far only really dealt with by modifying land use on slopes susceptible to such processes. It is not comprehensive however since it does not include major problems such as pollution. Sometimes there is a clear link between natural processes and human-induced problems. The Mindanao example is of this type, as would be landslips associated with quarry or mine waste.

To give a clearer picture of the problems associated with certain natural events, it is possible to draw a **hazard event profile** so that we can estimate the problems associated with the event (Fig. 1.21). This only refers to the natural aspects of the event, but it is possible to use a similar technique to describe the human impact as well.

10 Draw a grid similar to that in Fig. 1.21 and draw a 'hazard event profile' for a volcanic eruption.

11 Refer to the information on the dam burst at Stava and complete a human impact profile to compare with the one for the Colombian volcanic eruption (Nevado del Ruiz, 1985) already drawn for you (Fig. 1.22).

12 How would such diagrams help in planning for hazards in a drainage basin?

Frequency: Frequent
Rare
Duration: Long
Short
Areal Extent: Widespread
Limited
Speed of onset: Slow
Fast
Spatial dispersion: Diffuse
Concentrated
Temporal spacing: Regular
Random
Drought
Blizzard
Earthquake

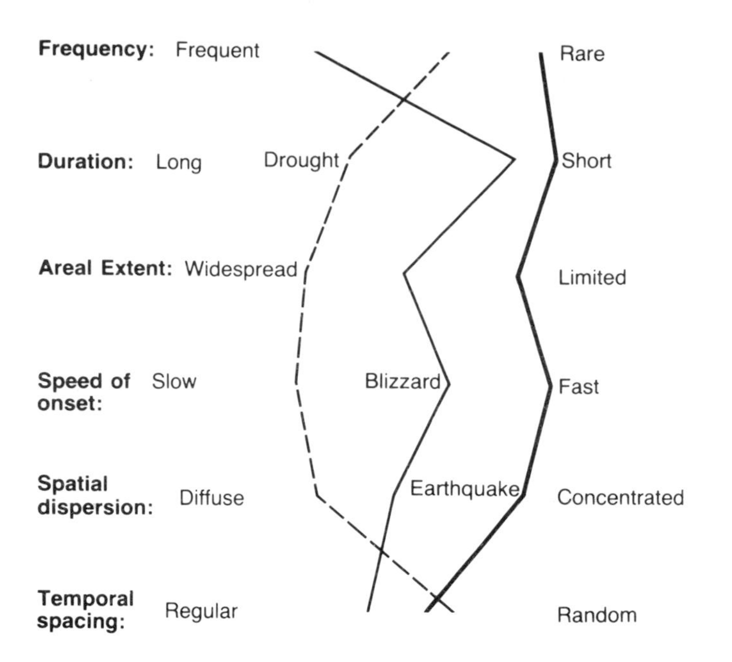

Fig. 1.21 Hazard event profiles for characteristic drought, blizzard, earthquake

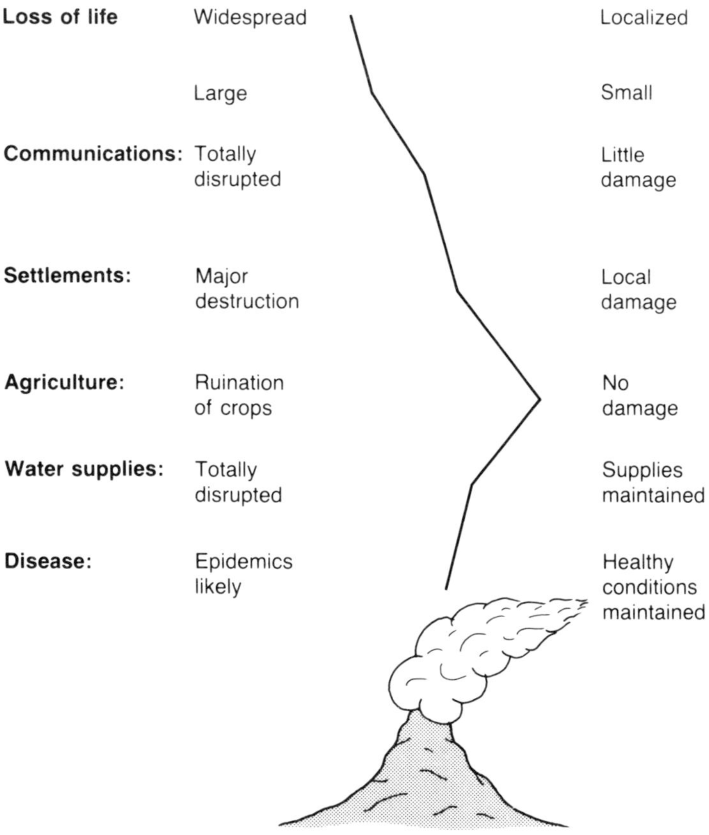

Loss of life
Widespread
Localized
Large
Small
Communications:
Totally disrupted
Little damage
Settlements:
Major destruction
Local damage
Agriculture:
Ruination of crops
No damage
Water supplies:
Totally disrupted
Supplies maintained
Disease:
Epidemics likely
Healthy conditions maintained

Fig. 1.22 A human impact profile for a natural hazard

Table 1.8 Types of adjustment to natural hazards

Types of hazards	Types of adjustment: Modify event	Modify vulnerability	Distribute losses
Avalanche	Artificial release	Snow shields	Emergency relief
Coastal erosion	Beach nourishment	Beach groynes	Flood insurance
Drought	Cloud seeding	Cropping pattern	Crop insurance
Earthquake	Earthquake reduction (theoretical)	Earthquake-resistant buildings	Emergency relief
Flood	Upstream water control	Flood-proofing	SBA loans
Frost	Orchard heating	Warning network	Crop insurance
Hail	Cloud seeding	Plant selection	Hail insurance
Hurricane	Cloud seeding	Land use pattern	Emergency relief
Landslide	–	Land use regulation	–
Lightning	Cloud seeding	Lightning conductors	Homeowners insurance
Tornado	–	Warning network	Emergency relief
Tsunami	–	Warning network	Emergency relief
Urban Snow	–	Snow removal preparations	Taxation for snow removal
Volcano	–	Land use regulations	Emergency relief
Windstorms	–	Mobile home design	Property insurance

(Source: White and Haas 1975, 86)

Decision-making Exercise B: The Management of an Urbanized Catchment

You are given a number of pieces of data on the East Brandywine Creek, a drainage basin about 50 km west of Philadelphia in the USA. The creek empties into the Delaware River estuary. It is on the fringe of the Philadelphia metropolitan area and is under imminent threat of urbanization. New settlements have been built since 1965 together with associated industrial parks and new roads. The locations of these are shown on maps (Figs. 1.25 and 1.26).

You are to produce a report from the Chester County Water Resources Authority stating the changes that are likely to occur in the channel of the creek in the section marked PQ on Fig 1.26 and how you suggest these changes are coped with. Your report should be illustrated wherever possible.

N.B. Since the data are derived from American sources, all figures are in imperial units, which are used in American scientific practice.

B

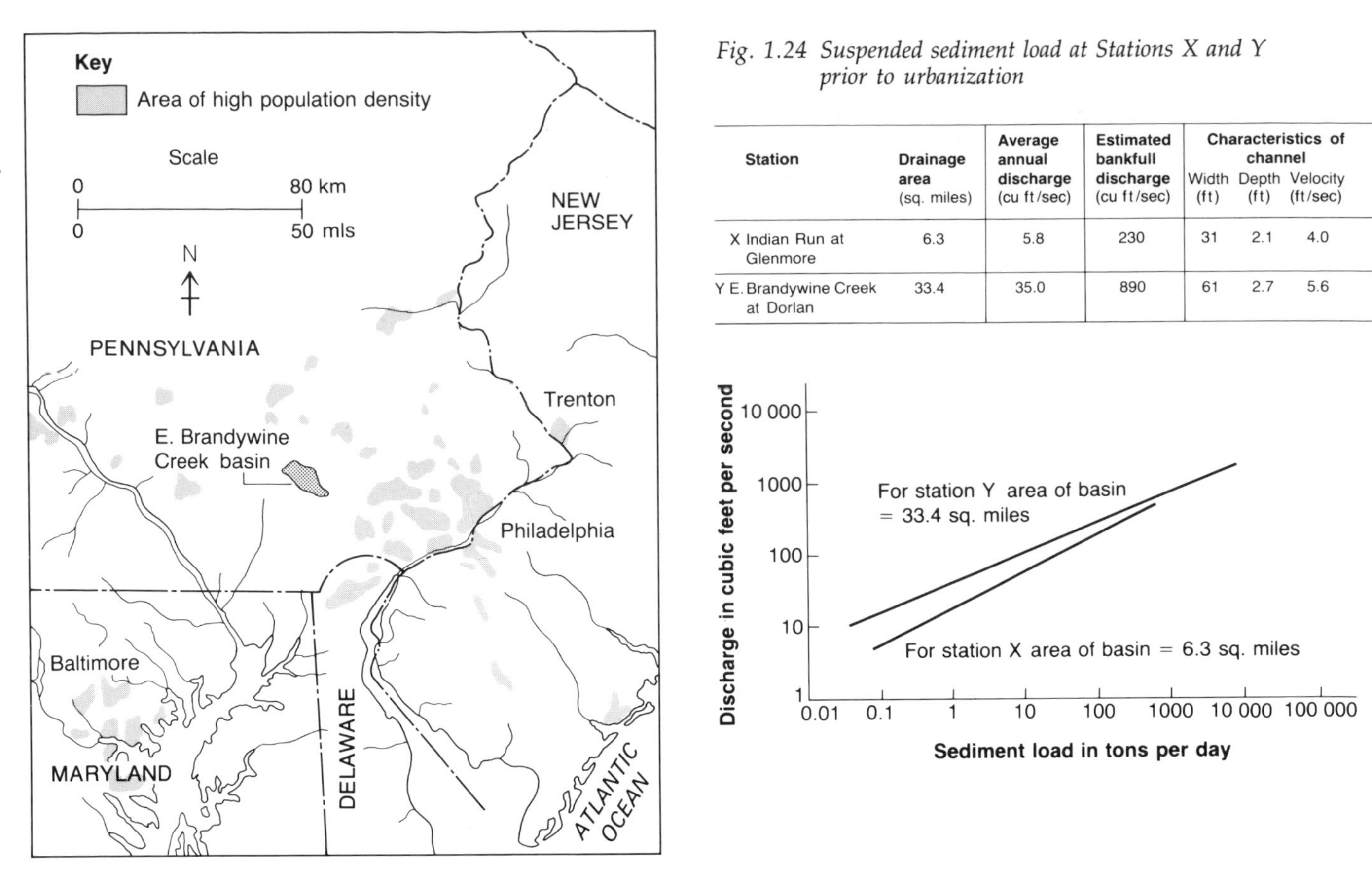

Fig. 1.24 Suspended sediment load at Stations X and Y prior to urbanization

Station	Drainage area (sq. miles)	Average annual discharge (cu ft/sec)	Estimated bankfull discharge (cu ft/sec)	Characteristics of channel		
				Width (ft)	Depth (ft)	Velocity (ft/sec)
X Indian Run at Glenmore	6.3	5.8	230	31	2.1	4.0
Y E. Brandywine Creek at Dorlan	33.4	35.0	890	61	2.7	5.6

Fig. 1.23 Location of East Brandywine Creek

Fig. 1.25 Location of new developments in the East Brandywine Creek basin

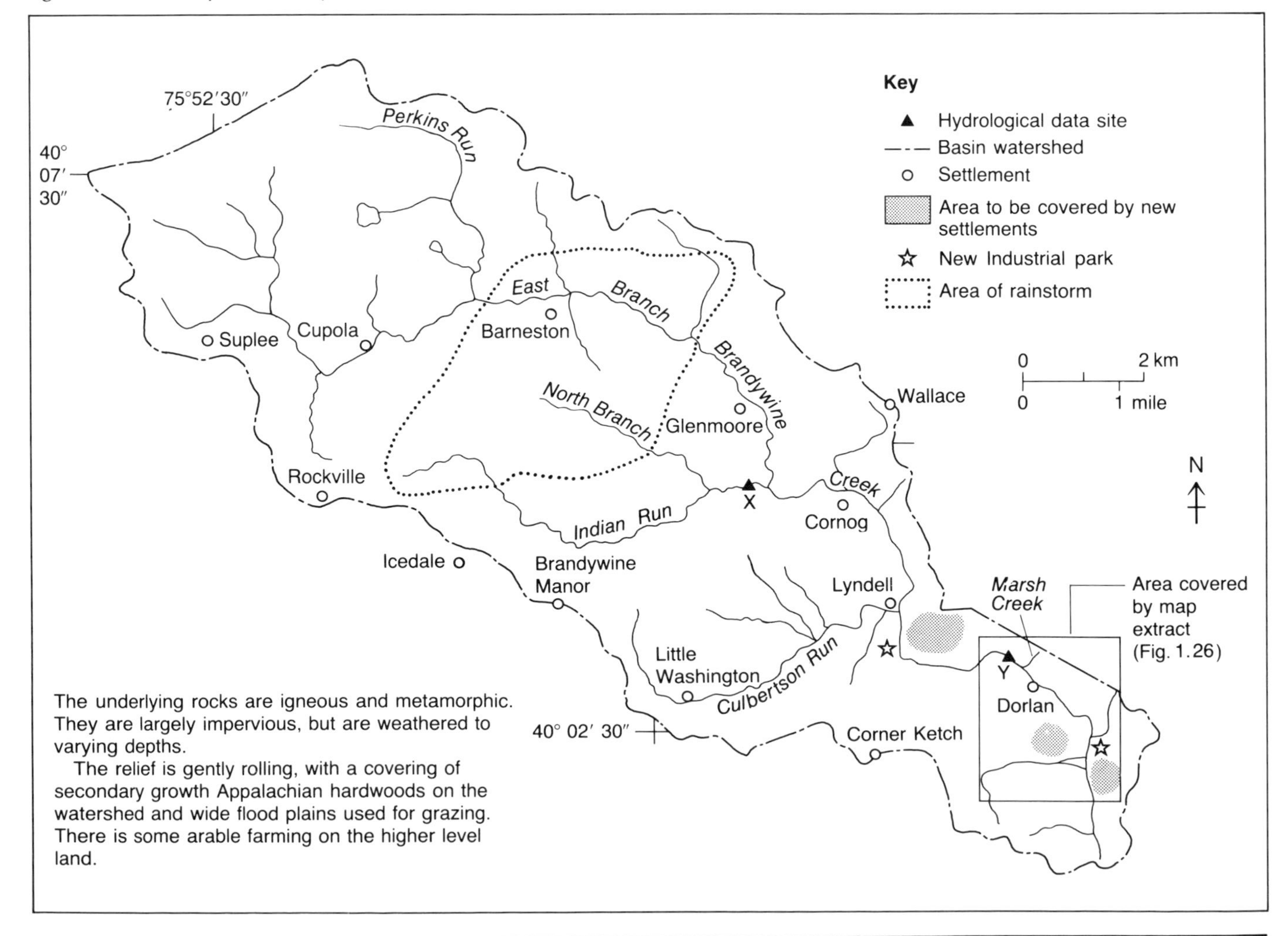

The underlying rocks are igneous and metamorphic. They are largely impervious, but are weathered to varying depths.

The relief is gently rolling, with a covering of secondary growth Appalachian hardwoods on the watershed and wide flood plains used for grazing. There is some arable farming on the higher level land.

B

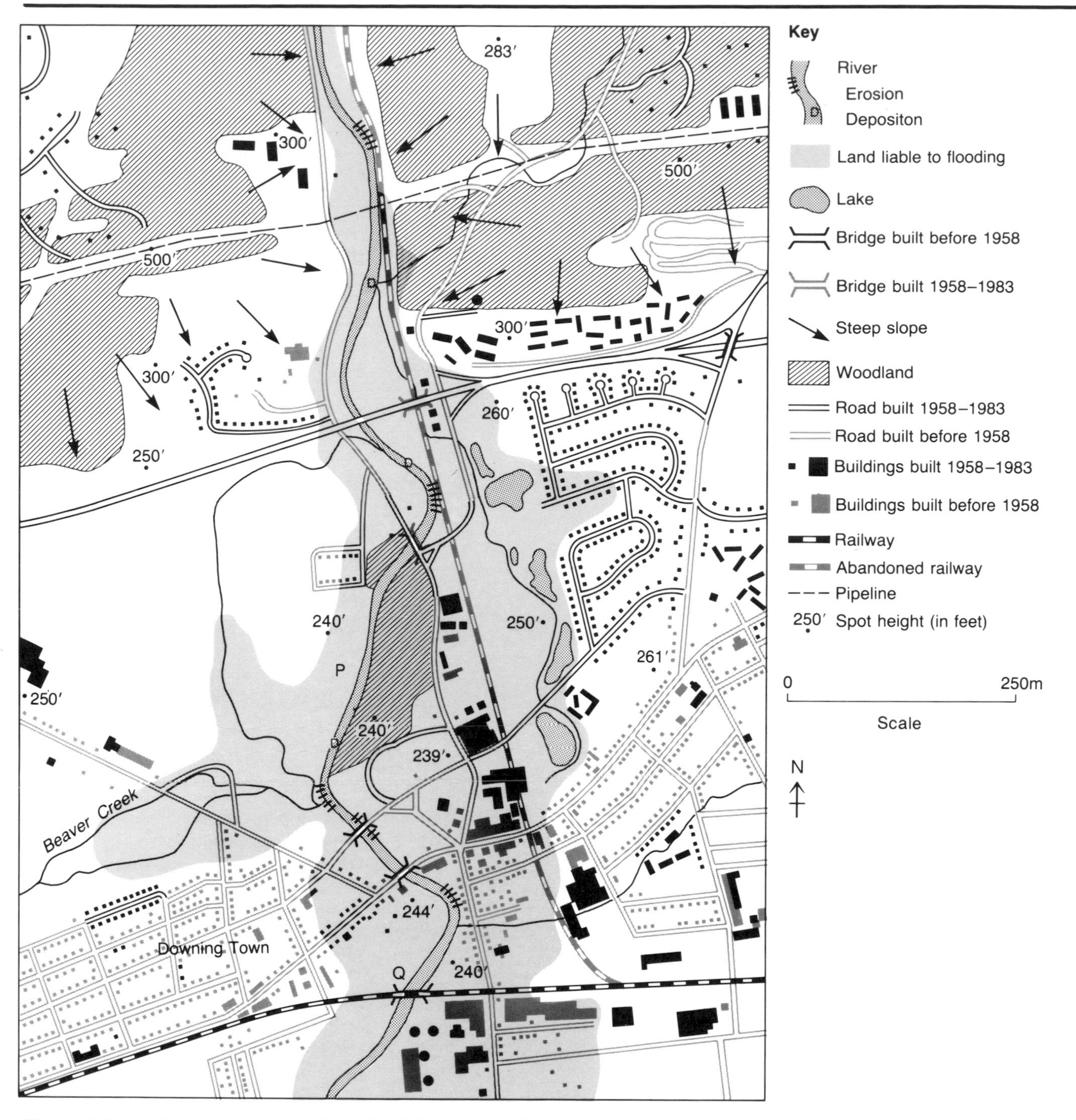

Fig. 1.26 Map of a recently urbanized stretch of the East Brandywine Creek basin

Fig. 1.27 Generalized hydrographs for Stations X and Y to show the response of the Indian Run and East Brandywine Creek to a heavy storm lasting for fifteen minutes (in the area demarcated with a stippled line on Fig. 1.26)

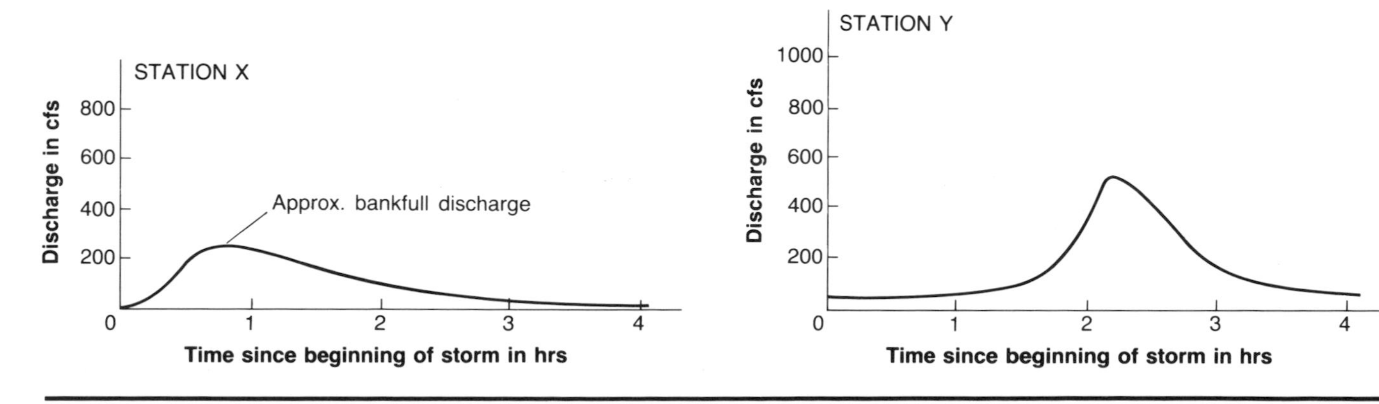

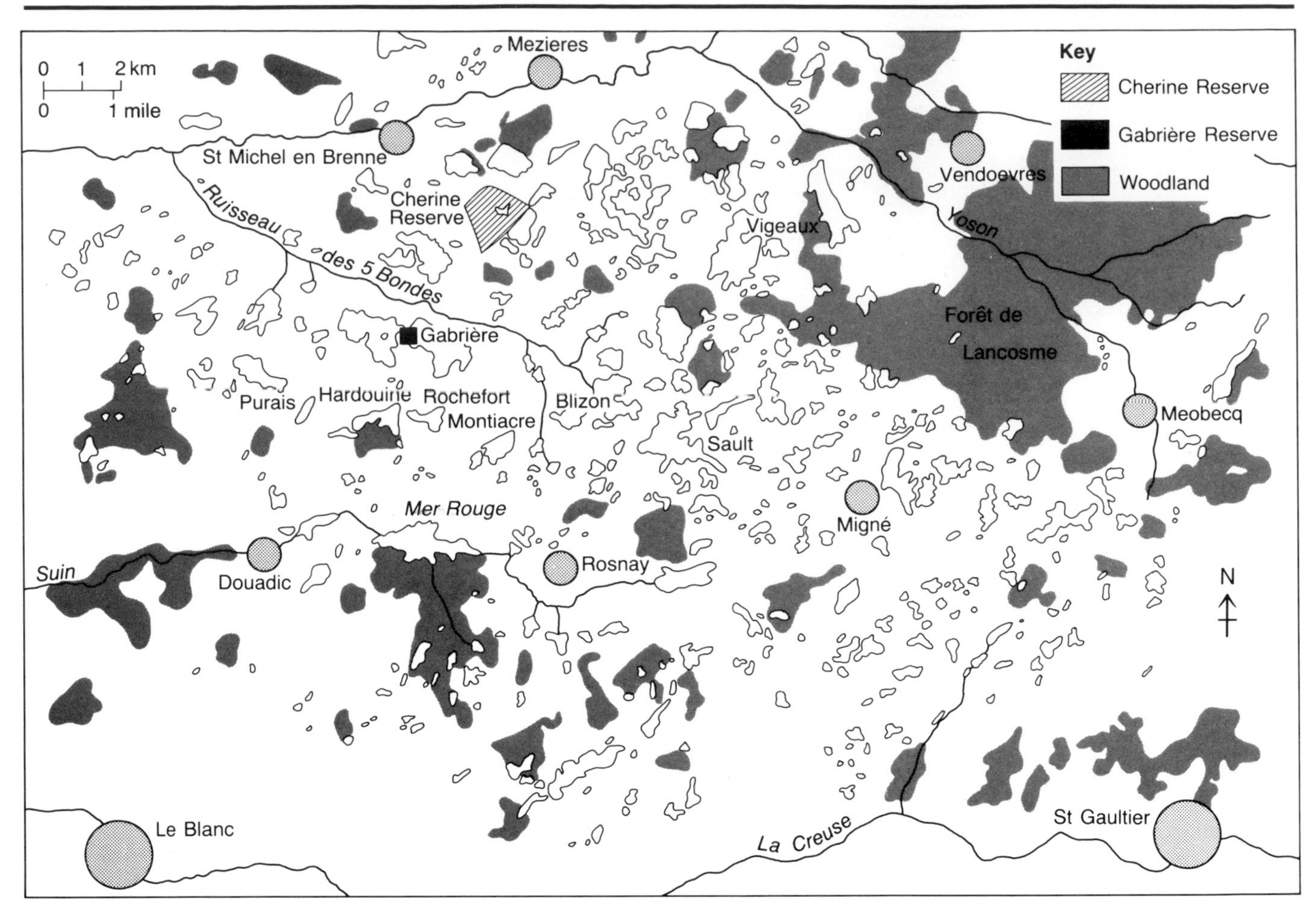

Fig. 1.28 The Brenne

Case-study: water management in the Brenne

The Brenne is an area of forests and man-made lakes covering about 540 km^2 between the towns of St Gaultier, Le Blanc and Mezières in the département of Indre, France (Fig. 1.28). It exhibits several interesting features of water management. The first is that the lakes were actually created in the first place to overcome a dietary problem experienced by the monastic communities in the area, and they remain mostly as fish farms.

In the eleventh century, the monks, established in large numbers at three local centres, needed to produce more from their harvests if they were to sustain their numbers, let alone allow them to increase. They were not permitted to eat meat for about 140 days in the year.

A campaign of deforestation began to enlarge their holdings and produce greater food supplies. The scheme was quickly found to be wanting, however, because the poor impermeable land led to winter quagmires and summer deserts. Harvests were poor. A supplement was needed. It had not escaped the monks' notice that water naturally built up in low-lying areas in the region and the monastic community had long been expert at fish rearing. Fish did not come into the category of the banned 'meat'. The monks of the Brenne, 200 km from the nearest sea, had to grow cereals and vegetables or raise freshwater fish to hold their vows.

The legacy of the monks is that there are now in the region of 840 lakes of varying size in the Brenne. All are still used for fish rearing and it is not unusual to see live carp on fishmonger's slabs in the region. Many of the fish are sold in eastern Europe. The method of using the lake basins is first to close a sluice gate in the shallow dam built across the valley. The basins are seldom deeper than three metres at the sluice, which is the deepest part. It only takes a matter of three or four days for a small lake to fill up, but the Gabrière may take up to two months. The basin may then be 'seeded' with carp fry from a hatchery near Le Blanc, or adult fish of several species may be returned as breeders. A mix of ages is returned to make the best use of the different food resources available. Three years later, the carp are substantial enough to be 'harvested'. Harvesting methods, which have not changed in centuries, consist of draining the lake until there is just one pool above the sluice gates. The fish are then netted from this pool (known as 'la pêche', Fig. 1.30). Tench, pike, roach, carp and others are then loaded on to lorries and taken alive to marketing centres, many of them in Germany and Poland. The lake is then filled up as quickly as possible, although, every seventh or eighth year, most lakes are left dry for repairing, cutting of vegetation, digging out the 'pêche' and to kill off any water-borne diseases, before the cycle is repeated. The annual production from the Brenne as a whole is some 900 tonnes, of which about 560 tonnes consists of carp.

The lakes are very attractive to wildlife, a fact which is further enhanced by the cycle of emptying and re-filling. There is a tremendous variety of duck, birds of prey, waders and other water birds in the area,

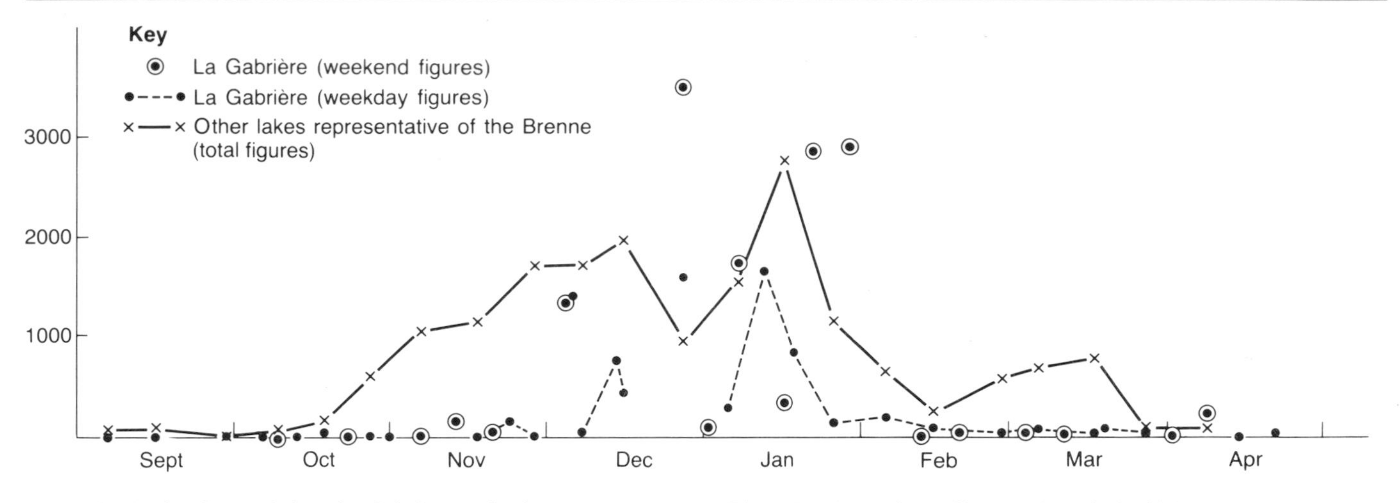

Fig. 1.29 Pochard recorded at the Gabrière and other Brenne lakes, 1983–1984

Fig. 1.30 An ecologically sound method of lake construction in the Brenne

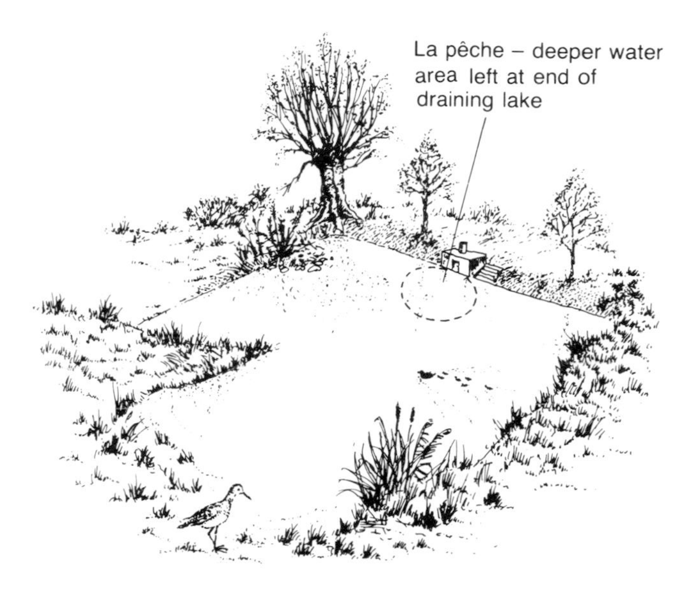

Fig. 1.31 An easy to manage, but sterile method of lake construction in the Brenne

each type exploiting a different niche in the rich **ecosystem**. Between the lakes often lie reed beds, woodland or, in some cases farmland, so the habitats are varied. Roe and red deer live in the woodlands, along with wild boar, and such a rich assemblage of fauna makes this an ideal area for the French to indulge in one of their favourite pastimes, 'la chasse' or hunting. The open season is for a four-week period, mainly in August, then from mid-September to the end of January. Nearly all of the lakes and the surrounding lands are privately owned and hunting rights are jealously guarded, but at any one time during the open season there are hunters using the area. There were 225 million shotgun cartridges sold in France in 1984!

It is against this background that a unique arrangement has been made for the Gabrière. This lake has been set aside as a reserve and a small information centre has been established to encourage people, especially from nearby towns, to take more interest in the natural resources of the area. It was formed by some local conservationists in collaboration with the Fédération Départementale des Chasseurs de l'Indre (FDCI), the hunters' association. Half the funding is from the FDCI and half from both local and national conservation bodies. In 1985, a substantial grant was given in addition by the World Wildlife Fund to support this very important step in the protection of European wetlands.

The enormity of this task is shown in Fig. 1.29, which shows the number of pochard, a migratory duck, both in the reserve and on all the other lakes in the Brenne. It appears that there are two peaks in the numbers. One is in December to January. It can also be noted from the graph that there is an extraordinary concentration of the species on the Gabrière at weekends. This is of course when the major incursion of hunters occurs. Then the numbers on the thirteen other lakes in the area are usually not as great as on the one reserve lake. In March, however, when a second peak in numbers is noted, numbers on the Gabrière are relatively low. This has been explained by the fact that the first wave, with its extraordinary numbers, has depleted the food resources for this particular species which therefore has to look elsewhere.

Another problem with water use in the Brenne is that it is now seeing a new type of lake being constructed. People who move into the area to build a weekend house or a holiday home are constructing private ornamental ponds which in view of the problems already mentioned are of dubious ecological value. The new lakes are not constructed as part of a chain and are not regularly emptied.

Many have ecologically less valuable trees, such as Weeping Willows, planted near them, and most have regular shapes which are of less value to wildfowl and wading birds. Other species of animal such as the European pond tortoise and rare dragonflies are also discouraged by this arrangement (Fig. 1.31). These new lakes are creating a fringe around the Brenne which has a superficially pleasing aspect but which, on closer examination, does not add to the environmental quality of the region.

The Brenne presents an interesting challenge. Should the individual landowners be allowed to develop their units of land and lakes without reference to the requirements of others and even without reference to total stocks of wildlife, which is most certainly an international resource? Would an overall management programme be more applicable to the Brenne than the present piecemeal development, or would it stifle a system that has proved its worth since the eleventh century?

The Brenne is an important economic resource. It provides agricultural land, commercial and sport fishing and tourist income for the growing population. It also provides hunting grounds for people from all over France.

13 What are the main conflicts in land use in the Brenne area?

14 The conflicts have been overcome to a degree by establishing the Gabrière reserve. Do you think this is an approach worth pursuing, or would you suggest an alternative management scheme? Explain your answer.

It is proposed to offer shares in a shooting syndicate at Sault (see Fig. 1.28). These shares will be offered to a 'new' sector of the market – industrial workers from Paris. Advertisements will be placed in company magazines offering the chance to shoot at Sault for six weekends throughout the year. Accommodation will be arranged in the village and guns and training will be provided as part of the package.

15 Write a paragraph for a company magazine in Paris outlining the benefits of the offer to the new hunters.

16 What problems could an influx of new hunters have in the area of Sault? You should refer to the map of the area (Fig. 1.28).

17 What arguments would *you* put forward for opposing such a scheme? Consider the international significance of the site as well as any reasons for local opposition.

Fig 1.32 The Brenne

Decision-making Exercise C: Planning a Water Park

C

You are provided with data on an area of recent gravel and sand workings on the outskirts of Reading, Berkshire. You are required to predict the demand for certain suggested uses from the population data provided. Subsequently, you should draw a plan of the park to show how you would manage the area using the base map (Fig. 1.34). This plan should include:

(i) Landscaping ideas, especially to segregate the park from the motorway section of the M329
(ii) Zonation of water uses, ensuring that conflicting uses are not placed too close to one another
(iii) Accommodation for clubs and development of an education and administrative centre for the park
(iv) A sound conservation scheme which protects certain types of habitat from recreational uses but still allows peripheral access.

The plan should include a map of the suggested use and a detailed explanation.

1. The history of gravel extraction on the site
Before 1965 the land in the area under consideration was farmland. Between 1965 and 1979 gravel was extracted from the site under three different contracts.

(a) In 1960, a company obtained permission to excavate sand and gravel from an adjoining site. They had built a gravel-washing and ready-mix concrete plant and by 1965 wanted to extend the workings. They were granted permission on condition that they restored the land to a level above the fifty-year flood level of the River Loddon. In 1969, work stopped in the area marked W in Fig. 1.35 since Berkshire County Council purchased the land for the construction of the A329(M) and the Woodley relief road. Another company then took over the gravel rights, excavating gravel for the construction of the two new roads. This excavation finished in 1973.

(b) The company which lost the mineral rights because of the building of the motorway was granted permission to excavate the area marked X to replace the land it had lost. A conveyor belt took material from this site to the plant. Working ceased in 1978. The land was restored, but several heaps of soil remained in the newly created lake.

(c) In 1970, application was made to Berkshire County Council to excavate material for the M4 and subsidiary roads. Despite the fact that the council's policy at the time was not to open new gravel pits, the application was seen to be in the national interest and since work would be over in two years and the potential recreational value of the site was seen to be very high, permission was granted on condition that landscaping was carried out by the contractors. The area marked Y was therefore excavated. Clayey material was found, however, and this was placed in the area marked Z, building it up above the original contour level. A layer of topsoil was placed on top of the clay. Finally, a weir was constructed at point P and the area was flooded.

2. Planning policy
Berkshire County Council's 1970 plan for the area had the aim of allowing 'the maximum reasonable extraction of gravel consistent with leaving the land to shapes and contours which can then play their part in the evolution of a valley park'.

Agreements were made on the final depth and extent of the workings and the location of dumped material so that the site could more easily be landscaped. The contractors were not to be allowed to fell any trees except with permission.

Wokingham District Council took responsibility for the area in 1974. General guidelines for land use were published. They were:

(a) that land use should meet the recreational demands of the public at large within the district

(b) that specialist interests should be catered for providing that public access was not prejudiced
(c) that a wide range of recreational facilities be provided (although these should be seen within the context of facilities in Wokingham District Council as a whole)
(d) that the visual amenities of the district should be improved
(e) that flora and fauna should be conserved
(f) that an element of flexibility should be built into the planning to allow for changes in recreational demand.

All these should be catered for in the light of the physical constraints imposed by the geography of the area.

It was only in 1978 that Wokingham District Council completed the purchase of the 568 hectares site and began development. Before that there had been five main land owners.

3. Local population structure

Population statistics for a huge new housing development, Lower Earley, which is immediately to the south-west of the area known as Dinton Pastures, were revealed by a survey in 1985 (Table 1.9). Predictions were made from these figures so that age groupings could be forecast until 2005, fifteen years after the assumed completion date of the housing development. Fig. 1.36 shows the population figures for the wider area surrounding Dinton Pastures in the form of a population pyramid for central Berkshire.

4. Damage to banks from waves

Certain gravel banks of lakes are susceptible to undercutting by waves. To overcome this problem, one of several stabilization methods could be used which have been tried elsewhere. Concrete-filled sand-bags have been used, but found to be unsuccessful as waves can wash out material between them and behind them. The laying of logs by the water's edge is more successful. The logs are pegged down and strengthened by the planting of willow and reeds. Other banks have been graded and the slopes protected with grass, while the base has been guarded by reeds. In yet others, old car tyres have been chained down as protection. Concrete or steel piling is also possible, but expensive. The places where banks are susceptible to erosion are marked on the base map.

Fig. 1.34 Base map of Dinton Pastures country park

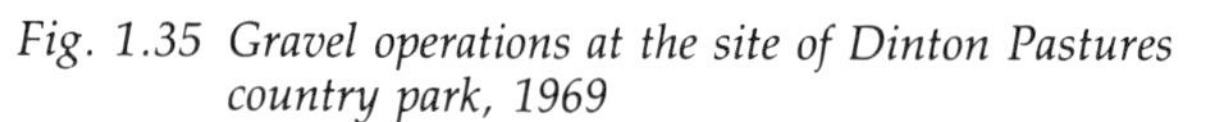
Fig. 1.35 Gravel operations at the site of Dinton Pastures country park, 1969

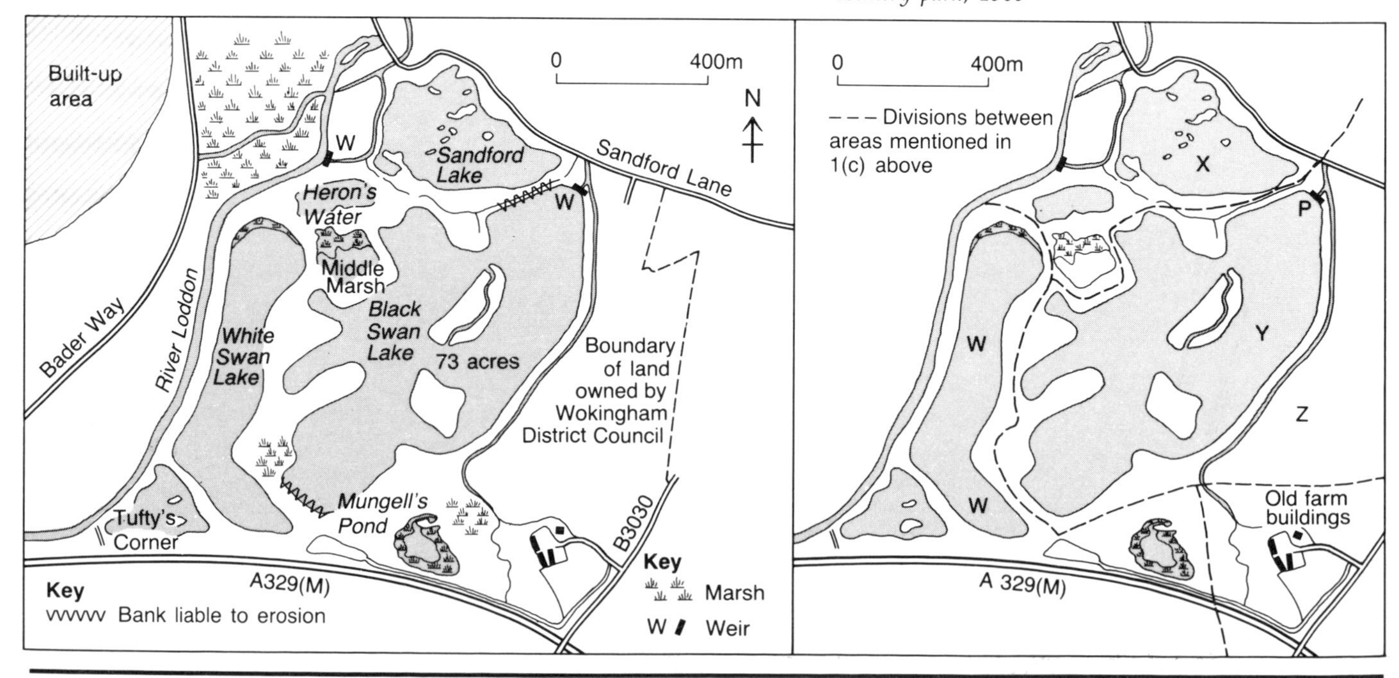

Table 1.9 Population figures for Lower Earley, 1985–2005

Age	Survey 1985	1990	Forecasts 1995	2000	2005
0–4	1 224	1 629	1 368	1 159	1 038
5–10	958	1 832	1 787	1 572	1 394
11–15	605	1 035	1 312	1 311	1 204
16–24	1 772	1 878	1 665	1 758	1 822
25–34	3 381	4 213	3 098	2 415	2 005
35–44	1 867	3 366	3 482	3 251	2 879
45–54	761	1 595	2 332	2 712	2 707
55–64	350	745	1 056	1 400	1 854
65–74	156	360	485	590	736
75+	71	154	205	275	332
Total	11 145	16 806	16 709	16 443	15 971

(Source: Berkshire County Council, *People in Lower Earley*, June 1985)

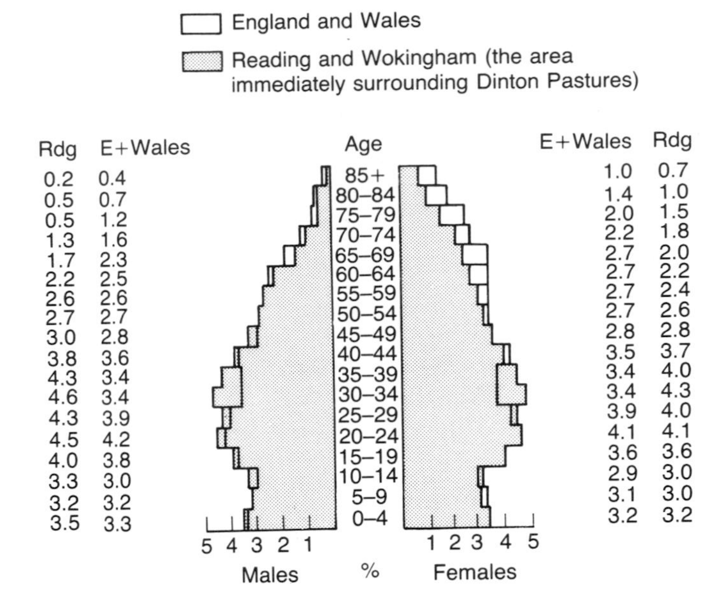

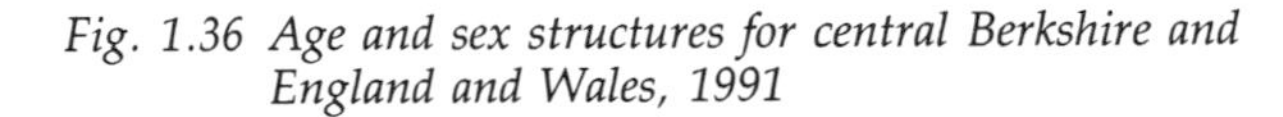

Fig. 1.36 Age and sex structures for central Berkshire and England and Wales, 1991

5. Compatibility of water uses

Table 1.10 Complementary and non-complementary water uses

	Fishing	Yachting	Wildfowl Attraction	Flora Conserv-ation	Power Boating	Sailboarding	Water Skiing	Canoeing
Fishing		x	√	√	x	x	x	x
Yachting	x		√	√	x	√	x	√
Wildfowl Attraction	√	√		√	x	√	x	√
Flora Conservation	√	√	√		x	√	x	√
Power Boating	x	x	x	x		x	√	x
Sailboarding	x	√	√	√	x		x	√
Water Skiing	x	x	x	x	√	x		x
Canoeing	x	√	√	√	x	√	x	

x = Non-complementary √ = Complementary

N.B. Boating is only practical on lakes over 125 hectares in area.

6. Suggested uses of the park

(a) *Footpaths*. These should be as dry as possible, not concentrating traffic in any particular area so as to minimize erosion problems. People should be able to start and complete walks at one point, preferably allowing access by car but avoiding 'backwards and forwards' routes wherever possible.

(b) *Boating activities*. These should be varied, but not conflicting. It would be useful to have a water sport centre somewhere in the park, allowing easy access from nearby roads and to the water's edge.

(c) *Fishing*. This should be kept away from conflicting uses, especially those aimed at wildfowl nesting areas and boating. Access to different types of fishing (river and lake) would be useful. Could one lake be used for a fish farm?

(d) *Habitat conservation*. As many different types of habitat should be conserved or created as possible, consistent with other uses planned.

(e) *Birdwatching*. Facilities for birdwatchers, such as hides, could be located wherever the habitat looks promising and where other uses do not make it pointless.

(f) *Picknicking*. Picnic tables should be provided at sufficiently close locations to car parks, but separated from them by trees, hills or some other feature.

In addition, the following could be provided or carried out:

(g) *An education centre/café*. This requires a large building with access and parking facilities.

(h) *Drainage*. Marshy areas in the route of suggested footpaths could be drained or filled with hardcore.

(i) *Bank stabilization*. The improvement of banks would help to stop erosion.

(j) *Bridges*. These could improve access to the park and access within it.

(k) *Fencing*. To separate conflicting land uses or to direct foot traffic, fences could be erected.

(l) *Screening*. Earth mounds could be built to separate different land uses or to screen noisy uses from more tranquil ones.

(m) *Treeplanting*. The following information shows which tree species would be useful from the ecological point of view. Those which would adapt easily to the water and soil conditions are marked with an asterisk.

(n) *Land-based sports facilities*. Suggested uses include a nine-hole golf course, a BMX track or a dry ski slope.

Fig 1.37a Views of Dinton Pastures

C

Fig 1.37(b) Dinton Pastures: Potential tree species to plant

Ranked value according to number of associated insect species	Ranked value according to number of bird species eating fruits
Oak 1	Rowan* 1
Willow* 2	Birch } 2
Birch 3	Pine } 2
Hawthorn* 4	Spruce 4
Poplar* 5	Oak 5
Apple 6	Hawthorn* 6
Pine 7	Beech } 7
Alder* 8	Yew } 7
Elm var. 9	Apple 9
Hazel* 10	Field Maple* } 10
Beech 11	Sycamore } 10
Ash* 12	Alder* 12
Spruce 13	Lime 13
Lime 14	Holly* } 14
Hornbeam } 15	Larch } 14
Rowan* } 15	Elm var. } 16
Field Maple* 17	Hornbeam } 16
Larch 18	Walnut } 16
Sycamore* 19	Hazel* 19
Holly* 20	Ash* 20

Case-study: the effects of large water management schemes

The decade 1975–1985 saw the completion of many major water control schemes as world industry demanded more energy. Schemes such as the Rio Paute hydro-electric power scheme in Ecuador, the Drakensberg pumped storage scheme in South Africa and the Vakhsk River scheme in the USSR have been completed to provide not only power but also water storage for irrigation purposes and domestic consumption. However, with the construction of these major schemes, there has been a small but growing body of opinion that they are not all that they should be. Do they take sufficiently into account the damage to the ecosystem likely to occur? Do they really encourage development in an area or are the huge schemes so much 'pie in the sky'? Are they the most sensible way to develop the resources of an area? In a major study called *The Social and Environmental Effects of Large Dams*, Edward Goldsmith said that 'one billion people will starve to death this century, and huge projects such as big dams will be one of the main causes'. Given that there will be 113 dams over 150 m in height by 1990, there is obviously a lot of momentum behind the 'big scheme' planners. Who is right?

In northern Quebec, a French-speaking province of Canada, there is a major water control scheme under construction. It centres on an inlet of the Hudson Bay known as James Bay. The rivers which flow into it drain an area larger than the total area of the British Isles and do so through rugged and desolate country peopled by 8500 or so Cree Indians and which is home to various species of wild animal. In 1971 the James Bay Project was initiated by premier Robert Bourassa. He hoped to see industrial expansion to the tune of 100 000 new jobs, the project acting as the impetus for a whole range of new industrial development.

The first thing that the Cree Indians knew of the plan was when one of their number, who had been educated in the south, read a report of it in the newspaper. The Cree lived in harmony with the taiga ecosystem, harvesting its animals at a rate dictated by the relative paucity of the environment and the provision of guns, ammunition and traps by the trading companies. The Inuit, or Eskimoes, also had a similar harmonious life style. Any development the size of that now reported to them would surely alter the balance of the ecosystem and call into question their way of life. Outraged, the Cree began to organize themselves. They found allies in a number of people dedicated to the protection of native rights in North America and the Defence Committee for James Bay was set up. They began to predict earthquakes, triggered off by the huge weight of impounded water and even a change in the climate, saying that the heat soaked up by the water would adversely affect both temperature and rainfall patterns. Perhaps as a panic reaction to such dire warnings, the developers set off almost at indecent haste, cutting a track at a rate of 1 km per day into the forest. This was even before the plan for the scheme had been finalised and its ecological impact had been assessed. When the report of that assessment was made in 1972 – based on evidence drawn from what little data was available from other areas, but from no field observations in the James Bay area – it concluded that the project would result in a functioning system probably just as satisfactory as the original natural system. It also recommended that this scheme should be used as a huge ecological laboratory for any future schemes in the sub-arctic regions, which was really an admission that very little was known about the ecological impact of such

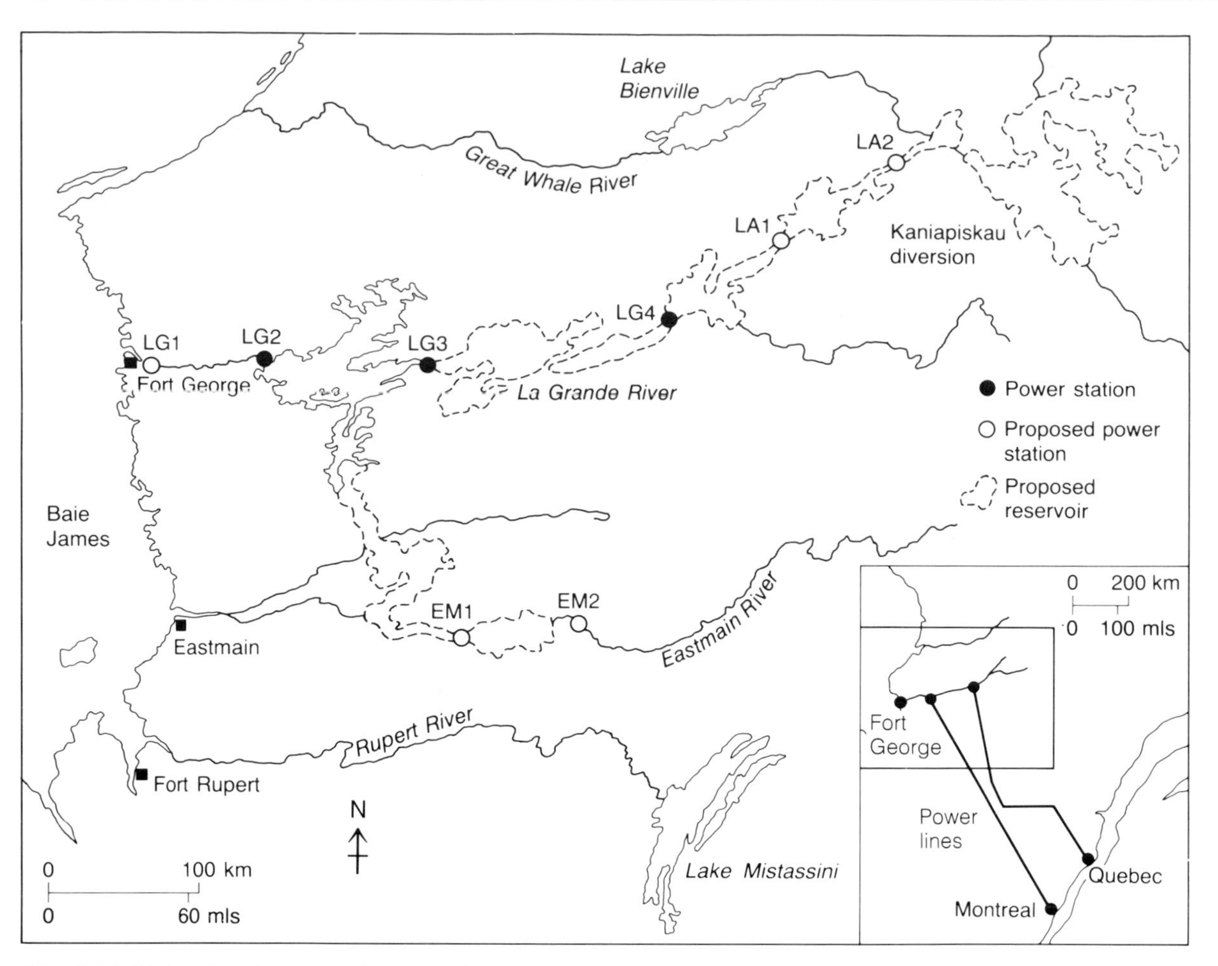

Fig. 1.38 Hydro-electric power schemes in the James Bay area

schemes. Its other conclusion was that the natives would suffer alterations to their lifestyle of potentially alarming proportions and significance. The natives hired some biologists who started to look for evidence of possible ecological changes to be brought by flooding and the creation of new channels. Their evidence was not encouraging. They said that the most important species to the Cree, the beaver, would be displaced and even drowned. Fish spawning places would be lost. Caribou migration routes would be flooded. In all, the food sources and source of furs would be decimated. Hearing this, a judge at the Quebec Superior Court found in favour of the Cree and suspended construction until the rights of the Indians to the land had been recognized and some settlement made. Construction in fact continued and on appeal the judgement was overturned. By 1975, however, the issues raised in the courtroom had led to an agreement between the developers and the Indians. The Cree gave up their claims to the land and the Province granted them native land reserves, exclusive trapping, fishing and hunting rights for those who spent more than six months of the year in the forest, and important rights to voice their opinion on future developments. These included a say in accepting or rejecting further developments based on the impact the present ones were likely to have on the environment and a say in the way that money set aside to carry out remedial work would be spent. They also got compensation of 135 million Canadian dollars. The Inuit also received 90 million Canadian dollars. The Cree were

Fig. 1.39 The ecosystem of the James Bay area before development

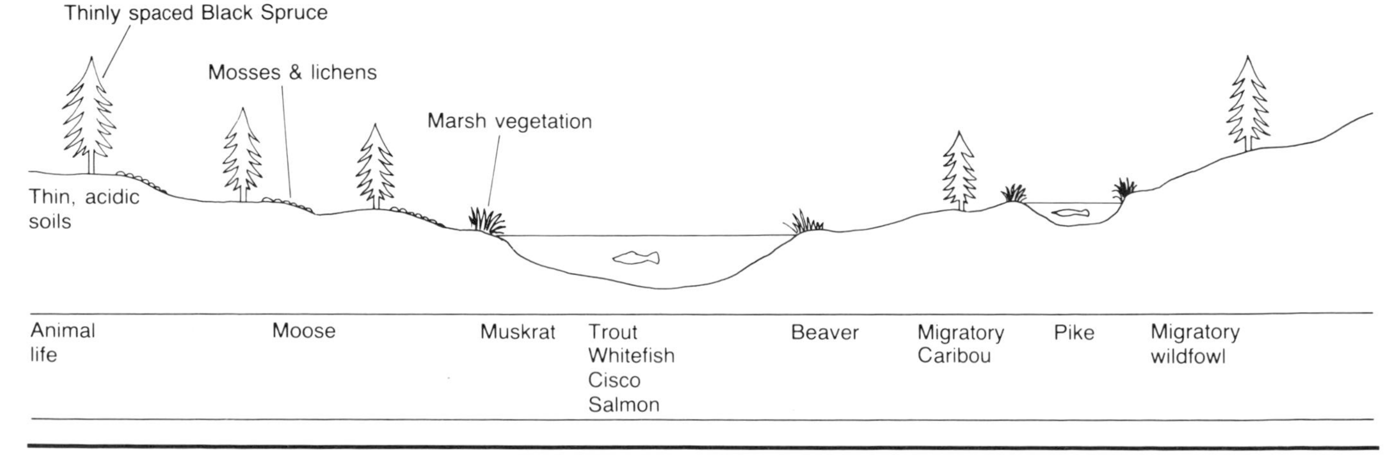

not popular amongst other Indian groups for signing away their rights to the land, but they said that if past developments were anything to go by, the James Bay Project would go ahead anyway. It was probably better for them to arrange to have a voice in its development than to stand by as it was built around them.

The present stage of development is shown in the map (Fig. 1.38). The three generating stations together are capable of producing 10 300 MW of electricity. To put this in perspective, it is about one sixth of the total generating capacity of the USA.

After ten years of construction, the first phase of the project has been completed. It is projected that the waters not only of La Grande Riviere, but also of the Great Whale River, the Eastmain and Rupert Rivers will be harnessed. In this way the generating capacity will rise to over 25 000 MW. So far, the expected industrial boom to make use of this electricity has not materialized and Quebec now has an estimated surplus of 5000 MW of electricity. Cheap electricity is being sold to the Americans so that plant can be utilized. The alternative is to let water pass over spillways without turning those expensive turbine blades. In fact, Quebec has $15 000 million tied up in the project.

Meanwhile, scientific studies have progressed on the ecology of the area after inundation. It had been expected that the slow-growing nature of the trees and fish would be greatly upset by the new soil and water conditions. On the whole, however, what the scientists have found so far is encouraging. It suggests that the ecosystem is much less fragile than had been expected. Beaver have built long lodges to utilize their preferred habitat on the water's edge where lake levels sometimes fluctuate greatly. They seem to have adapted well to the new conditions. Some fish have even benefitted from the changes. When riverside plants were inundated as the lakes filled, they decayed and released their nutrients into the water. This led to an abundance of food and a fish population explosion in some cases. On the other hand, flooding also led to a build-up of mercury in the waters, leached from surrounding rocks. This was taken in by the fish and those at the top of the food chain, such as pike, concentrated it. Pike now have an average of 2 mg of mercury per kg, four times the level that the Canadian government allows in fish for human consumption.

It seems that the ecosystem is not as fragile as was thought by some before the James Bay Project got underway. In the 6000 years since glaciers left the area at the end of the Ice Age, plants and animals had come to adapt to the environment in a very efficient manner, one tolerant of the rapid changes such as discharges and temperature fluctuations accepted as normal in this environment. There are those who say that the work of the bulldozer and construction gangs is only a scratch on the surface

Fig. 1.40 A decision-making sequence for the James Bay area

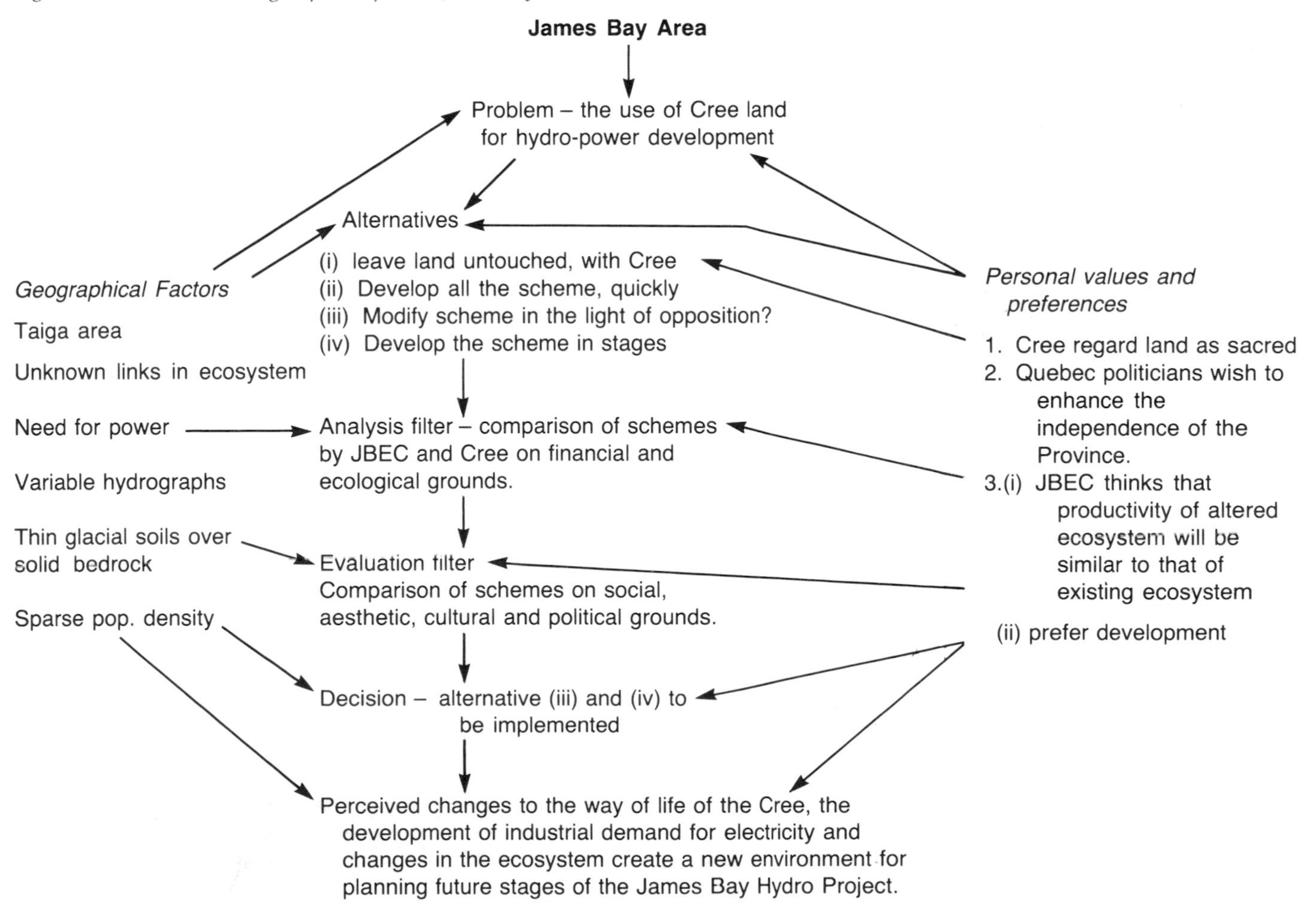

compared to the work of the ice.

The decision-making sequence for this particular case is summarized by Fig. 1.40 (previous page).

18 The opposition to the James Bay scheme suggested that earthquakes and climatic change were likely to occur as a consequence of the scheme. In your view, what would be more realistic claims to make against it?

19 Redraw Fig. 1.39 to show the possible effects on the ecosystem of the changes predicted by the biologists hired by the Indians. Add the information found since the scheme has been completed in a different colour.

20 Set out a list of reasons why the Cree agreed to the scheme.

21 Do you think that the scheme will have long term benefits or not? Explain your answer.

Fig. 1.43 An aerial view of one of the La Grande dams

Fig. 1.41 Part of the LG2 township in summer

Fig. 1.42 LG2 in winter

Fig. 1.44 Roads must be built to serve 176 000 km²

Case Study: the Rio Chico, central Luzon, Philippines

Location

The map (Fig. 1.45) shows the Rio Chico, a left-bank tributary of the Rio Grande de Cagayan in north central Luzon, the largest island in the Philippine archipelago. The area is one of high relief. The Chico cuts through a series of deep gorges in its passage to the wide, fertile plains of the Cagayan valley. Rainfall in the Central Cordilleras is extremely seasonal, being in the Monsoonal circulation. From November to April the totals are less than 500 mm, but then from May to October there is more than 1500 mm of rainfall. This leads to a great change in the discharge of the Rio Chico so that its once peaceful channel, a mere 50 m across, becomes a raging torrent, one which carries both large quantities of water and huge amounts of silt.

History

In 1973, the government announced its plan to build four hydro-electric power dams and one irrigation dam in central Luzon. The World Bank supported the plans. When they got to learn of the massive extent of the project, tribes who live in the affected area reacted very strongly against it. The upland people (the Bontoc) entered into a mutual defence pact with the lowland people (the Kalinga) to defend their rights. At this point the central government sent in PANAMIN (the Presidential Adviser on National Minorities) to persuade the Kalinga that the idea was a beneficial one and to some extent the peace pact broke down. The Kalinga signed agreements for the scheme to go ahead, although there is considerable doubt that the agreements were arrived at fairly. Still the resistance of the Bontoc was unbroken, so in 1978 PANAMIN was withdrawn and the Philippines constabulary was drafted in to deal with local resistance which had by this time become partly an armed struggle. In 1976 the first signs of the guerilla army, the New People's Army, had been seen in the valley. The struggle became violent, so that by 1981, when the Philippines five-year National Energy Programme was published, no mention was made of the Chico scheme. There is no doubt that the plan has not been dropped, however. The World Bank has suspended its support, but not cancelled it. Plans revealed in the five-year plan show that in the next twenty years, forty similar schemes are projected for the Philippines.

Tribal information

The hill peoples of Luzon, the Bontoc, are fiercely independent agricultural peoples who have a distrust of central government. They are even traditionally isolated from the lowland tribes, the Kalinga. The Bontoc, one of the tribes known collectively as the 'Igorot' or 'people of the mountains', became independent first of the Spanish when they ruled the islands, then of the Americans, and now of the Philippines government. Few changes have occurred affecting their simple subsistence way of life, although elements of the market economy have been reaching them in recent years. The agriculture consists of rice farming on

Fig. 1.45 The location of the Chico scheme

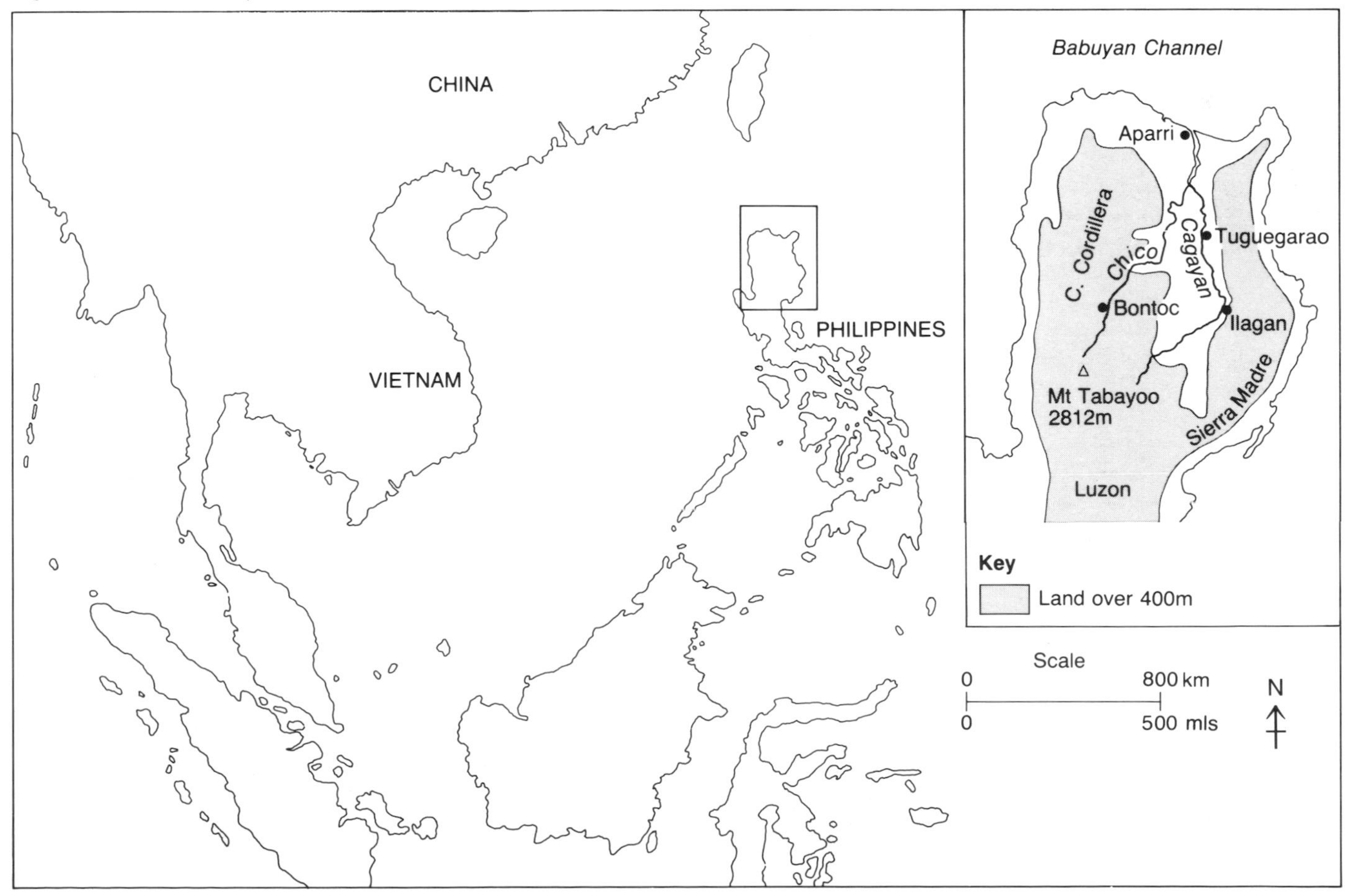

painstakingly constructed terraces. These rise in ten-metre steps from the floor of the steep-sided valley to 1000 m above the river bed in places. The tribes see themselves as being at one with the land. Theirs is a spiritual attachment. They have their fields, their burial grounds and their sacred groves. In the area affected by the plan there live some 90 000 Igorot in sixteen villages. Over 2750 ha of land would be either destroyed or partially destroyed by inundation.

The economy

The Philippines is a poor nation. It has survived through strong foreign support, especially from the USA, which has supported the government against the communist threat from the New People's Army. It depends for 95% of its energy on imported sources, especially oil. Its exports include sugar and other foodstuffs.

Table 1.11 Philippines trade for 1981

Total $USm	Primary products (excl. oil) %	Oil %	Machinery %	Other manufactured goods %	Miscel-laneous %
Imports 6737	11.4	18.5	28.3	30.0	11.8
Exports 5722	60.0	0.5	2.4	25.7	11.4

The World Bank supported the Philippines in its plan to spend 1000 million US dollars (at 1983 prices) on the Chico dams.

The World Bank, or IBRD (International Bank for Reconstruction and Development) to give it its proper name, was set up in 1947. It is the sister organization to the IMF (International Monetary Fund) and has the same member countries. It was originally set up to help repair the war-ravaged economies of Europe. The IMF was set up at the same time, but it is the World Bank which allocates funds for development projects. It does not have unlimited funds, although the amounts it deals with are enormous. Its funds come from three sources. Quotas, 90% of which are promised as a guarantee against the Bank's loans, are received from member countries. The Bank also sells bonds on the world's financial markets and lastly receives interest so that it has an income. It has now turned its attention from a relatively prosperous Europe to the world's less developed countries (LDCs) and is involved with development projects in these countries. It is not just a source of funds. It also has a research department which offers advice on many development issues. It set up the International Development Association in 1960, an agency which makes loans over longer periods than did the IBRD so that it has earned the nickname of the 'soft loan window'. The Bank is subject to strong political pressures, and has to be assured of the political stability of a country before it will support projects there.

By 1973, the Philippines had ten hydro-electric power stations operating, the largest of which was a 212 MW station. The Chico dams were projected to be of massive capacity:

Dam	Capacity
Chico I	100 MW
Chico II	360 MW
Chico III	100 MW
Chico IV	450 MW
Irrigation	50 000 ha

An analysis of present electricity consumption per sector is shown in Table 1.12.

Table 1.12 Electricity consumption by sector (mKWH), 1981

Sector	Manila	Rest of country	Total
Residential	3047	976	4028
Commercial	3396	392	3791
Industrial	3995	532	10 529
Others	72	145	1203
Utilities and losses	55	241	4301
Totals:	10 566	13 286	23 852

The industrial use is principally for primary industry such as logging and mining. The majority of the companies involved in these exploits are foreign owned, such as the largest enterprise on Mindanao, the Kawasaki sintering plant.

Similarly, the largest farming enterprise on the Cagayan plains is that of the Cagayan Sugar Corporation. It stands to gain most from the irrigation of the valley. This would probably put the small farmer under great pressure to sell his land once irrigation becomes available. In the past, land has even been expropriated in the Philippines under similar circumstances. IRRI (the International Rice Research Institute) in Manila has recently turned more to the study of alternative crops to rice since it has found that its new varieties of rice, such as IR8, need very large amounts of both fertilizer and water. This inevitably selects the wealthy landowners as those most likely to benefit from improvements in plant breeding and techniques, because they are the only ones who can afford them. The flood of grain on local markets from the new high-yielding varieties tends to depress prices, so that small farmers become less and less able to keep pace. Once they have sold out to the wealthier farmers, prices invariably go up again.

Despite the postponement of the Chico hydro-electric power dams, a scaled-down version of the irrigation dam is still being built. It will probably irrigate about 19 000 ha, being a run-of-the-river dam instead of relying as was originally planned on the impounding of water by the Chico IV dam.

Ecological information

The rain forest of the Central Cordillera is under threat. Logging often results in massive and irreversible loss of soil. One example comes from Benguet province, in the north of Luzon, where in 1952 the Ambuklao Dam was built. This was supposed to be of 75 MW capacity, but logging of the watershed and the dumping of mining tailings have meant that it is no longer fully operational. The stage of development reached by much of Filipino industry is that of the robber economy. There is very little replanting of forest trees once they have been logged and there is precious little watershed control so that the mining industry, run as it is on a small scale by individuals and small groups, has few checks on it. The Igorot see themselves as guardians of the forest since their irrigated farming techniques use the lower slopes and rely on a protected watershed so that the balance of sedimentation is not upset. (For further information, see the section on Tropical rain forest, pp. 58).

Political information

The regime of President Ferdinand E. Marcos had been under considerable pressure from internal sources and internationally to change its ways. There was a history of opposition to the government being combated by the gun. For example, when the exiled leader of the opposition returned to the islands in 1983 he was gunned down at the airport. The government was widely held to be responsible for this and similar acts, despite denials and court action against individuals accused of the murder. 'There was deep suspicion about President Marcos' own financial dealings and civil unrest grew partly as a result of this. On 25 February 1986, Marcos was forced to flee the country and was eventually given refuge in Hawaii. After his death, his wife and close colleagues were brought to trial on charges of racketeering, though they were cleared of this in the United States. Meanwhile, the wife of the assassinated leader of the opposition, 'Cory' Aquino, became President. She also found it difficult to reconcile the demands of the people and the aspirations of the multinational companies. Serious civil unrest continued.'

Charles Drucker, who wrote the article in *The Ecologist* from which much of the present information comes, states, 'Large hydro-power projects, like the Chico Dams are not true instruments of national development. They constitute, rather, part of an elite strategy to convert the Philippines' natural and human resources into corporate profits, through the medium of international trade.'

Fig. 1.46 A farmer working rice fields in Luzon

Decision-making Exercise D: The Chico Dams

D

Stage One
Using the information supplied here and with research into other available sources, convene a meeting run by PANAMIN. This meeting will take place in the regional capital of Bontoc and will take statements from the following people:

(i) the tribal chief of the Bontoc, who would speak up for traditional rights, especially religious

(ii) a leader from the Kalinga, who would be supportive of the present way of life and against major changes, having had second thoughts after the agreement they signed

(iii) a government adviser on hydro-power development, who would think of the economics of power production as his first priority

(iv) a Bontoc farmers' leader who would be intent on the preservation of the present way of life

(v) an American adviser on ecological matters who would be against the introduction of a 'robber' economy

(vi) an official of the World Bank, who would be interested in the economic stability of the country

(vii) a member of the Philippines constabulary, who would be interested in the maintenance of law and order, no matter what decision is made

(viii) a written statement should be received and read from the New People's Army, expressing their support for the people's struggle against the big scheme.

The meeting is run by an official from PANAMIN. (This person should decide on the order in which people are to be heard.) He or she should sum up the points made, taking into account that he or she is a representative of the central government.

There should be two written reports of this meeting, prepared by a journalist from a magazine interested in the support of minority groups and small-scale industry and by a journalist from the pro-government newspaper.

Stage Two
A meeting of the Bontoc and Kalinga should be convened to react to the PANAMIN meeting and to plan a future strategy of opposition.

Stage Three
A government decision should be made about whether the plan to dam the Chico should go ahead or not. Important elements in the decision might be:

(i) Is it in the interests of tribal harmony to go forward?

(ii) Is it in the interests of national security?

(iii) What are the economic implications of building the dams and of not building them?

(iv) What are the alternatives?

Conclusion
Attempt to set out a decision-making sequence diagram to explain the decision that has been made. Does your decision concur with the actual one of not proceeding with the dams at this moment?

The president of the World Bank stated that he could only 'worship at the altar of economic pragmatism' in making his decision on which major projects to support. This would mean that no developments should take into account cultural, social, aesthetic, spiritual or ecological factors. They should only maximise profits. Do you agree with him? Explain your answer.

Systems diagrams

To show flows of anything, including such things as electricity, finances and water, we can draw a systems diagram. This enables us to see direction of movement and inter-relationships between factors affecting that movement.

Let us consider a very simple flow to begin with. To do the washing up you use water from the tap (an input). This is poured into a sink (a store) and eventually when the washing up is finished, the plug is taken out and the water goes down the wastepipe (output). In a systems diagram this would be shown as:

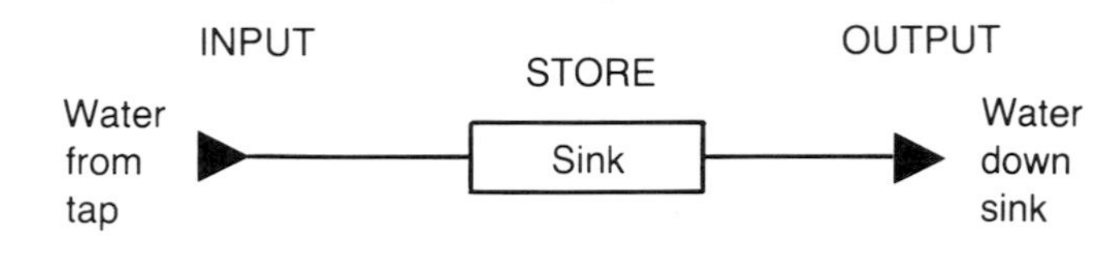

In fact, this is a simplified version of reality because not all of the water that enters the sink finally goes down the plug hole. Some of the hot water evaporates directly from the sink. When the washing up is finished, some of the water may be used to wash down the draining board. If a choice is involved, either a conscious one, as in this case, or a divergence of path as in sunlight either getting through the clouds or being reflected back off them, a symbol is used to denote this. The diamond signifies a decision, a switch or a regulator of some kind. The water used to wash down the draining board flows back into the sink and is shown as a feedback loop. The whole diagram would be shown as follows:

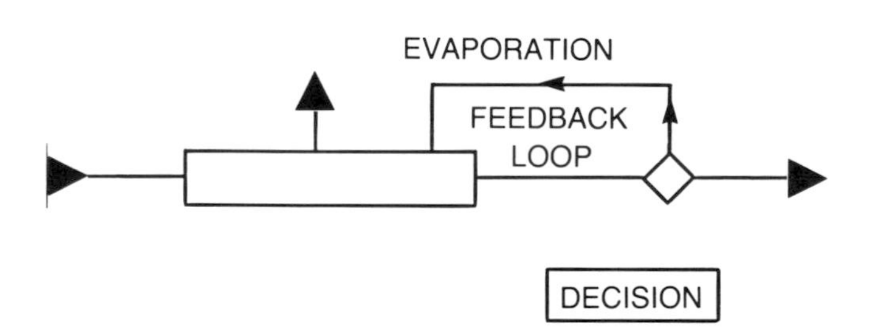

It may be necessary to simplify a systems diagram if the inter-relationships are very complex. This can be done by isolating part of the diagram and showing it as a sub-system. If details of that sub-system are not essential to the study at hand, it may be shown as a 'black box' within which the details are omitted. The part of the diagram shown in detail is known as a 'white box'. Thus, for one system, the hydrological cycle, the diagram could be shown as follows:

ATMOSPHERIC SUB-SYSTEM

Evaporation

Condensation

Water stored in atmosphere

WHITE BOX

INPUT

Precipitation

OUTPUT

INPUT

Evaporation from seas etc.

GROUND SUB-SYSTEM

BLACK BOX

River flow

If it is necessary to make lines cross in such a diagram and the two lines do not inter-relate, a bridge may be used. Thus the symbols for systems diagrams are:

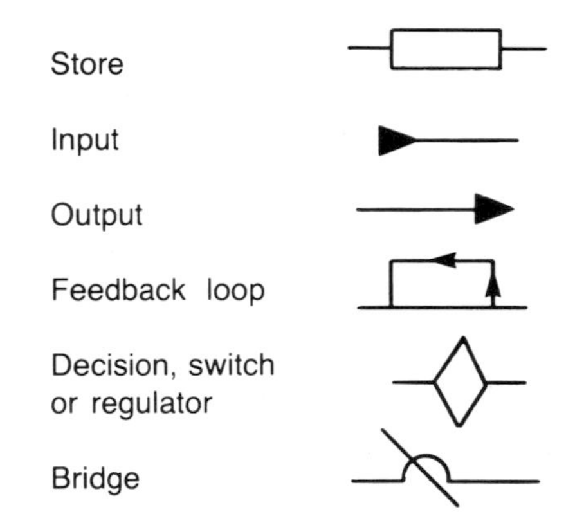

1 Using the information on this page, attempt to draw systems diagrams to illustrate the following:

(i) your income, savings and expenditure for one week

(ii) the manufacture of a cardboard container

(iii) a central heating system.

Fieldwork techniques for stream measurement

(a) Velocity of flow

An idea of velocity relationships in a channel can be gained by looking at Fig. T1.1. The **isovels** are lines of equal velocity and it can be seen that the greatest velocities in this particular channel are away from the banks which create friction with the water and away from the surface with its friction with the air. Any measurement technique must therefore take into account these variations in speed of flow. A suggested procedure is as follows:

(i) Select a stretch of stream that is both accessible and possible to wade across
(ii) Measure out a suitable length, perhaps 10 metres, along one bank. Place a marker at the beginning and end of this section
(iii) Select a float which will be at least partly submerged, such as an orange or a heavy ball, and which offers little resistance to the wind. Place the float at the beginning of the measured stretch and allow it to float naturally between the two markers. Time how long it takes for the float to pass from one marker to the other
(iv) Perform the timing a number of times, perhaps ten, and average out the result
(v) If the stream is shallow, multiply your answer by 0.85, since the water will be moving faster at the surface than at depth in such a channel
(vi) Divide the distance by the average time to find an average velocity in metres per second.

(b) Channel cross section

The part of the channel in contact with water is known as the **wetted perimeter**. An estimate of the length and shape of this wetted perimeter can be made by the following method.

(i) Stretch a tape measure across the channel from one side of the stream to the other.
(ii) At regular intervals across the channel measure the depth with a metre rule.
(iii) Draw a diagram similar to Figure T1.2 for this particular channel and work out the total cross sectional area as shown. Note that since channel shape changes very quickly downstream, it is advisable to measure several sections and average them out over a given distance. This could be done at the beginning, middle and end of the stretch marked out for measuring velocity of flow.

(c) Discharge of stream

If the average cross sectional area of the stream is measured in square metres and the velocity of flow is measured in metres per second, a multiplication of the two would give an answer in cubic metres per second. This is a measure known as the **discharge of the stream**, the amount passing a given point in a given time. Discharge can be expressed in cubic feet per second (cusecs), litres per second or cubic metres per second (cumecs) as in the example above.

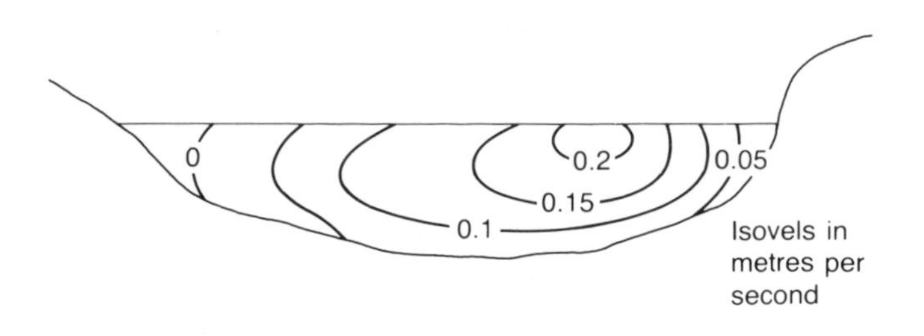

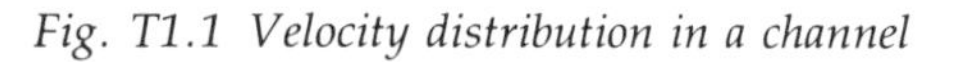
Fig. T1.1 Velocity distribution in a channel

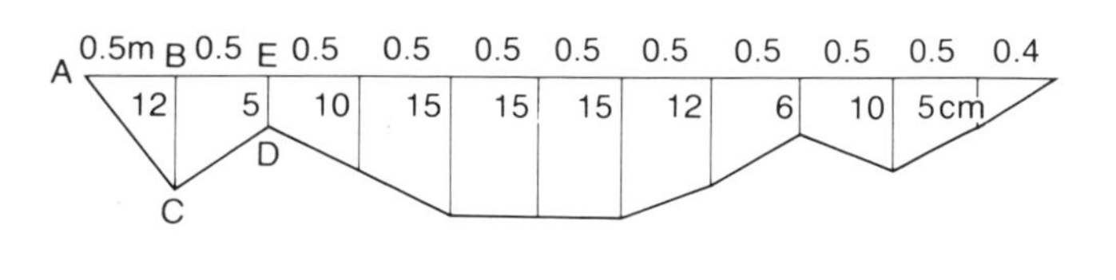

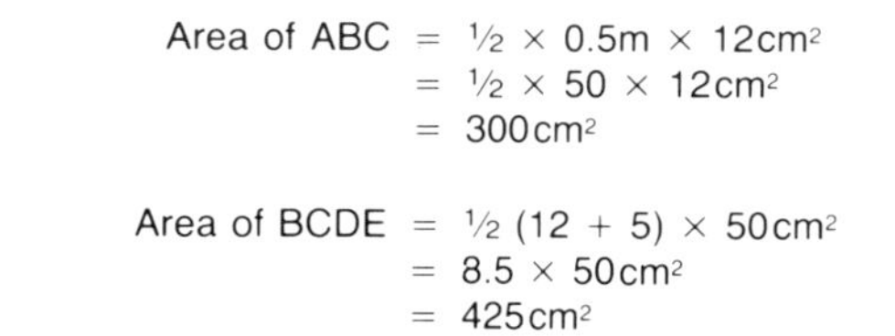

Total area is found by adding the total areas of the two triangles and the nine trapezia. To convert the area in cm² to an answer in m² divide the total by 10 000.

Fig. T1.2 Method used to estimate the cross-sectional area of a stream

(i) What is the cross sectional area in square metres of the stream shown in Figure T1.1?
(ii) If the average velocity of this stream were 0.2 m per sec, what would be its discharge?
(iii) What experimental or operator errors are possible using any of these techniques?
(iv) An alternative way to measure velocity of flow would be to use food colouring. A few drops of this are still visible in the water after it has flowed several metres. Under what circumstances would this be better than using a float?
(v) A further alternative would be to use a **current meter**. This is a piece of apparatus with a small propellor fixed on a shaft. The depth of the propellor is set 0.6 of the channel depth (at which depth the average velocity occurs). Revolutions of the propellor are recorded automatically and the velocity found as a function of the number of revolutions per second. In what ways is this method more useful than using a float? Does the method have any drawbacks?

The use and interpretation of hydrographs

Hydrographs are diagrams which show the relationship between precipitation and run-off. They may be drawn from a storm, for a longer period of time or for the 'water year'.

The **storm hydrograph** has a number of important features. Any stream which does not periodically dry up is said to have a **base flow**. This is the water being released to the channel from groundwater sources. On top of this for the storm hydrograph comes the amount of water being directed to the channel by run-off, throughflow and overland flow. If the soil is dry at the start of the storm, the first rain that falls will be largely used for recharging the soil water table. In other words, the first process to take place will be infiltration. At this stage, vegetation may well use a good proportion of the water. As the soil fills up and reaches **field capacity** (when it can take no more) the water may start to percolate further into the bedrock, if it is porous or permeable. It may also begin to be transferred down the slope by throughflow, through the soil. Eventually, if neither of these processes is able to remove the water as fast as it is being added to by precipitation, overland flow will begin. In all, these processes lead to a peak in discharge of the stream, which may be a considerable time after the precipitation began. The time between the two events is known as the **lag time** and is dependent on a number of factors such as the intensity of the precipitation, the steepness of the gradients in the drainage basin, the amount of vegetation, the depth and porosity of the soil as well as any actions taken by the inhabitants of the area to speed or hold back drainage (Fig. T1.3).

Storm hydrographs for different shaped catchment areas have distinctly different shapes themselves. Some examples are shown in Fig. T1.4. The steepness of the profile of the main stream, the drainage density and the bifurcation ratio may also affect the hydrograph shape. **Bifurcation ratio (Rb)** is obtained by expressing the number of one order streams as a fraction of those in the next order. A first order stream does not have any tributaries. A second order stream occurs when two first order streams meet, a third order stream occurs when two second order streams meet and so on. Thus $Rb = \frac{Nu}{Nu+1}$. The drainage density is also an important determinant of hydrograph shape. Examples of the possible effects on hydrograph shape for all of these phenomena are shown in Fig. T1.4.

The annual hydrograph for a basin often shows features in addition to those already mentioned. These relate to features of the basin itself, such as bedrock porosity, amount of surface vegetation and degree of urbanization and to features of the climate. If snow falls in the area there may well be a seasonal snow melt showing as a peak in the hydrograph. If summer temperatures are high and plant growth luxuriant, run-off will decrease in those periods. Seasonal rainfall, such as monsoonal rains, will also result in peak flows.

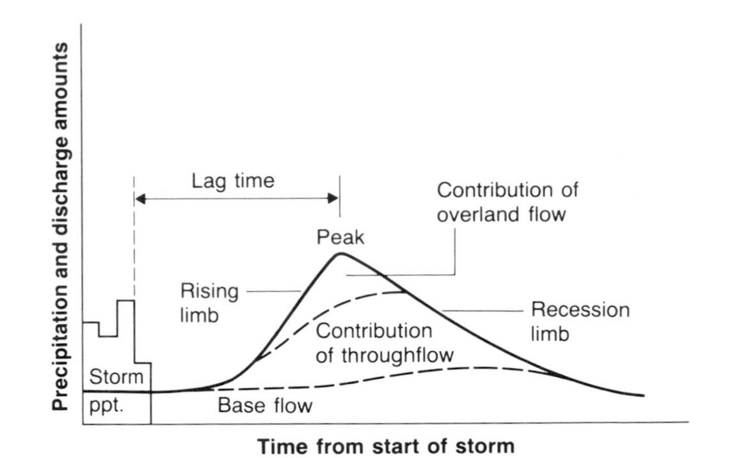

Fig. T1.3 The storm hydrograph

Fig. T1.4 The relationship of hydrograph shape to certain basin characteristics

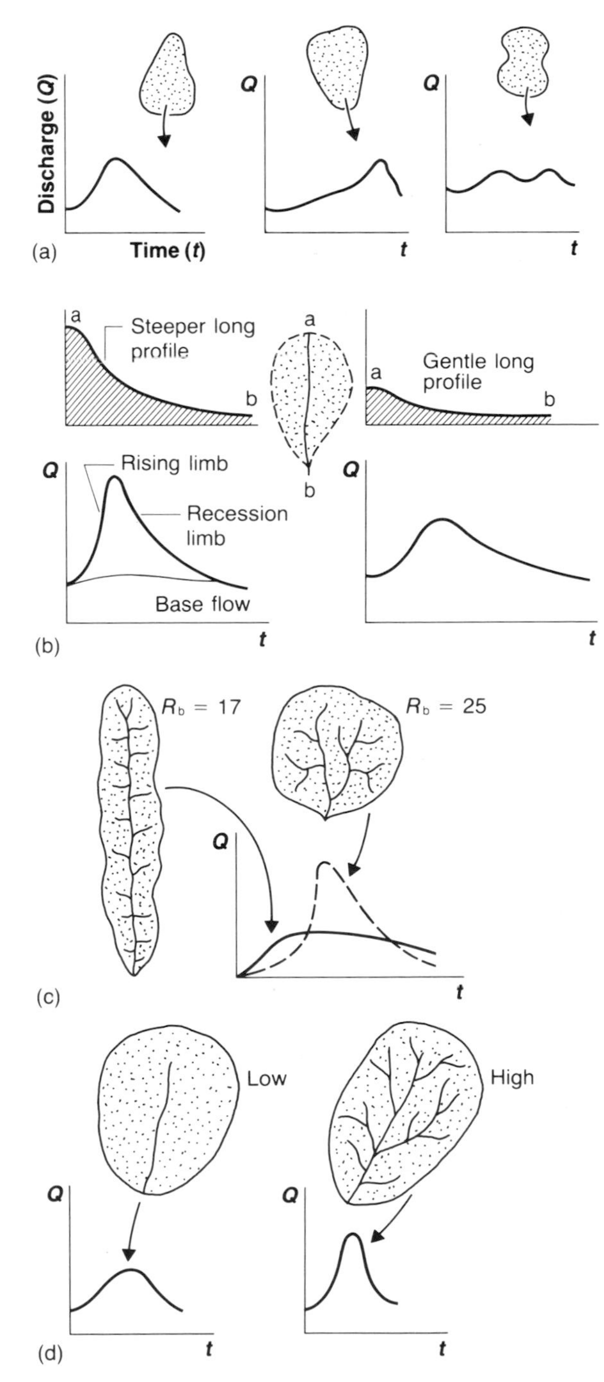

Fig. T1.5 Two Yorkshire hydrographs

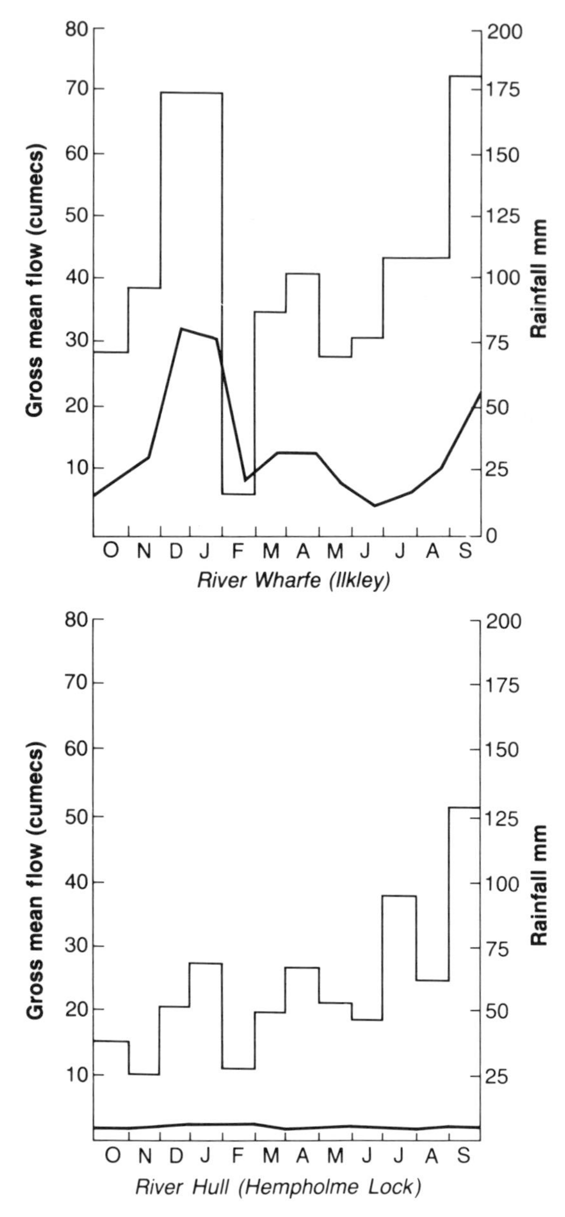

Hydrographs are used especially by water engineers to predict and accommodate flood surges in a channel. The land use of the flood plain can be planned if it is known that the river is a moderate one, only reacting slowly to rainfall events. If, on the other hand, the reaction of the river is violent, other decisions about land use might be made.

1 Attempt to draw generalized hydrographs for a storm falling in the Afon Cwmllan and Afon Cwmpenamnen basins (Fig. 1.6, p. 17). Label your hydrographs to show how they are related to basin characteristics.

2 Study the hydrograph of Yorkshire rivers (Fig. T1.5) together with the simplified geology map (Fig. T1.6).

(a) Describe the relationship between precipitation and discharge for the two rivers.

(b) What other factors must be taken into account in explaining the hydrographs?

(c) Complete two summary diagrams by copying each hydrograph at the centre of a sheet of plain paper. Then label each point on the discharge curve for which you have obtained an explanation. You should then have two annotated hydrographs.

3 Study the hydrograph for the River Nile and its tributaries (Fig. T1.7). With reference to an atlas and climatic statistics, explain the important features of that hydrograph.

4 Refer to Fig. T1.8 and an atlas map of Kenya. The river Tana flows to the sea near Garsen, starting near Mt Kenya. Comment on the fluctuations shown. How do they come about? What effects would they have on life in the area?

Fig. T1.6 The simplified geology of part of Yorkshire

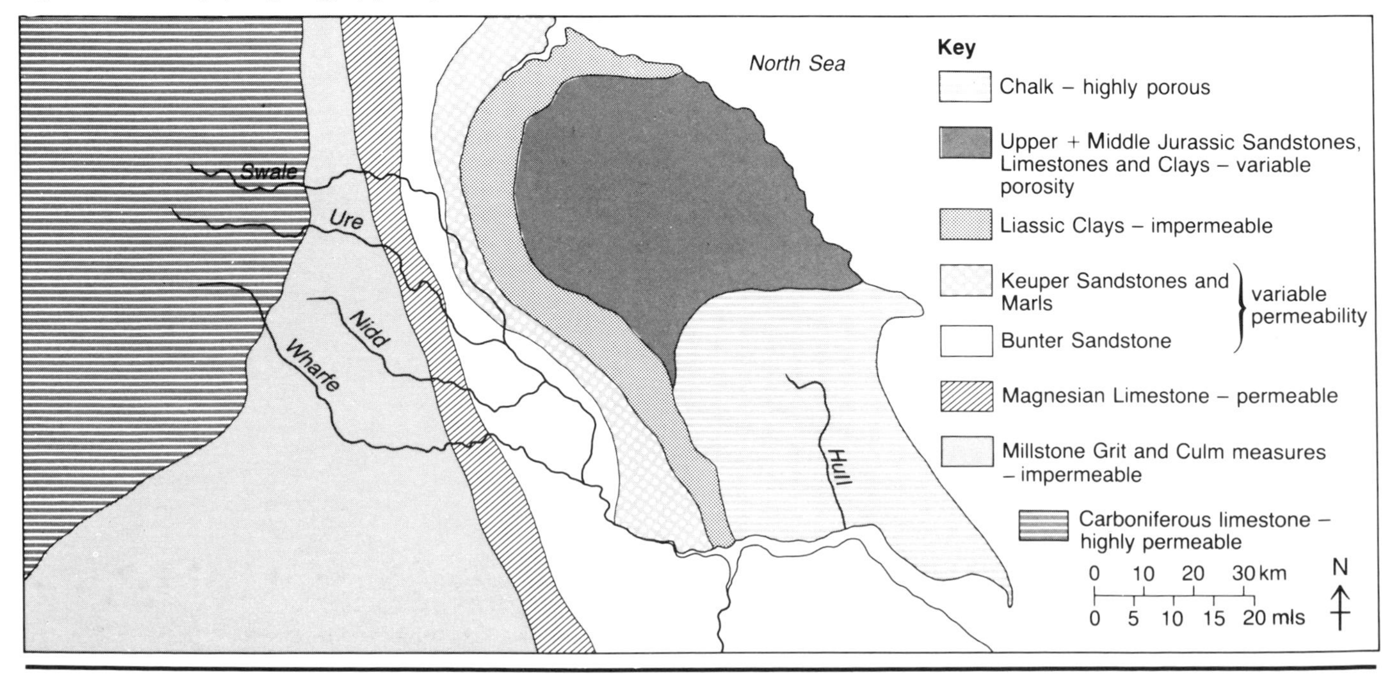

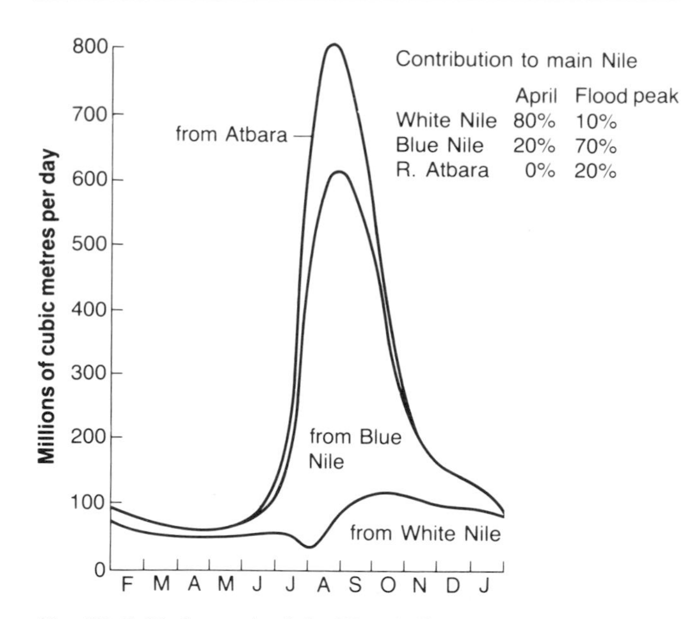

Fig. T1.7 Hydrograph of the River Nile

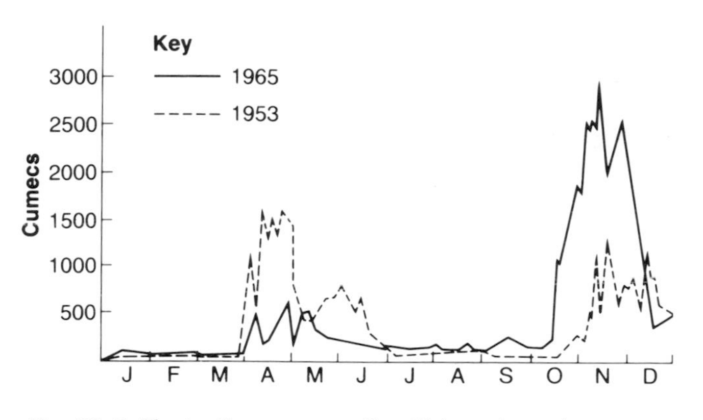

Fig. T1.8 Fluctuations on a smaller African river, the Tana, in Kenya. The two most variable years, 1953 and 1965, are compared.

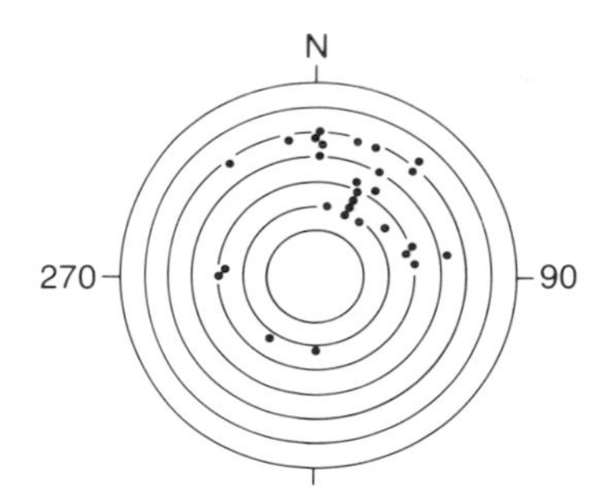

Fig. T1.9 Polar diagram showing the relation between corrie floor altitude (concentric circles) and the preferred north-easterly aspect (radii). Northern Nain-Okak part of Labrador, circles 0–915 m

Fig. T1.10 Polar diagrams, as Fig. T1.9

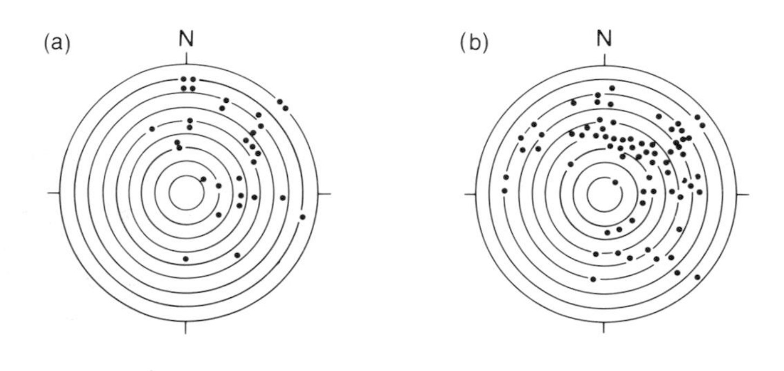

(a) North Caernarvonshire, North Wales, circles 270–750
(b) West-central Lake District, circles 300–780

Polar graphs

Graphs may be drawn on circular or polar graph paper to show the effect of orientation on distribution. The polar paper on page iv may be copied to provide you with paper on which to complete this exercise and the one associated with the orientation of landslips in Wellington, New Zealand.

The concentric circles on the paper may be used to show height. They should be labelled after finding the maximum and minimum heights concerned in the data. The orientation is plotted using an angular bearing. Thus if something is facing due south, it has a bearing of 180 and is plotted accordingly. Other less straightforward bearings are plotted using a protractor.

Figure T1.9 shows a completed polar graph for corrie floor height and orientation in part of Labrador. The distribution shown clearly picks out the preferred orientation to be north-easterly. This is explained by the fact that the sun would have been most effective in melting the snow from the south-facing and, especially, the south-west facing slopes during the Ice Age. The prevailing winds at this time were south-westerly ones, so that they would have swept the snow from the south-west facing slopes, leaving it to collect on the north-east facing slopes. The processes of ice and snow slipping which lead to corrie formation would therefore be more active on the north-east facing slopes.

1. Complete a polar graph to illustrate the following data concerned with the height and orientation of vineyards in the upper Rhone valley near Sion, Switzerland.
2. Explain the distribution you have found.

Height of vineyard (metres)	Orientation of vineyard (x°)
800	140
600	300
750	150
550	160
600	330
720	155
775	180
730	145
480	150
830	160
510	340
540	135
600	120
600	130

Theme 2 *Managing Ecosystems*

The Workings of an Ecosystem

The **ecosystem** is all the living organisms of a defined area together with their inter-relationships with the environment. Thus the study of ecosystems (ecology) is a highly complex one. In order to come to some understanding of the scope of the study, we can look at a few of its component parts.

The ecosystem is often represented as having a number of compartments (Fig. 2.1). The living components are described as belonging to different **trophic levels**; grass would belong to trophic level one, grazing animals to level two and so on. The whole system is powered by the sun, but successive levels of the ecosystem obtain their energy in different ways. Representatives of trophic level one produce their energy from inorganic sources and are called **producers**. Other animals graze on these plants and are **consumers**. When members of either group die, they are broken down by the **decomposers** and their nutrients are returned to the soil to be recycled through the producers again. The consumer group may be further subdivided into **herbivores**, **carnivores** and **omnivores**.

The vegetation in the ecosystem is also known as the **primary production** of that ecosystem. The amounts of this vary greatly from one ecosystem to another, as shown in Table 2.1. It is the vegetation which, as we have seen, provides the foundation upon which the rest of the biotic part of the ecosystem depends.

Fig. 2.1 The components of an ecosystem

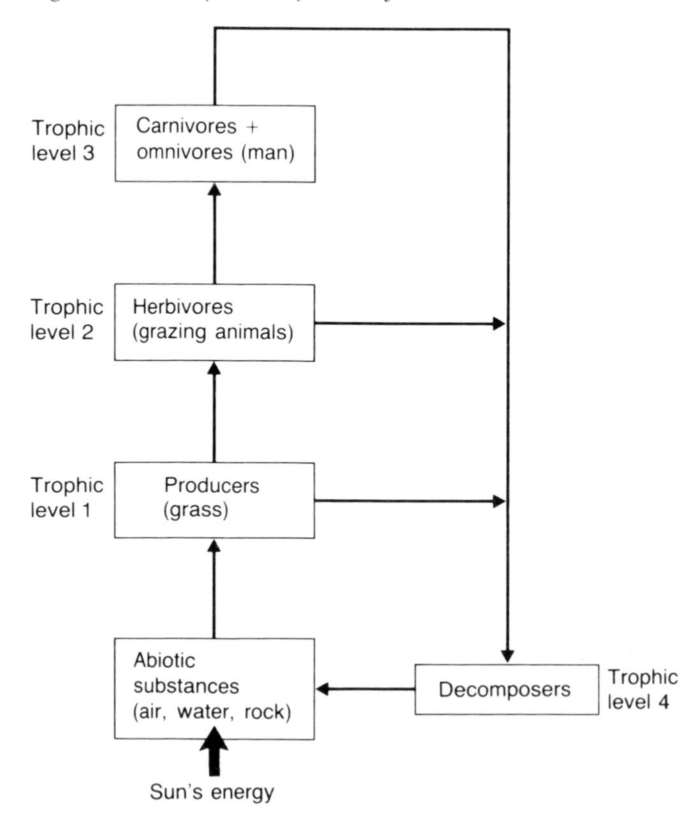

1 Fig. 2.2 shows a diagrammatic representation of inter-relationships in a tropical grassland (savannna) ecosystem. Redraw the diagram as a systems diagram (see Techniques page 47).

2 In what ways might the growth of a human population in a savanna area affect the ecosystem? Add any modifications you suggest to your systems diagram in a different colour.

3 Answer the following from Table 2.1.
(i) Which ecosystem covers the largest area?
(ii) What is the net primary production of that ecosystem per unit area?
(iii) How much biomass does it produce per unit area?
(iv) In terms of world biomass production how important is it?
Now repeat questions (ii), (iii) and (iv) for the land-based ecosystems which cover the next six largest areas of the world.

4 From your answers to question 3, is it true to say that climate plays an important part in determining the productivity of an ecosystem? Justify your answer.

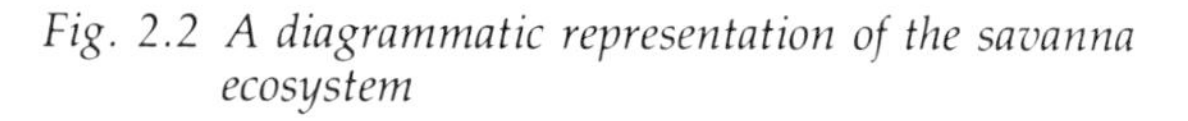

Fig. 2.2 A diagrammatic representation of the savanna ecosystem

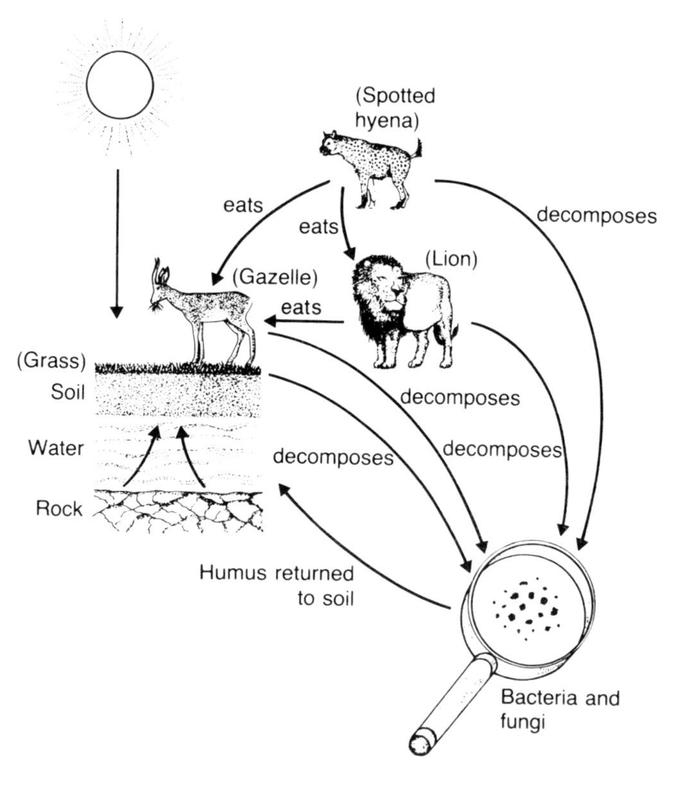

Table 2.1 Net primary production in the world's major ecosystems

Major ecosystem	Area (10^6 km^2)	Net primary production/unit area (dry g/m^2 year)		World net primary production (10^9 dry tons/ year)	Biomass/unit area (dry kg/m^2)		World biomass (10^9 dry tons)
		Normal range	Mean		Normal range	Mean	
Lake, stream	2	100–1500	500	1.0	0–0.1	0.02	0.04
Swamp, marsh	2	800–4000	2000	4.0	3–50	12	24
Tropical forest	20	1000–5000	2000	40.0	6–80	45	900
Temperate forest	18	600–2500	1300	23.4	6–200	30	540
Boreal forest	12	400–2000	800	9.6	6.40	20	240
Woodland, shrubland	7	200–1200	600	4.2	2–20	6	42
Savanna	15	200–2000	700	10.5	0.2–15	4	60
Temperate grassland	9	150–1500	500	4.5	0.2–5	1.5	14
Tundra, alpine	8	10–400	140	1.1	0.1–3	0.6	5
Desert scrub	18	10–250	70	1.3	0.1–4	0.7	13
Extreme desert, rock, ice	24	0–10	3	0.07	0–0.2	0.02	0.5
Agricultural land	14	100–4000	650	9.1	0.4–12	1	14
Total Land	149		730	109.0		12.5	1852.0
Open ocean	332	2–400	125	41.5	0–0.005	0.003	1.0
Continental shelf	27	200–600	350	9.5	0.001–0.04	0.01	0.3
Attached algae, estuaries	2	500–4000	2000	4.0	0.04–4	1	2.0
Total Ocean	361		155	55.0		0.009	3.3
Total Earth	510		320	164.0		3.6	1855.0

(Source: Whitaker in M. Bradshaw, *Earth, the Living Planet*, Hodder & Stoughton)

The biotic components of an ecosystem are very susceptible to change through the influence of people. They may be used to provide products, as they are in a hunter-gatherer society, in which case the seeds of the fruits collected may germinate near to human habitation, leading to a concentration of particular species in that new area. This may lead naturally to the domestication of certain plants, which may give way to the development of forest clearings as in the shifting agriculture system. Eventually, full-blown arable agriculture may develop. The same sort of sequence may well occur with animal domestication. In these ways, the ecosystem may change from being totally natural in the first instance to totally managed in the final stage. The numbers of plant and animal species and the ways in which they inter-relate may alter dramatically through this sequence.

The natural ecosystem of an area is the result of gradual changes over a very long time. The vegetation has slowly altered through the ages to take advantage of changing environmental conditions. For example, in the United Kingdom, we might take as our starting point the end of the great glaciation of the Ice Age, which finally uncovered most of the country about 10 000 years ago. At this stage there were only poor, thin soils, many of them composed of ground rocks. The first plants to colonize them were lichens, which as they grew gave a humus content to the soil that was not there before. They also helped to break down the debris further, gradually creating a true soil. In these soils, mosses and herbs could at last take root. As the humus content of the soil increased, other, deeper-rooted, plants invaded and eventually an oak forest emerged as the dominant vegetation. This process is known as **plant succession** and the final, stable form of vegetation is known as the **climax vegetation** (Fig. 2.3). Any interruption of the succession by human interference, changing climate or soil conditions, for instance, will lead to the development of a **sub-climax vegetation**.

5 Fig. 2.4 represents the vegetation in a natural ecosystem in the humid temperate belt (of the kind which used to cover the United Kingdom). Draw a similar diagram for the vegetation in one area of the country today that is farmed on an arable basis. What major changes have there been in (a) the variety of species, and (b) the numbers of individual species?

6 What other changes are there likely to be in the ecosystem between the natural oak woodland ecosystem and the farmland?

Fig. 2.3 Plant succession and climatic climax

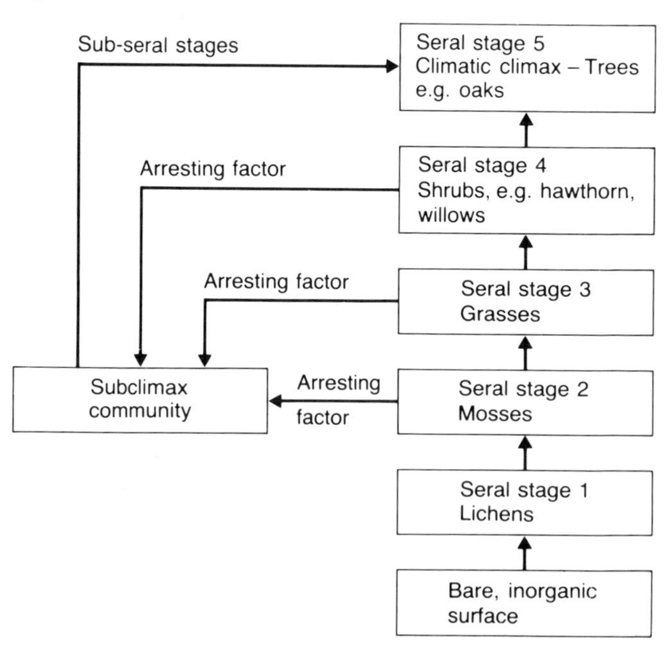

Fig. 2.4 A section through the climatic climax vegetation for the United Kingdom

Case-study: the sand dune environment – Morfa Harlech

One of the best examples of plant succession is found in the sand dune environments. The correct term for a succession in a sandy environment is **psammosere**. A good example of this is found in the Morfa Harlech sand dunes in North Wales. The extent of this system is shown in Fig. 2.5 and an indication of the rate of growth of the system is given by the fact that the castle at Harlech was constructed in the thirteenth century, and it was built on the shoreline.

The sand dune material comes from two sources. Some of it is derived from glacial moraine deposited in Cardigan Bay to the south following the melting of the ice sheets when the last glacial period ended. The remainder comes from material eroded by the sea further south. The waves and the wind combine

Fig. 2.5 Map showing the extent of dune system at Morfa Harlech and the use of the area

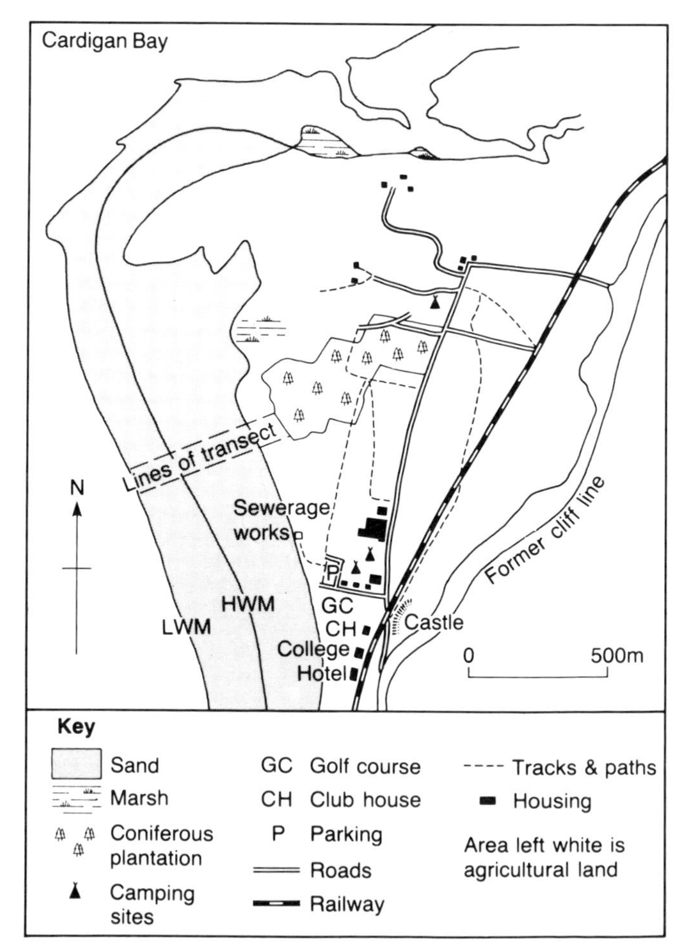

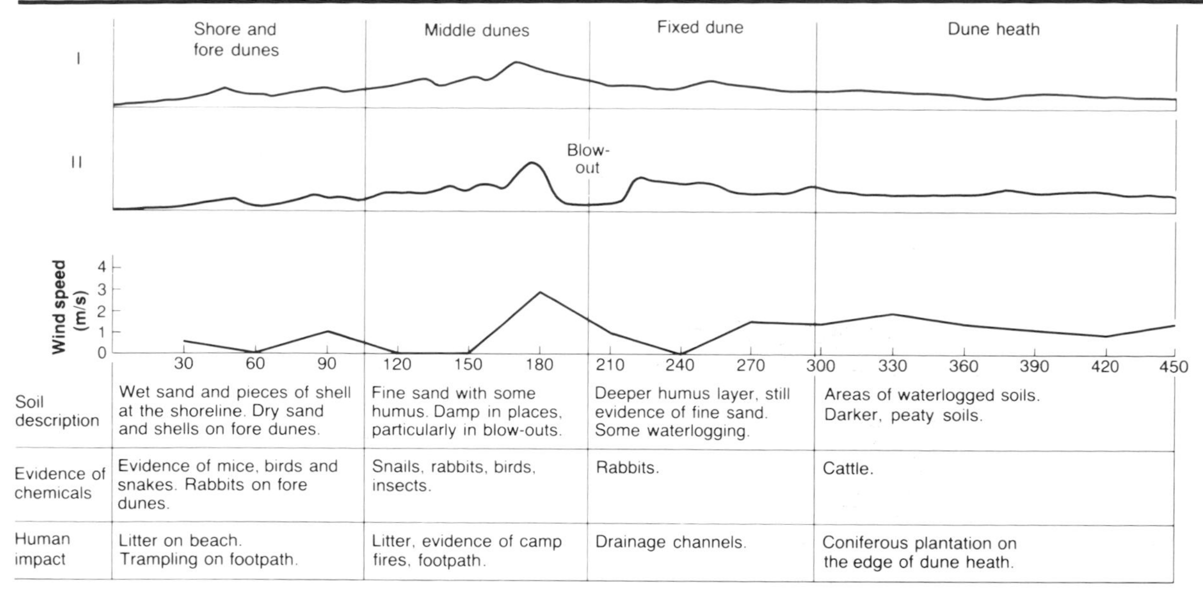

Soil description	Wet sand and pieces of shell at the shoreline. Dry sand and shells on fore dunes.	Fine sand with some humus. Damp in places, particularly in blow-outs.	Deeper humus layer, still evidence of fine sand. Some waterlogging.	Areas of waterlogged soils. Darker, peaty soils.
Evidence of chemicals	Evidence of mice, birds and snakes. Rabbits on fore dunes.	Snails, rabbits, birds, insects.	Rabbits.	Cattle.
Human impact	Litter on beach. Trampling on footpath.	Litter, evidence of camp fires, footpath.	Drainage channels.	Coniferous plantation on the edge of dune heath.

Fig. 2.6 Results from fieldwork transects

to move the material by means of longshore drift in a northerly direction with deposition of the material at the point where the coastline changes direction. Dunes will then form as surface obstacles obstruct the wind and, in slowing it down, sand is deposited particularly on the landward side of the obstacles. These may be vegetation such as sea sandwort which is tolerant of salty conditions and will establish itself on the strand line. This process will continue until the characteristic shape of the dune is formed, as shown in the foredunes in the transect in Fig. 2.6.

This part of North Wales is also recovering from the huge weight of ice which covered most of the area during the last glacial period. These ice sheets had the effect of depressing the land and, with the melting and disappearance of the glaciers, the land is now rising relative to sea level. This return movement of the earth is known as **isostatic readjustment**. This has the effect of helping the deposition of material by the sea as a large, nearly flat beach is formed. This provides a good supply of material for the wind to whip up and transport to the dune system. Hence the dune system is moving towards the sea with the oldest dunes further inland. The oldest dunes will be subjected to the normal processes of weathering, mass movement and erosion and will be lowered and hence slightly undulating in form. The largest dunes will be the mobile dunes as shown in Fig. 2.6. Areas of salt water trapped behind the dunes will become salt marsh and as the access to the sea is cut off this will become freshwater marsh which in time will silt up. These areas may become marshy again during times when the water table is much higher and are known as **slacks**.

The dunes in their newly formed state are very unstable and can be easily destroyed by high winds. Vegetation tends to stabilize these features. Colonization of the dunes is a chance event in such an unstable system. The first plant to colonize will be sea couch grass and then as the dunes become higher the marram grass will arrive. Marram grass is well adapted to this environment, being very hardy, with tough leaves which roll in on themselves to limit transpiration. Its sharp, spiky leaves reduce its attraction to grazers and its deep root system gives it stability in very rough conditions. The root system is effective in binding the sand particles together and slowing down the movement of the dunes.

As one moves along the transect line (Fig. 2.6) so the number of colonizers increases and marram eventually gives way. The undulating nature of the dune shape gives shelter from the wind, making the establishment of plants easier. Animals such as the striped snail and nematode worms appear to aid the growing depth of humus and a shallow soil starts to develop. Eventually bushes and trees will grow, providing habitats for birds, while the marshy slacks will provide aquatic habitats for toads and larger birds such as heron. An example of a food chain in the dune ecosystem could start with the snail-consuming plants which are eaten in turn by the thrush, with the buzzard being the top carnivore. The buzzard is not likely to live in the sand dune ecosystem but could live in the bleaker high moorland of the interior.

7 Draw a systems diagram to show the inputs and interaction within the sand dune environment.

8 What will the effect be on the sand dune environment of the following developments:
(i) construction of a large scale groyne system reducing the supply of sand?
(ii) draining and improving the pasture on the dune heath?
(iii) the siting of a large camp site just inland of the fixed dunes?

Some of the features of the sand dune ecosystem are shown in Figs. 2.6 and 2.7. This is information which was collected by a group of sixth form geographers. They used a series of transects through the dunes as a means of sampling them and took information at fixed points and by means of quadrats. These techniques are explained on Techniques page 79.

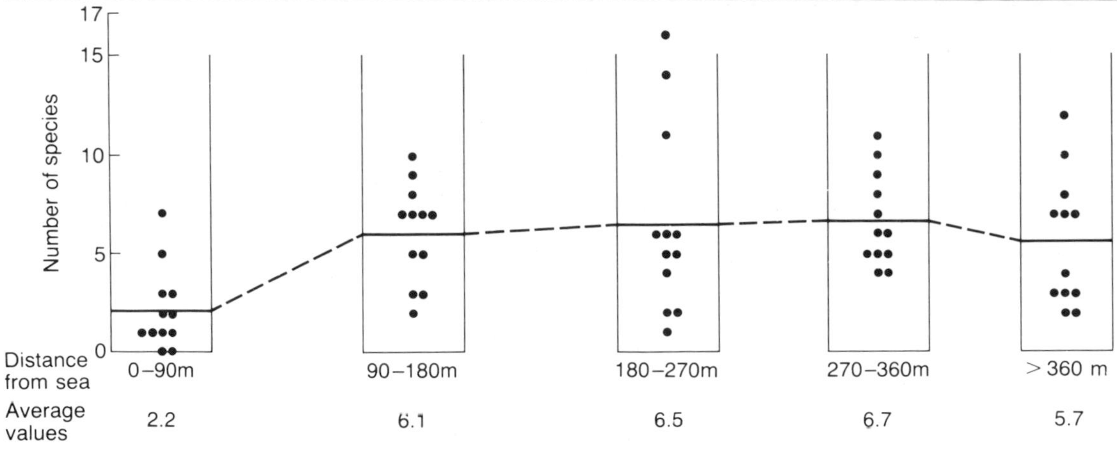

Fig. 2.7 Dispersion graphs to show the number of plant species and distance from the sea

9 Use the information to describe any correlations in the fieldwork results shown in Figs. 2.6 and 2.7. You will notice the deep blow-out on the second transect. Try to put forward reasons for this occurring.

A useful method of analysing data is to use the **chi² test**. Techniques page 78 has the method and explanation.

10 Apply the chi² test to the information provided in Table 2.2 (p.53) on distance from the sea against the pH of the soil and the amount of bare cover. What would you expect the relationships to be? What do the results tell you? Can you explain them?

11 How could the data collection be improved to get better results?

Human impact on the dune system

Dune systems can be very popular with a variety of people for a variety of reasons. The transect information gives some indication of the use of the system and some of the impacts such as litter. However, the dune system can be sought after for more formal uses.

12 Study Fig. 2.5 and list the number of human uses of Morfa Harlech, taking the area from the old cliff line to the low water mark. For each of the uses identified, describe the impact they make on the natural system. Are there any of the uses which are incompatible with any others?

Trampling is one of the main problems in the sand dune system. The vegetation can be removed by the effects of walking and climbing on the dunes. Many of the early colonizers are easily disturbed and removed. This can upset the delicate balance between the rate of accumulation and erosion of sand. This will affect the formation of soil and the animal populations. The blow-out hollow in transect II was formed by the removal of vegetation in one spot which left the area open to erosion by the wind. Once a hollow has been formed, then the eddying nature of the wind will continue to remove sand until the damper sand at the water table has been reached.

Table 2.2 Observed values for chi² testing

1. Distance and % bare cover

		% Bare Cover				
		0–20	**21–40**	**41–60**	**61–80**	**81–100**
Distance from Sea	0–90m	3	1	2	1	5
	91–180m	7	2	1	0	1
	181–270m	10	1	0	0	1
	More than 270m	25	0	1	0	0

2. Distance and pH values of soil

		pH values of soil				
		6.5	**7.0**	**7.5**	**8.0**	**8.5**
Distance from Sea	0–90m	0	0	2	2	3
	91–180m	0	0	1	2	3
	181–270m	0	2	2	3	2
	More than 270m	2	4	4	2	0

Fig 2.8 (a) *A view across the mobile dunes at Morfa Harlech.*

(b) *Students collecting data at Morfa Harlech*

Decision-making Exercise E: Using the Dunes

Fig. 2.9 Map of Morfa Harlech dune systems showing vulnerable areas and a possible camping site

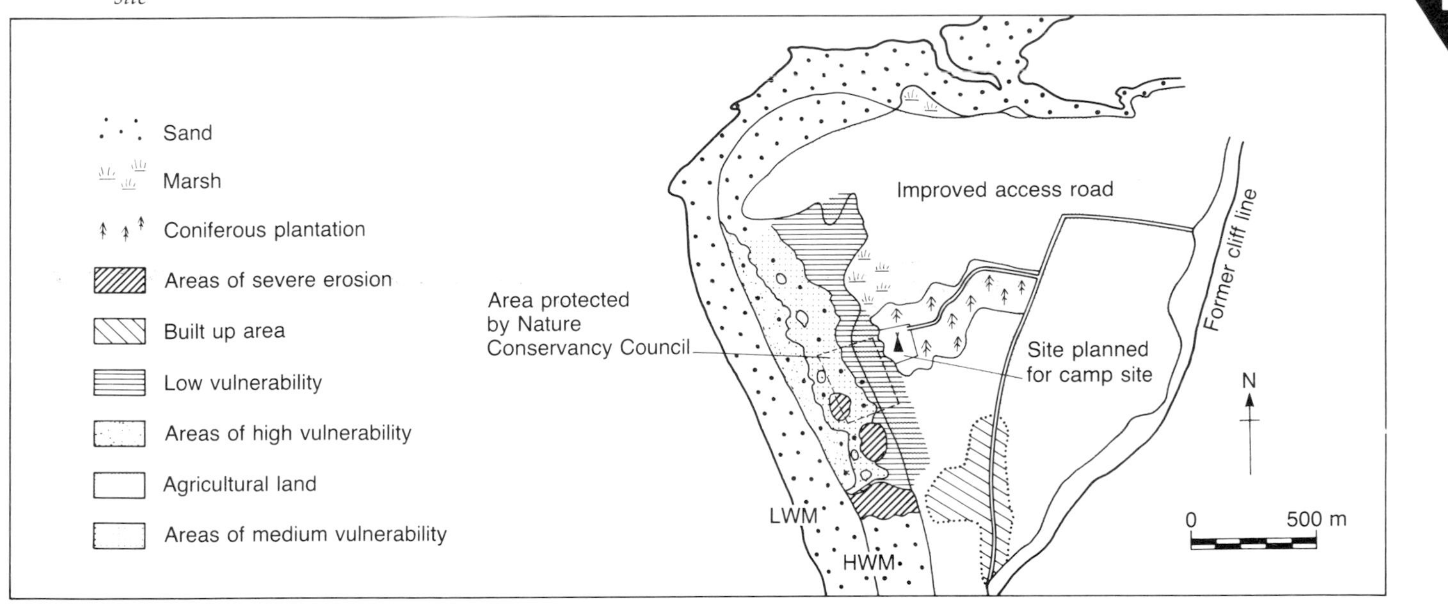

For the purposes of this decision-making exercise, assume that the local council is planning, along with the Forestry Commission, to build a new camp site as shown in Fig. 2.9 on Morfa Harlech. The plantation track is to be improved for access by private cars and the site would be on the edge of the wooded area.

The map (Fig. 2.9) shows areas of the dune system which have been eroded already and areas which are considered by the Nature Conservancy Council as being vulnerable to trampling. In areas of high vulnerability small changes to the ecosystem can unbalance it and lead to its destruction.

Your task is to take on the rôle of the planner and you have narrowed the options down to three possibilities:

(a) You accept that damage will occur, but by concentrating the impact you hope to retain some of the system in its natural state. You therefore propose to enlarge the NCC area and strengthen the fencing around the area to stop people gaining access. Notices will be posted and strong action taken against offenders. You will construct an aggregated (loose stone) path across the

E

dune heath and continue the path across the dunes by using duckboards. People will be strongly advised to keep to this path.

(b) This option attempts to spread the impact over a wide area and will plan to use the same sort of path to the edge of the dunes and then allow people to use informal paths over the dunes. Information will be given to the public on the vulnerability of the footpaths and the dune system in general. It is hoped that better public awareness of the problem will be all that is needed.

(c) Keeping to the already eroded areas. This option will use similar paths to cross the dune heath and will take the path south to the area of maximum erosion. The path will run along the back of the fixed dunes and link up with the existing path through the dunes. There will be notices posted advising the public to keep out of the dune system.

1 You are to take the role of a university lecturer employed by the local planning authority to explain to the local people the structure and functioning of the sand dune ecosystem. They need to know the effects of the three options on the ecosystem. You will therefore need to show:
   (a) the structure and functioning of a sand dune ecosystem which has not been affected by human interference, and
   (b) the effect on this structure by each of the three options in turn. You will need to use the correct terminology of an ecologist but must be careful to avoid using terminology without full explanation. You should refer to biogeography textbooks to give you more information.

2 Taking one of the following roles explain your preference for one of the options:
   (a) a representative of a local conservation group wanting human interference kept to a minimum,
   (b) local farmers wanting to improve the pasture on the dune heath and graze cattle,
   (c) local residents interested in improving tourist potential and encouraging the local council to build more houses on the dune heath.

3 All the plans involve a certain degree of public information. Design the leaflet that you would make available to the campers giving instructions and explaining the need to conserve and care for this environment.

The Tropical Forest – deforestation and its consequences

Considerable public alarm has been shown over recent years concerning the possible consequences of deforestation, particularly in the Tropical belt. This has not stopped the process, which shows every sign of increasing in extent. There is some debate as to the rate of deforestation in tropical areas, but one group, Earthlife Foundation, suggested that there may be no such forest left in forty years. Others are far more conservative, but whatever the truth, the process is a very worrying one. Many scientists are now engaged in studying the inter-relationships that occur in tropical rain forest areas in an attempt to quantify the use that can safely be made of them and what limits should be imposed on their development.

Tropical rain forest is concentrated in three areas of the world. It is found in South America, particularly in the Amazon Basin, in Central Africa, especially in Zaire, and in South East Asia, particularly in Indonesia. The probable extent in these three areas is shown in Fig. 2.10. Any figures concerning extent are to some degree tentative since even satellite images find it difficult to separate primary from secondary forest. Similarly figures to show deforestation must be treated with some caution.

The ecosystem of the tropical rain forest is a highly complex one in which the myriads of component pieces relate to one another in a multitude of ways. Any description of the inter-relationships in tropical forests must necessarily be a simplification. Research is being stepped up on this relatively poorly understood ecosystem in an attempt to be able to predict the consequences of the changes at present being carried out there. Nevertheless, some broad generalizations can be drawn on the ecosystem as a whole.

Fig. 2.11 shows the principal stores of nutrients and the ways in which they are cycled in a tropical rain forest ecosystem. The most remarkable thing about the diagram is that it shows a paucity of nutrients stored in the soil and in the litter. At any one time, the vast majority of nutrients is stored in the biomass, which is the plants and animals in the ecosystem. To outward appearances, the ecosystem is a very rich one. In it live literally millions of species of plants, insects, birds and other animals. Half of the earth's species at present depend on these forests. In fact, this amazingly varied ecosystem is

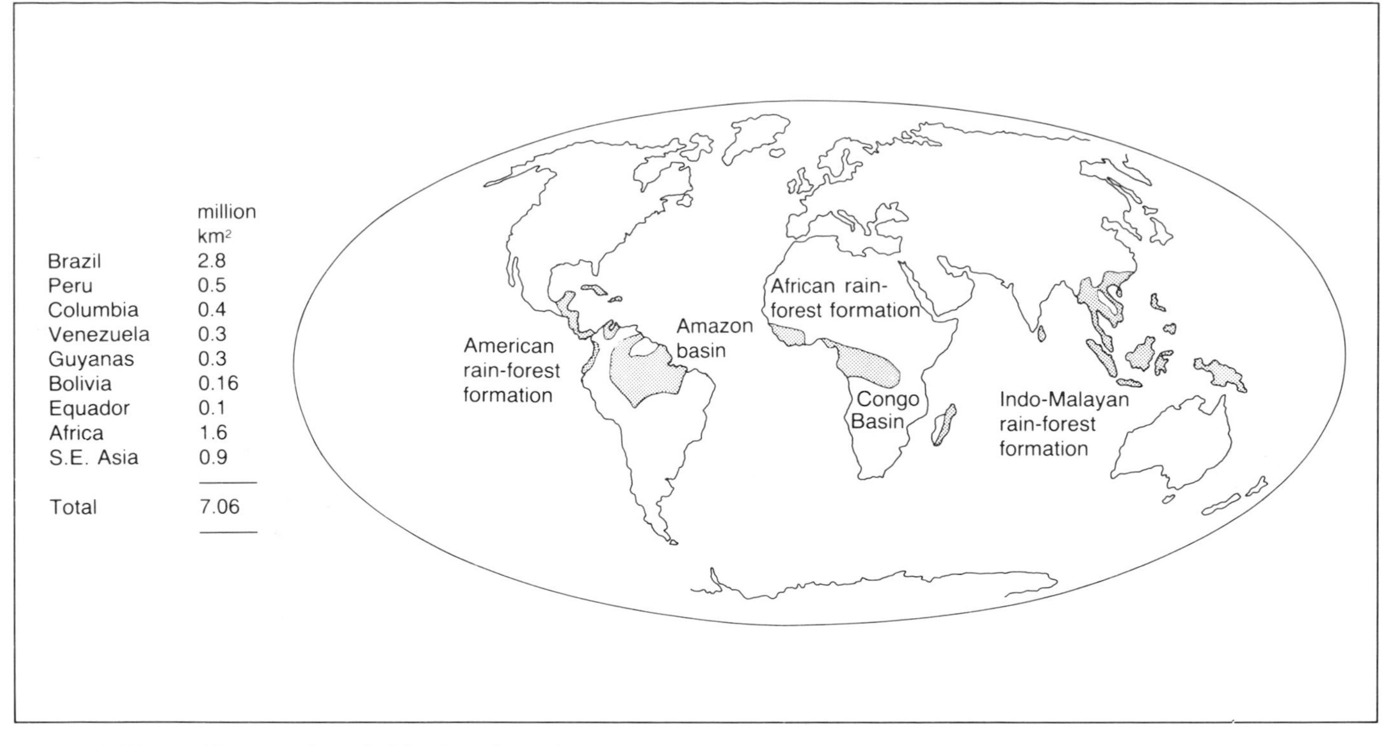

	million km²
Brazil	2.8
Peru	0.5
Columbia	0.4
Venezuela	0.3
Guyanas	0.3
Bolivia	0.16
Equador	0.1
Africa	1.6
S.E. Asia	0.9
Total	7.06

Fig. 2.10 The world extent of tropical hardwood forest

supported by the efficiency of the recycling process rather than by the soil itself.

When a tree falls, it is quickly attacked by fungi and bacteria which also thrive in the hot, humid climate. Whereas minerals are leached from the decaying tree, the root mat of all the other vegetation in the area is concentrated into the top layer of the soil to filter out the percolating minerals. As much as 90% of the root area may be concentrated into the top 20 cm of soil. The soil only holds a small proportion of the minerals in the ecosystem. As much as 90% of the potassium, 92% of the magnesium, 74% of the calcium, 66% of the phosphorus and more than 60% of the nitrogen in the system is held in the plant biomass at any one time according to experimental evidence from the Venezuelan forest. The poorer the soil, the higher is the percentage of the total biomass taken up by the root system, because it needs to be that much more efficient at filtering out the minerals. In a relatively rich tropical forest soil, the root system might make up about 10% of the total plant biomass, whereas in a very poor soil it might rise to 60%. Decaying matter on the forest floor is thus virtually recycled directly back to the living biomass without much relation to the soil.

Fig. 2.11 Nutrient transfer in tropical rain forest

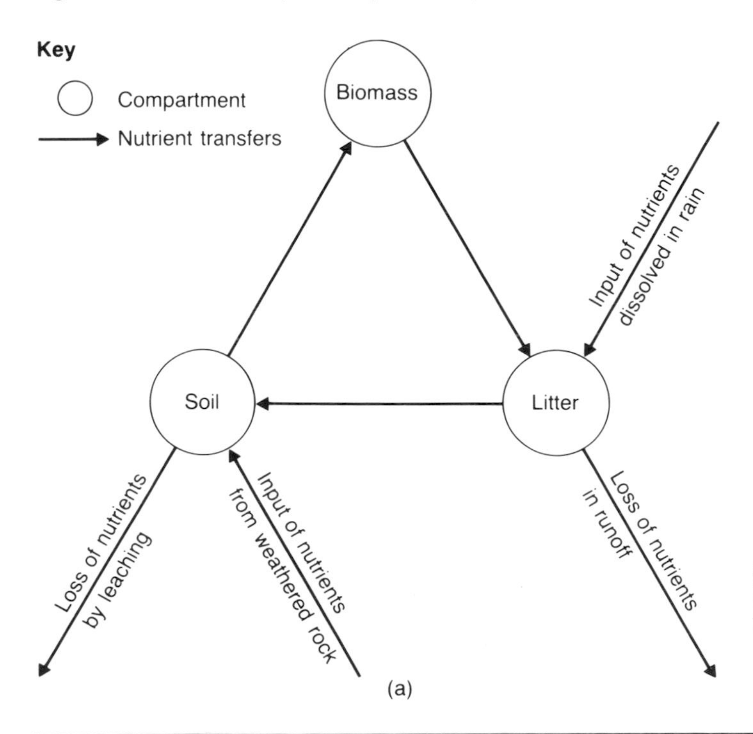

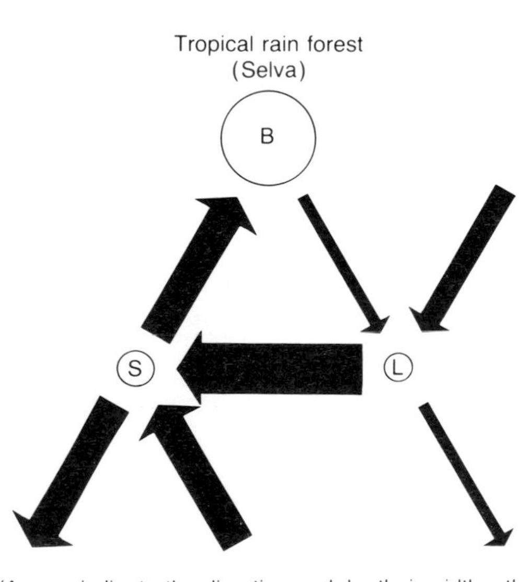

Fig. 2.12 Climate figures for Manaos

Station	Latitude	Height above sea level (m)		J	F	M	A	M	J	J	A	S	O	N	D
Manaos, Brazil	3°S	50	*T* (°C)	26.7	26.7	26.7	26.7	26.7	26.7	27.2	27.8	28.3	28.3	27.8	27.2
			R (mm)	233	228	244	216	177	92	55	35	51	104	139	195

13 Describe the climate of the tropical forest as shown in Fig. 2.12. You should refer to:
(a) the annual variation in temperature
(b) the total rainfall
(c) the variation in rainfall. (Can you pick out any rainfall seasons?)
(d) the fact that most of the rainfall occurs in heavy storms in the afternoons and that the humidity is always very high.

14 With reference to Fig. 2.13 state what effect this climate will have on the weathering of the bedrock underlying the forest? Does this explain why tropical forest soils are usually developed over deep sandy or clay material?

15 In Fig. 2.11, comment on the following:
(a) the uptake of nutrients by the biomass
(b) the relatively small amount of nutrient transfer from the biomass to the litter
(c) a relatively large flow of nutrients from the litter to the soil despite a small store of nutrients in the litter
(d) the balance between the amount of nutrients coming into the system from rainfall and that leaving it through leaching, runoff and throughflow.

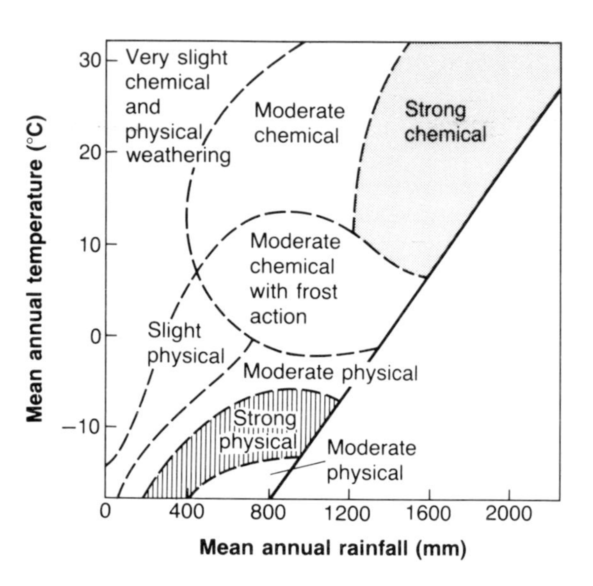

Fig. 2.13 Weathering related to climate

The tropical forest appears to the traveller to be a tangled mass of vegetation without much order or organization. This is certainly not the case, as closer inspection of field data reveals (Fig. 2.14). A layer of tall trees (up to 60 m in places) forms a **canopy**. These canopy trees are of many different species and two trees of the same species may be several kilometres apart. Below this is another layer of trees, some of similar species to those making up the canopy, and some totally different types. Below this again is a third layer of trees, many being the saplings of those in the other layers waiting until the fall of one of the forest giants lets in enough light to promote rapid growth. The lowest layer of trees also contains species that are adapted to live in low light conditions.

On the floor of the forest is what is known as the **undergrowth**. This is often thought of as being dense, since it is so along the banks of the many rivers that thread their way through the forest. On the contrary, the undergrowth is normally very sparse away from the rivers since the light conditions are so poor.

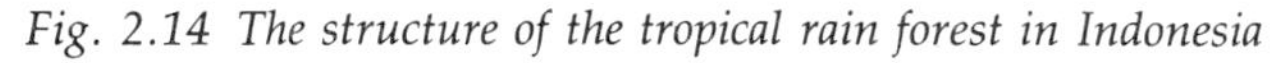

Fig. 2.14 The structure of the tropical rain forest in Indonesia

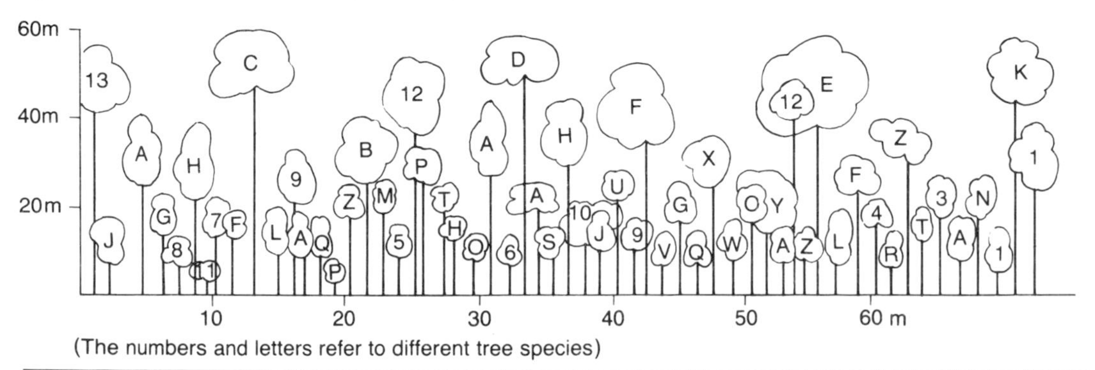

There are many remarkable plants in the tropical forest. Again as a reaction to the light conditions, plants called lianas grow up the trunk of one tree, hang down from the canopy, and grow up the next tree again. They may be 100 m long, with their flowering parts up in the canopy. Other strategies for reaching the light are exhibited by the epiphytes. These small shrubs live in the higher reaches of their host tree and suspend a web of tightly packed fine roots to catch falling leaves and moisture. Yet another way to reach the light is used by the parasites, which penetrate their hosts to suck nutrients from them and thus are able to rest in the high branches, well away from the nutrients of the forest floor. Stranglers start life in the same way as the epiphytes but then they send their roots down to the forest floor, surrounding and gradually killing their host. In all, there are some 40 000 species of flowering plant in Brazil, and some 20 000 different species in West Africa.

If the plant life is varied, then so is the animal life that is harboured by it. We are used to seeing pictures of the African savanna with its vast herds of antelope and other animals. In fact, the tropical forest holds even greater stocks of animals than this. In Brazil, there are monkeys in the highest branches, sloths in the middle layers, and water loving animals such as the cayman, on the floor (Fig. 2.15). The bird species are legion, and snakes, mammals, insects and reptiles of various sorts complete the picture. It is a very safe bet that there are as yet many species in several of these categories that have not been discovered. There is even a strong rumour that in the central African country of Burundi lives the only remaining member of the dinosaur family, closely resembling a brontosaurus! If it is not possible to prove this incorrect, you will see just how dense and unyielding the forest can be.

The tropical forest is under threat because of three main pressures. Consumption of hardwoods in the northern countries, many of which have very strong connections with the forest countries through former colonial ties, has increased markedly. In 1950, 4.2×10^6 m^3 of hardwood were used compared to 8.0×10^7 m^3 that were used in 1980 and a predicted 1.1×10^8 m^3 that will be used in the year 2000. The second pressure is the demand by the rural poor for land on which to grow their food. A third, closely related demand, is for land on which to grow products for export to alleviate the balance of payments problems that are inherent in many countries which also have tropical forest within their boundaries. It is not just the forest itself that is under threat but the people who inhabit it, and perhaps even the climate of the world.

There are perhaps about a thousand tribes in the world's rain forest. It is impossible to be more accurate because, for instance, one new tribe was contacted per year, on average, as Brazil's Trans-Amazonian highway was driven into the rain forest. In 1500 there were probably between six and nine million Indian inhabitants of what is now the Brazilian rain forest, but this figure has already collapsed to less than 200 000. The same pattern is repeated around the world. Many of the indigenous peoples have the potential to help the rest of the world's inhabitants. They often use products from the rain forest which could be developed into important drugs, as many already have. The rosy periwinkle, a seemingly insignificant weed from Madagascar and other areas, has been developed into such a useful anti-leukaemia drug that, whereas in 1960 four out of every five children who contracted leukaemia died, now only one out of five dies. How many more such helpful materials exist in the forest? Some Indians, for instance the Mbuti

Fig. 2.15 A Brazilian tropical rain forest food web

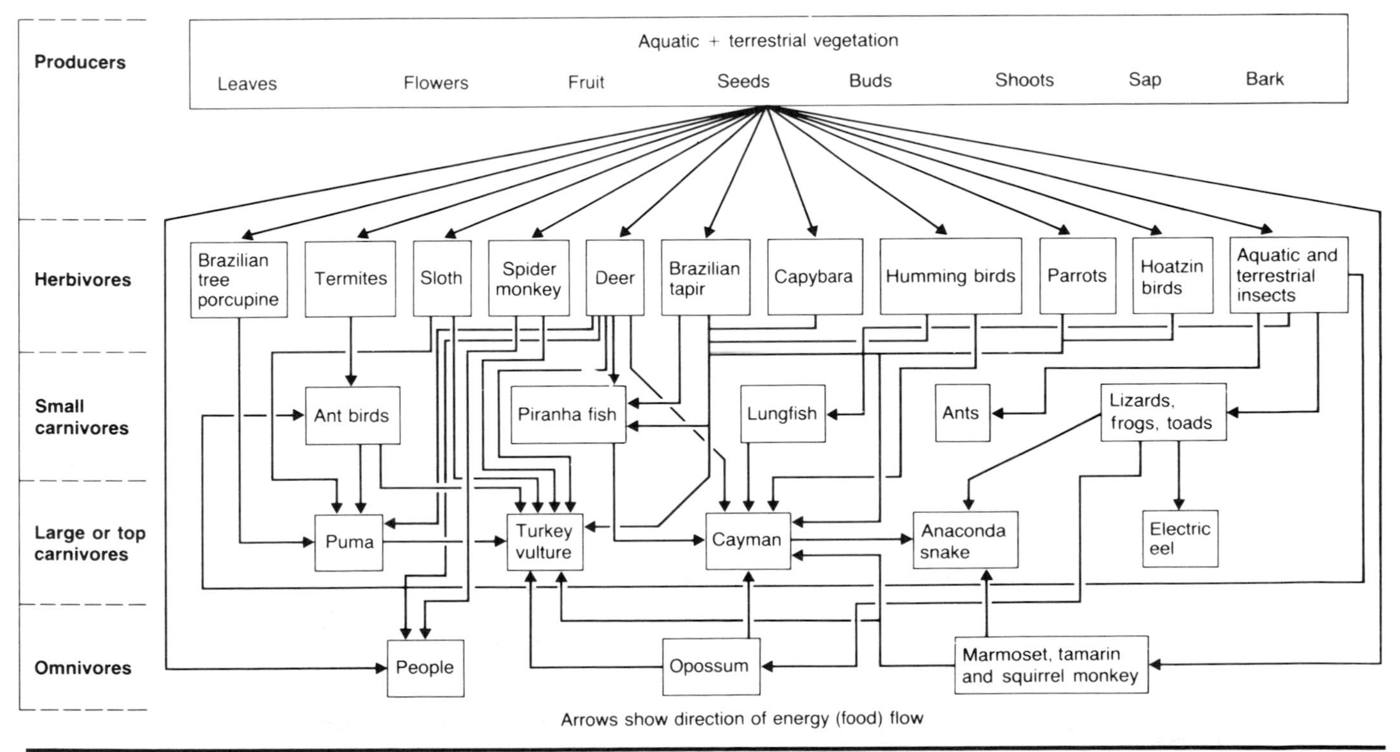

pygmies of the Zaire–Uganda border, have developed remarkable, self-sustaining forms of agriculture that provide all their requirements in a way that many farming systems transplanted from other environments cannot do. It is beyond dispute that such people could help us in our search for better ways to exploit our environment if only we give the forests time to yield up their secrets. Read Fig. 2.16 below.

Case-study: Transmigration in Indonesia

Indonesia is an island state lying between the South China Sea and the Indian Ocean. It comprises 13 667 islands in total and stretches some 5440 km from west to east. The outlying islands retain a dense cover of tropical rain forest, but the original vegetation has all but disappeared from the two central islands of Java and Madura. The population density map (Fig. 2.17) shows the reason for this. Over 60% of the total population of 162 million people live on the two central islands. On Java, the population density is higher than that of Bangladesh and among the highest in the world at 700 people per km^2 on average and 2000 per km^2 in irrigated areas. On the other hand, some 12 675 islands are unpopulated. It is against this background that a major drive to relocate people has taken place in Indonesia. Before independence, the islands formed the Dutch East Indies and it was the Dutch who initiated the relocation policy, known as Transmigration, in 1905. There has been a steady influx of people into the outlying islands as a result of the transmigration policy, but the process has been much more dramatic since 1975. Between 1979 and 1984, over 3.6 million people had been relocated from the three islands of Java, Madura and Bali to places such as West Irian. There are contrasting arguments to explain what has been happening.

The Indonesian government has long had problems keeping political control of its far-flung outposts. There are independence movements on several of the islands. Its stated aims in promoting transmigration are to promote national unity, national security, an equal distribution of the population, national development, the preservation of nature, helping the peasant farmers and improving the lot of local peoples. In the words of the World Bank report on the project, the objectives of Indonesia's policy are to . . .

'(a) improve living standards and employment opportunities of impoverished and landless labourers from Java, Madura, Bali and Lombok
(b) alleviate population pressures on these islands, especially in environmentally vulnerable areas
(c) promote balanced population distribution and regional development throughout Indonesia
(d) increase food and tree crop production.'

It argues that 'Transmigration will solve the overcrowding problems of the central islands'. In the words of the same report, 'the low population densities in some areas . . . impede regional and economic growth. These facts have been so striking

Fig. 2.16 An article from the Observer

Death of a life-cycle

Our land, compared with what it was, is like the skeleton of a body wasted by disease. The plump, soft parts have vanished and all that remains is the bare carcass.'

So lamented Plato, contemplating the destruction of the once plentiful forests of Attica. As the trees came down springs dried up and the land became barren. Over two millennia later the ground is still rocky and infertile.

Now the wasting sickness has spread with our civilisation to the most vulnerable, and vital, organs of the living earth.

Rapid destruction of the great tropical rain forests threatens to make a million species of plants and animals extinct by the end of the century. It is endangering water supplies on which a 1,000 million people depend for food, and may alter the world's climate.

The World Wildlife Fund has just launched its most important campaign, to try to save the forests. On Thursday a British organisation, Earthlife, will announce its contribution to the campaign, aimed at raising £3 million to help save three important forests in Cameroon.

The rain forests are the most exuberant celebration of life ever to have existed on earth. They cover only six per cent of its land surface, but are home to almost half of all living species. Yet they grow on some of the planet's poorest soils.

They achieve this biological miracle by constructing an intricate and delicate web of life, spread like a damp green blanket over almost sterile ground. Trees, 150 feet high, virtually perch on the surface thrusting most of their roots flat out across the forest floor into the thick, nourishing layer of debris that covers the ground. In the twilight, almost windless world between this rich carpet and the tree canopy is the most extraordinary biological powerhouse on earth.

Yet we know no more than the barest outline of life in the forests. We have not even discovered more than 15 per cent of their four to five million species, let alone studied them.

Experts say the largest area of the Brazilian Amazon forest for which there is a reliable tree survey is a mere 13½ acres. Only recently we discovered an unknown major tributary of the Amazon, and realised that several mountain ranges are in the wrong place on our maps.

Indeed, we don't even know for sure how much we are cutting down. Conservative estimates are that about half of them have already gone, that 18 million acres are going every year and another 11 million are severely damaged.

On present trends Malaysia, Nigeria, the Ivory Coast and all Central America will be bare by the end of the century.

As the trees come down there is nothing to protect the precious nutrients from the tropical rains which can strip 75 tons of soil from an acre in just one storm.

Much of the forest is cut down for hardwoods, like mahogany. Only six per cent of the wood taken from forests is exported, but it is still worth over £4,000 million on world markets. As Asian forests are logged out, the companies are turning to Latin America.

Their tracks often open up the jungle to agriculture, the greatest source of destruction.

Poor people, denied access to land elsewhere (93 per cent of Latin America's farmland is owned by just 7 per cent of its people), hack out plots in a desperate attempt to find somewhere to grow food.

Magnificent

But farmers find that the apparent fertility of the forests is an illusion. After a year or two of good harvests the soil reverts to its original sterility. When the land is destroyed they move farther into the forest and start again.

Rich farmers do the same damage on a grander scale. They are, as Dr Gerardo Budowski, a leading authority puts it, 'turning forests into hamburgers.'

Cattle ranching is the major cause of destruction in Central America and the Brazilian Amazon. It produces cheap, lean beef suitable for fast food chains and almost all of it goes north to the United States.

Within seven years or so, the land has become useless, but the ranchers have made their profits, and they move on to flatten more forest.

Costa Rica's beef production has trebled over the past 20 years. But it has gone for export; the local people now eat half as much as they did—less meat than a domestic cat in the United States. And the country's magnificent rain forests, which contain more bird species than the whole of North America, have been slashed by a third.

A US presidential report warned that a million rain forest species may become extinct by the year 2000. That would be, as one leading expert put it, 'an evolutionary Rubicon whose crossing *homo sapiens* would do well to avoid.'

Even the death of a single species diminishes us all, because each is a storehouse of substances that may well prove invaluable to man. The humblest bacterium, it is said, can synthesise in its brief life more organic compounds than all the world's chemists combined.

Already half the medicines in the world's pharmacies are derived at least partly from rain forest products, and new sources of drugs are being found all the time in these natural laboratories.

Two chemicals from the rosy periwinkle, from West Indian rain forests, have quadrupled the chances of successfully treating Hodgkin's disease, given lymphocytic leukaemia patients an excellent chance of recovery, and proved effective in fighting other cancers.

Not least, the disappearance of the forests leads to the extinction of men. There were over six million Brazilian Indians when the Europeans first arrived in Latin America, now there are only 200,000.

When the forests go, rainfall drops, and fresh water supplies are disrupted. Even worse, their disappearance adds to the 'greenhouse effect' which most scientists expect will change the world's climate in the next century. One result could be that the Soviet Union rather than the United States becomes the world's leading grain producer, with enormous geopolitical consequences.

This spring an international attempt by three UN agencies to draw up a global rescue plan for the forest collapsed when key countries like Brazil, Zaire, Columbia, Venezuela and Burma failed even to turn up.

GEOFFREY LEAN, our Environment Correspondent, warns that rapid destruction of tropical rain forests is endangering vital food and water supplies

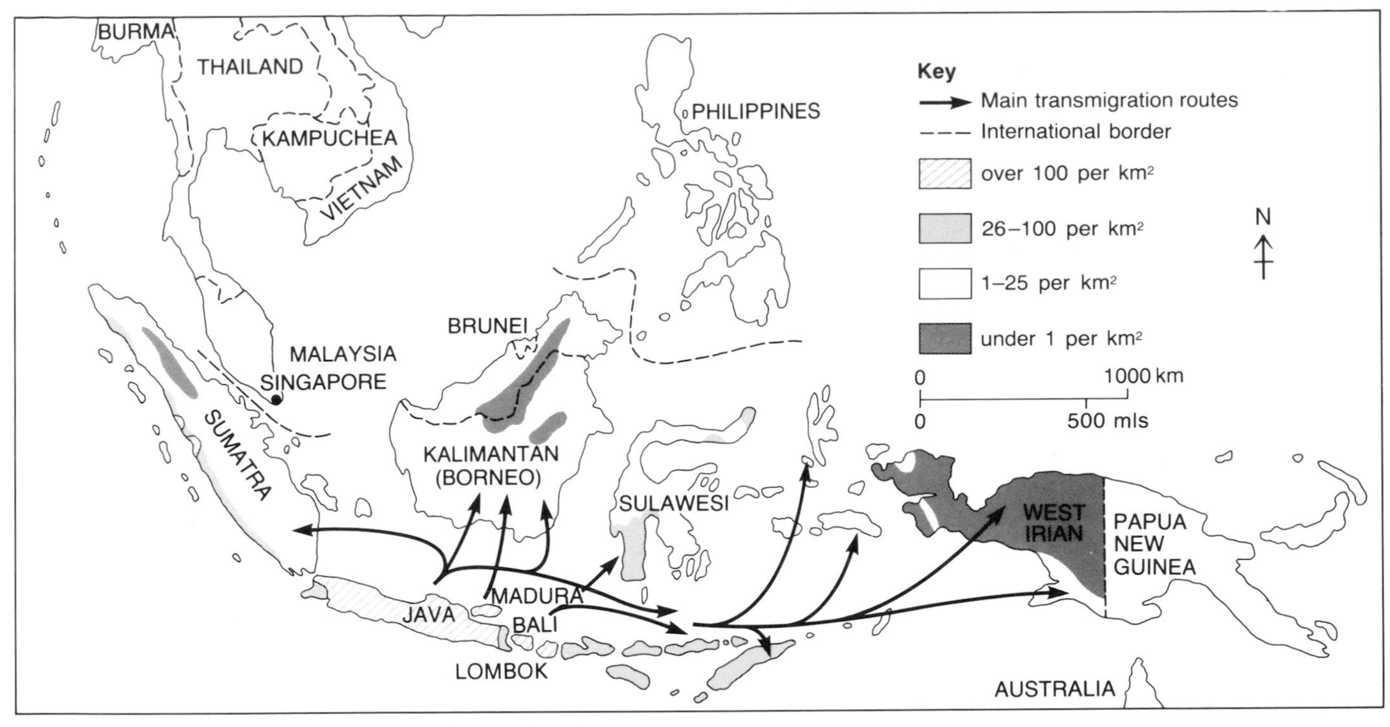

Fig. 2.17 Indonesia: population density and transmigration routes

for so long that programmes to wed the underutilized labour of Java with the underutilized land of the other islands have been carried out since the beginning of the century.' Opposition to Transmigration comes from those who say that it is really a 'Javanization' policy. They point out that, despite the rapid increase in the numbers involved in movement, population pressure at the centre has not declined. In fact, just as the poorest move out, the richer students and entrepreneurs move in to be nearer the 'core'. The net effect is 'the export of poverty'.

Dire consequences are being felt in the forests. In Kalimantan, 200 000 ha per year are felled in response to the demand for land from Transmigration. Top soil is being lost at an increasing rate as the bulldozers clear the forest. The settlers who learnt their farming techniques on the rich volcanic soils of Java, Madura, Lombok or Bali, have little idea how to cope with the poorer lateritic soils of the other islands. They often experience rapidly declining yields after a short while and total crop failure after five years. After this, they may abandon their plots and migrate to the nearest urban area, or continue to farm by becoming shifting agriculturalists. The ecosystem is under great stress as the forest becomes fragmented. There is an attempt to fit the settlements and agriculture to the environment in that settlements are not sited on slopes of more than 8% for food crops and more than 25% for the tree crops. In the World Bank report, it states that forest which contains rare flora and fauna as well as areas of 'particular ecological importance' should be avoided.

Fig. 2.18 Transmigration settlements in West Irian

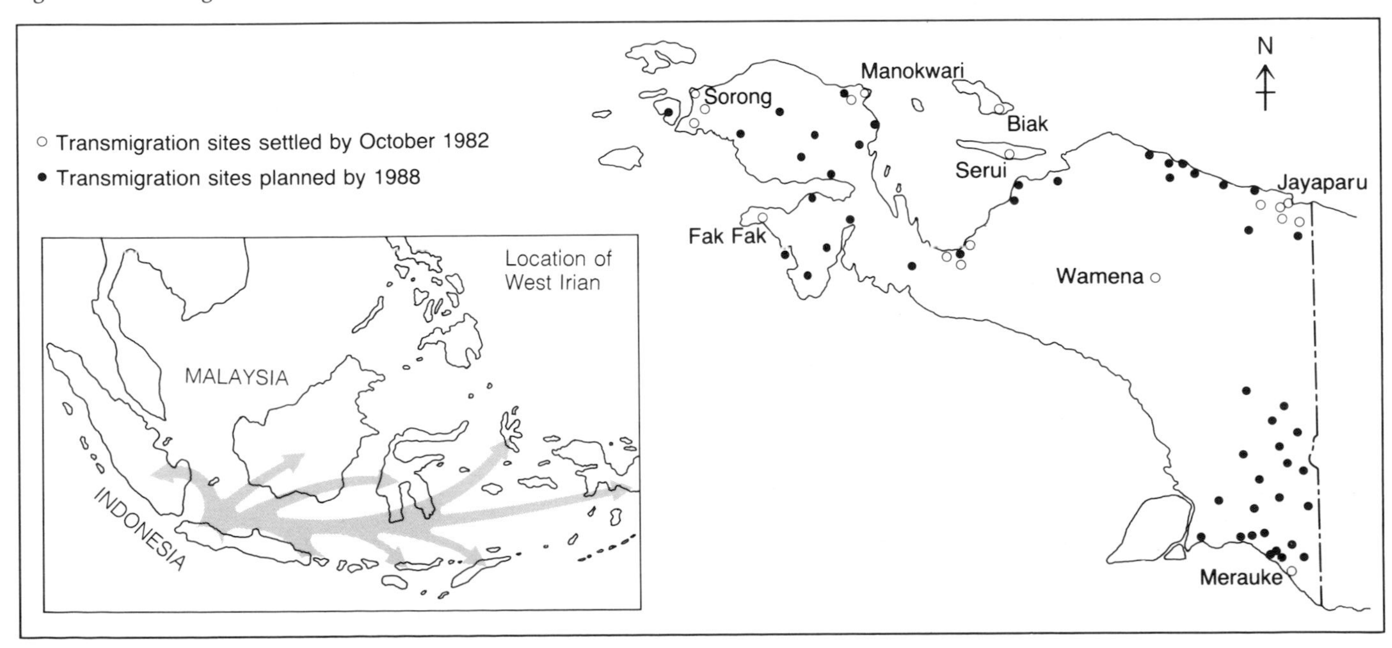

Table 2.3 Transmigration programmes (1969–89) (from a World Bank report)

No. of families	Repelita I (1969/74)	Repelita II (1974/79)	Repelita III (1979/84)	Repelita IV (1984/89)/*a*	Total
Sponsored	46 000	83 000	366 000	450 000	929 000
Spontaneous /*b*	17 000	35 000	169 000	300 000	529 000
Total	63 000	118 000	535 000	750 000	1 458 000

/*a* Targets
/*b* Families with limited or no government assistance

Notes: Repelita = five year plan
Sponsored migrants = landless agricultural labourers or smallholders living at near subsistence level. They receive considerable government support during the first five years in the form of transport, land, housing, social services, and a subsistence and agricultural package for the first few years after settlement.

Fig. 2.18 shows the settlement pattern being established in West Irian. The state was only annexed by Indonesia in 1969, yet it is now one of the main targets for population movement. Here the soils are particularly fragile and poor, but the government has recently published figures which show that in a five year period it intends to settle 55 000 people in a country where the total population is only 1.2 million (Table 2.3). Estimates of the real extent of settlement, as opposed to government figures, put the number to be settled at anything between 500 000 and 1 million. West Irian has already lost 700 000 ha in the Transmigration programme. The influx, working on the unofficial figures, would lead to the further loss of some 3.6 million ha of forest by 1989. The report does say that two sites in West Irian will be monitored to see what social and environmental impacts there will be. In one area in the south-east of the country, however, two thirds of the tribal lands have been set aside for transmigrants. This is as a result of the Indonesian law not recognizing hunting, gathering or shifting agriculture lands. The consequence of this is that, as well as the ecosystem being fragmented to the point where it can no longer function properly, the inhabitants who lived in harmony with the forest are driven out under the influx of the transmigrants. The displaced people have been heading in ever greater numbers (10–12 000 people in 1984) to Papua New Guinea, where no such restrictions exist.

The World Bank report says that 'there has been a remarkable receptiveness to settlers in most parts of Indonesia'. In reality, the problems of conflicting land claims are acknowledged by the fact that in West Irian, 50% of the transmigration plots are made available to local people, although it is further acknowledged that there is not the demand from locals to fulfill this percentage. Despite the problems, transmigration remains a major plank of the Indonesian government's policy which is funded by the international community (Table 2.4). The World Bank report on this project nevertheless does not include one direct reference to the effect of settlements and agriculture on the tropical forest ecosystem.

Table 2.4 International funding for transmigration

World Bank	$400.3 m.
United Nations Development Programme	$ 4.9 m.
World Food Programme	$ 56.7 m.
European Economic Community	$ 11.5 m.
Asian Development Bank	$ 34.3 m.
Islamic Development Bank	$ 10.0 m.
Federal Republic of Germany	DM 88.5 m.
France	$ 1.9 m.
Netherlands	FF 22.2 m.
United States of America	$ 15.0 m.
Others, including NGOs	$ 5.1 m.
TOTAL (approx.)	$600.0 m.

16 Write down what *you* think are the most likely reasons for the Transmigration policy.

17 With reference to Fig. 2.14 (p.60), what do you think will be the main effects on this ecosystem from Transmigration?

18 What are likely to be the responses to Transmigration from:
(a) a political leader from West Irian?
(b) a tribal chief whose land is being settled?
(c) a member of the Islamic Development Bank?
(d) a landless peasant from Java who has become a government-sponsored migrant?

Decision-making Exercise F: Opening Up the Forest in Rondônia

F

Brazil is divided into states, one of which is the newly-created state of Rondônia. This occupies the south-western corner of the Amazonia region next to the border with Bolivia. It is being opened up to settlers through the creation of an infrastructure, especially new roads. There is considerable controversy about whether this is good for the country as a whole or for Rondônia in particular.

You are provided with data concerning the development, and with a series of statements expressing the views of people concerned with the plan. (The statements are mock-ups of the views and not actually made by any individual.)

You are to complete the following:

1. Write a further statement made by a landless peasant from the south-west of Brazil. You recently lost your small plot of land when it was bought up by a government agency which had the intention of starting a soya bean farm on the estate, which includes your land. You now find that your labour is not required on the estate and the capital you received from the sale of your land is running out. You should use the same format as that employed for each of the statements on the following pages. State in particular whether you support the opening up of Rondônia, and what are your alternatives to moving to the area.

2. Write a newspaper article for the government newspaper summarizing the arguments for the plan.

3. Make a summary of the decision-making sequence used by the Brazilian government to open up Rondônia.

4. Write a letter to the prime minister of Brazil setting out your own personal position on the project.

A statement from an ecologist
The proposed Polonroeste Project in Rondônia carries certain major threats for the environment:

1. The ecosystem is very fragile and the inter-relationships that exist within it are not as yet fully understood. To develop an area which until recently was totally rain forest is therefore premature and ill-judged. In the most detailed study ever of the forest, in an area 100 km to the north of Manaos, investigations are being made of five different sized areas of forest left when clearance has been made of large areas for cattle ranching. These areas of forest are squares left in their original state in agreement with the developers. So far it has been found that the exposure to wind around the edges of the forest has led to desiccation of some species and their subsequent death. A troupe of howler monkeys in one isolated patch of forest has died, probably due to the forest not being large enough to support them. On the other hand, another species of fruit-eating monkey has also died out in the same area despite having sufficient space and food. The reason is not so far understood, but it could be related to subtle changes in species mix, both in plants and animals. The lowering of humidity levels within the forest due to increased wind action has led to the development of this different species mix, both in plants and animals. The lowering of humidity levels within the forest due to increased wind action has led to the development of this different species mix. The rain forest ecosystem is a giant machine. It has not yet been shown how many cogs can be lost before the whole thing grinds to a halt.

2. The soils underlying the forest are again most fragile. Most of the nutrients in the system are stored in the biomass and the shallow lateritic soils are easily degraded once they are cleared. Acid, infertile soils, with a pH value of 4 or below, cover nearly 75% of Amazonia. Only just over 10% of Amazonia has moderately fertile or very fertile soils. A comparative study of the changes in soils in an area near Manaos has shown that several major changes happen to the soil if it is managed unwisely. On a slash and burn plot, very little surface erosion was evident, but on a similar sized plot cleared by bulldozer, a surface erosion figure of some 6 cm was soon noted. Equally, although the slash and burn plot retained some nutrients after clearing, the bulldozed plot lost nearly all of its nutrients through rapid leaching very soon after clearing. Compaction

was much greater in the bulldozed plot. Roots can only penetrate the soil if it has a penetration resistance figure of below about 9 kg per cm^2. A figure of 9.5 kg per m^2 was reached at a fairly shallow depth in the bulldozed soil, rendering it unsuitable for most crops. A substantial reduction in infiltration rate was noted in the bulldozed plot, though not sufficient to cause surface run-off.

3. It is not known how large a patch of forest has to be before it can retain all its complex inter-relationships. In the absence of firm information, we would do well to avoid large scale developments in such areas. The Tropical Forestry Working Group, Washington DC, has produced a clear statement to this effect and it speaks for all the principal global environmental and conservation organizations and some 200 scientists, technicians and resource specialists of several nationalities dedicated to the good management of the world's tropical forests.

An idea of the complexity of the ecosystem is given in the diagram of the food web in part of the rain forest in Amazonia (Fig. 2.15, p.61).

Fig. 2.19 (a) From the Observer

Brazil buys dream of American billionaire

from JAN ROCHA in Sao Paulo

TWENTY-THREE of Brazil's leading bankers and businessmen will meet President Figueiredo in Brasilia tomorrow to celebrate the transfer of control of American billionaire Daniel Ludwig's Amazon forest empire to Brazilian hands.

The new Jari company that will administer the wood pulp and rice production enterprise, abandoned by Ludwig after innumerable setbacks, is a consortium of the country's biggest banks, insurance companies and engineering, construction and mining firms, supported by $180 million from the Brazilian Government. Control of the company will be exercised by manganese mining magnate and close friend of Ludwig's, Augusto Trajano de Azevedo Antunes.

Most of the 22 businessmen and bankers who agreed to back the deal worked out by Antunes show little optimism about the chances of Jari becoming a lucrative investment but bowed to government pressure to participate for 'patriotic reasons.'

For many Brazilian economists and members of opposition parties, however, the whole Jari episode is a scandal. Congressman Horacio Ortiz has called for a parliamentary inquiry to investigate the application of government money 'practically double what many Brazilian States have to dispose of to meet the needs of millions of people.'

The Jari project, covering up to nearly nine million acres of Amazon land, began in 1967.

Ludwig, banking on a forthcoming world shortage of pulp, looked around the world for a piece of land large enough to plant millions of fast-growing gmelina trees. Wooed by Brazil's military Government, he bought a vast tract of Amazon forest for a song and set about clearing the jungle.

The project included a pulp mill floated across the world from Japan, a huge rice paddy on the banks of the River Amazon, kaolin quarries and cattle herds. Thousands of Brazilian peasants were recruited to work in Jari, and a shanty-town housing 35,000 people sprang up.

But Ludwig's high-handed manners, his refusal to allow Brazilian journalists to visit the project, the absence of any government supervision of his activities and the allegation that Jari was claiming more land than it legitimately held title to, evoked accusations by congressmen and Press of an unwelcome 'foreign enclave'.

When the Government refused to act as guarantor for the import of a second floating pulp mill from Japan, Ludwig bluntly threatened to withdraw from the project unless the Government took over all infrastructure costs.

But growing Brazilian hostility to the Ludwig empire was probably only a minor part of his decision to pull out. The real problem was the 'Amazon factor,' a combination of climate, distance, soil and pests, that increases the odds against any project, let alone one of the gigantic proportions of Jari.

Ambitious production targets could not be met: only 31,000 tons a year instead of 170,000 tons of rice were produced, and the tree plantations ran into endless problems with the thin Amazon soil.

Ludwig's imperial style of management did not help: he would make sudden trips to Jari, making impulsive decisions to change plans, endlessly firing managers and technicians.

Fourteen years after Ludwig's Brazil honeymoon began, he finally washed his hands of the project, leaving the Brazilian Government to pick up the bill for $180 million worth of debts, including $29 million to Lloyd's Bank of London.

Fig. 2.19 (b) Deep regolith in latosol, near Manaus

Fig. 2.19 (c) Tropical rain forest, Upper Rio Negro - Amazonas

Fig. 2.19 (d) Buttress roots

A statement from a member of the National Indian Foundation (FUNAI)
Rondônia is home to several indigenous peoples, who need to be protected from incursions made into the area by settlers attracted by the Polonroeste Project. As a means of protecting these people, special reserves have been set up where they can carry on their traditional way of life, unaffected by the new settlers.

Indian Posts have been set up throughout the area. These offer health care to the Indians using funds supplied directly by central government. A new regional office has been opened in the small town of Vilhena, where a clinic staffed by a doctor, four nurses and a medical technician has been set up. The posts allow close contact with the Indian groups and ensure that the plans are not monolithic and over-centralized. They can therefore react more sensitively to the needs of the local people. The loan agreement for the first phase of the Polonroeste Project states (in section 4.05):

'The Borrower and the Bank agree that the strengthening of the measures to protect the indigenous Amerindian population in the programme area is essential to the successful carrying out of the project. To this end, the Borrower shall take all necessary measures to put into effect promptly the special project for protecting the interests of the Amerindian communities located in the programme area.' Our response to this clause has been the special funding of the Indian Posts and stipulating that any development should have a 'Certidao Negativo' or 'no objection' clause before it can go ahead.

The Indian Foundation is the Successor to the Indian Protection Service, which was disbanded when atrocities against the Indian came to light. One hundred and thirty-four people from that agency were charged with offences and the new Foundation came into being, since when there has been much more careful supervision of its activities. It is now truly representative of the indigenous peoples. It carries on research into the territories of Indian groups of whom very little is known at the moment, such as the Uru-eu-wau-wau, to ensure that they do not suffer from settlement schemes like the Polonroeste Project.

Fig. 2.20 Map of Rondônia, showing Indian posts and reserves, from The Ecologist, *Vol. 15, 1985*

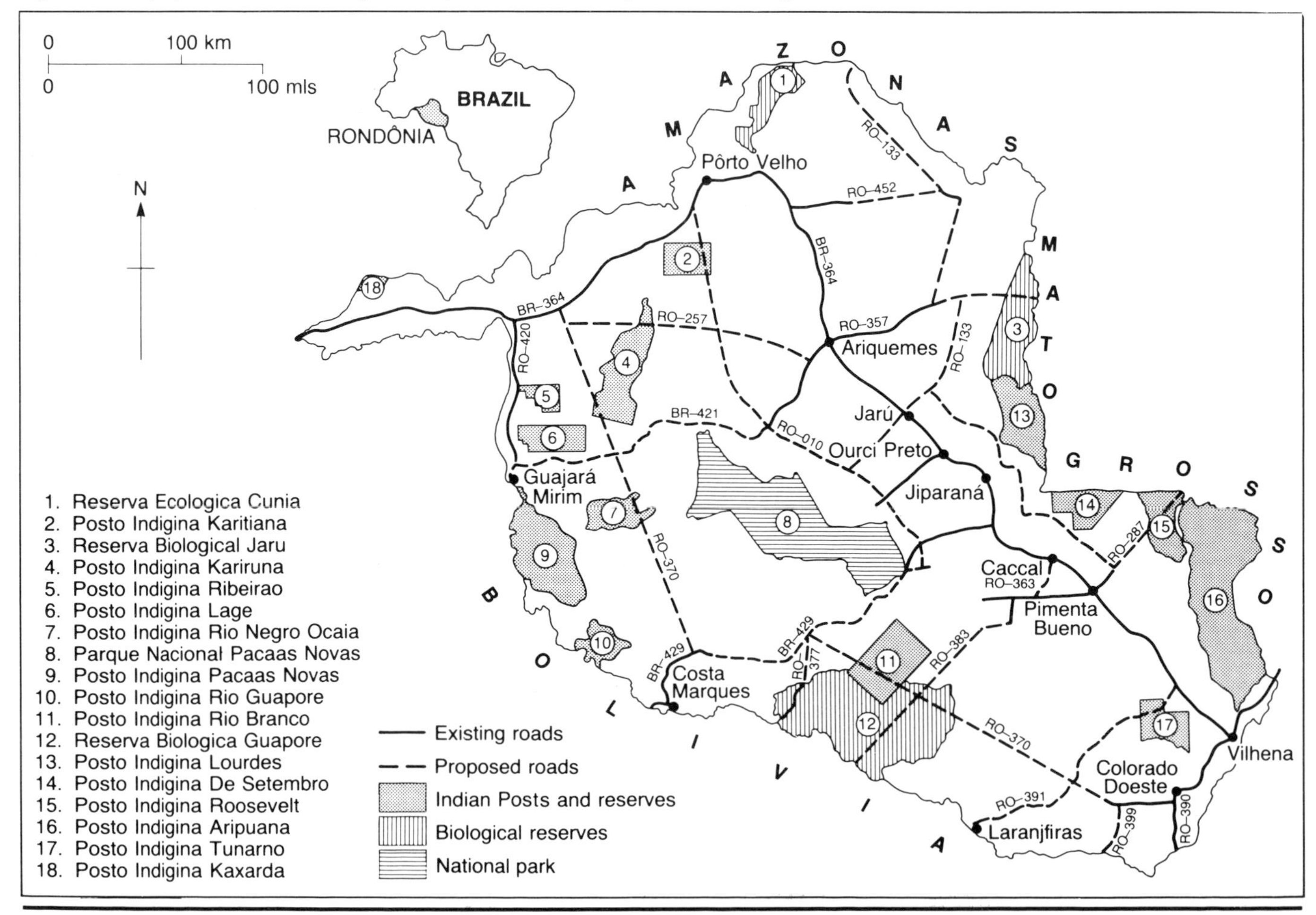

F

A statement from an anthropologist working in the Polonroeste area
There are many groups of indigenous peoples in Amazonia, many of them only recently contacted and their way of life very poorly understood. These peoples are highly susceptible to projects such as Polonroeste. The government has paid lip service to them by setting up FUNAI, but this agency has several very major shortcomings as far as the protection of the Indians is concerned.

Another government agency, INCRA, the land agency, is active in advertising the land and prospects of Rondônia. Most of the people it entices into the area come from the south of Brazil. There the soils are rich, and a sustainable peasant agriculture is possible. When these people reach Rondônia, they find poor organization waiting for them. Sometimes they are allowed to take over land which has been designated as Indian reserve land. One example of this is along the southern borders of the Guapore Biological Reserve, where the relatively fertile lands have been settled by incomers despite an earlier designation of the area as an Indian reserve. Where they settle land that has poor, infertile soils, they soon find that agriculture is much more difficult than they had imagined. In order to make enough from the land to survive, they often go beyond the area that has been allocated to them, sometimes coming into conflict with the Indians as they do so. There have been several incidents of settlers being murdered by Indians as a result.

There is much that could be learnt from the Indians. They have a detailed knowledge of the ecosystem. They are hunter-gatherers who are able to live in the environment with few of the traumas experienced by the new settlers. Their store of knowledge is in jeopardy if we are not more careful with our 'progress'.

Fig. 2.21 (a) Newly cut forest road, 25 km north of Manaus

F

The Polonroeste Project involves the building of a 1000-mile, all-weather road into Rondônia, together with feeder roads and measures to improve the agricultural productivity of the area. The 158 000 square miles covered by the project contains many Indian groups, whose lands are being cut through by the roads, thus cutting their territories into fragments. This has a disastrous effect on their way of life.

New settlers are supposed to have clearance that there are no Indians on their land before they start developing it. The 'Certidao Negativo' has however been given out with gay abandon and the training of new Indian Agents has been stopped, ostensibly for lack of funds. This does nothing to show that FUNAI really means business. At a time when FUNAI still contains some previous members of the discredited Indian Foundation, this can do nothing to instill confidence in it.

There is in Rondônia a tribe known as the Uru-eu-wau-wau. Little is known about them, except that they are fiercely independent and at the moment unwilling to cooperate with any development plans. In the absence of any firm information about the extent of their territory, developments have gone ahead which break Brazilian and international law. Their land has undoubtedly been settled and cut by roads. They have had no say in this, and their reaction will be a violent one, which will inevitably turn other people against them. The Polonroeste Project is based on inadequate information and is to be conducted by incompetent agencies, with consequent stresses and problems for both the environment and the people. It should be curtailed until these vital aspects have been investigated further.

Fig. 2.21 (b) Native peoples' confrontations in Latin America from New Internationalist, *Jan. 1986*

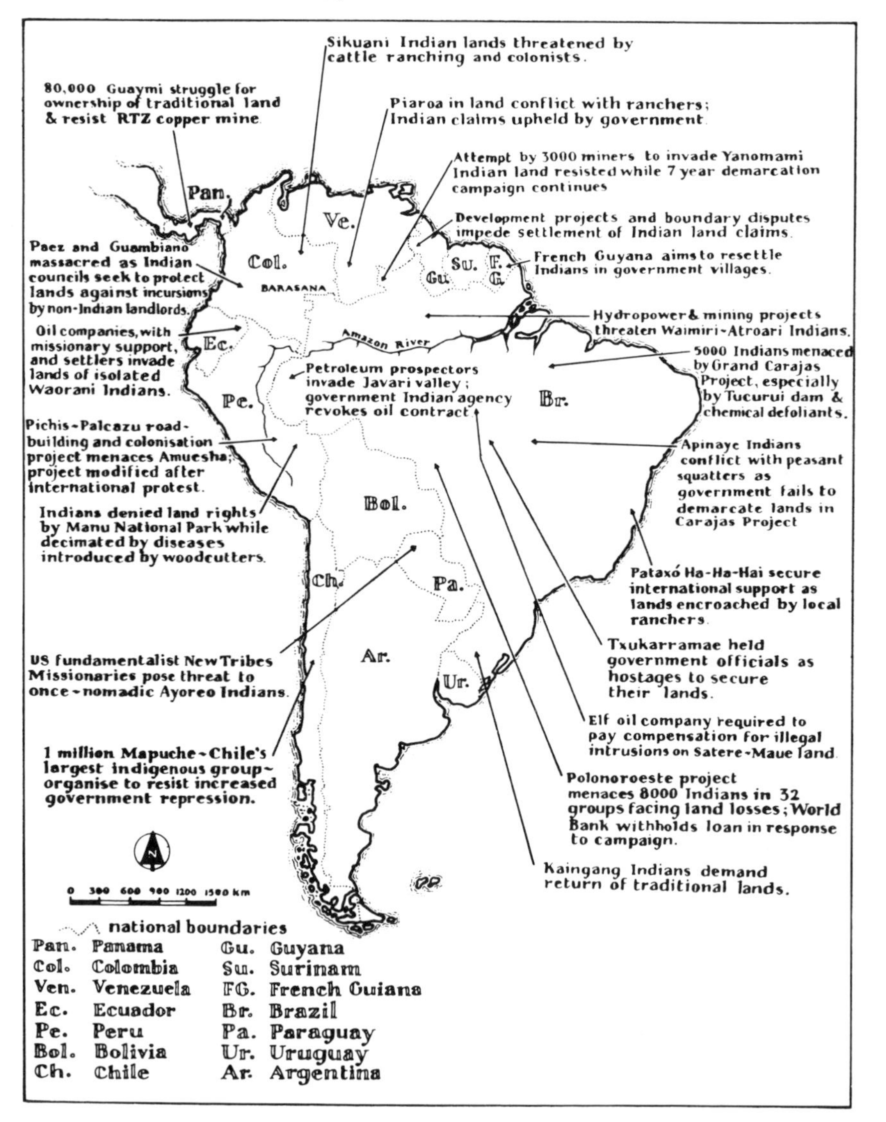

A statement from a member of INCRA

INCRA is the Brazilian National Institute for Colonization and Agrarian Reform. Our task is to coordinate the settlement schemes in Rondônia with the overall development plan and to ensure that the development fits in with the demands made by FUNAI.

There are many landless peasants in Brazil. For many years they have been migrating to the cities when life becomes intolerable in their rural areas. São Paulo, Rio de Janeiro, Belo Horizonte and many other Brazilian cities are among the fastest growing cities in the world as a result of this process. There they live in squalid conditions, often greatly resenting the move they have made but being unable to reverse it.

INCRA is offering this army of peasants new hope. It is granting them title to land in Rondônia and helping them to set up in the new environment. Already many new settlements have grown up and agricultural production is beginning. We are proud to be making probably the largest agricultural reform in the world. By the time we have completed the project, we will have made a considerable dent in the estimated 2 500 000 landless poor in Brazil today.

The main aim of the Polonroeste Project is to ensure growth of production in harmony with preoccupation for the preservation of ecosystems and natural resources. We are succeeding in this aim by selecting the migrants and by working closely with FUNAI. We have divided up the land so that each family has sufficient to produce not only for themselves but also room to expand in the future. Already the area is producing maize, bananas, vegetables and animal products, especially beef. The future looks bright for agriculture in Rondônia.

Fig. 2.22 (a) Cocoa estate in Reconcavia, Bahia

Fig. 2.22 (b) Buildings on the edge of Manaus

Fig. 2.22 (c) Recently cleared patch for slash and burn in forest 60 km north of Manaus

F

A statement from an agronomist
The Polonroeste Project is an unmitigated disaster for Brazil. I make this statement after considering the evidence provided by several exhaustive studies of agricultural potential and the way it has been exploited in Rondônia.

The Brazilian government has conceived the plan so that it can avoid addressing the real problems of agriculture, especially as they exist in the south. There the land has been taken from the poor, by fair means or foul, in order to make room for the huge, government-encouraged estates to grow export crops, especially soya beans. The resulting social injustice is prevented from boiling over by the offer of large holdings in Rondônia. It is a good job, from the peasant's point of view, that the holdings are large, because the land is so poor compared with that which they have been forced to leave. They are often forced to use the land by a modern form of shifting cultivation, since the plot they cleared in the first year becomes almost non-productive in the second. The system which drove them off the land in the south is also catching up with them here. Already, several peasants have sold up their holdings to large companies which have amalgamated the plots into large units for the running of cattle, an alarmingly inefficient way of using the forest.

The designation of plots has been done in an incredibly insensitive way. I came across farmers whose land was crossed seven or eight times by the meanders of a river. This made parts of their plots totally inaccessible to them. Some people's land is almost pure sand; other people's is on rock outcroppings. Some plots are so steep that they cannot be safely ploughed. Some may not be near water. This is the result of somebody sitting in an office, dividing the land up into squares by using a ruler! This strategy may have worked to some extent in the prairies, but it cannot possibly work in such a sensitive environment as the rain forest.

Fig. 2.23 (a) Peach palm and breadfruit trees, on the outskirts of Manaus

The help given to new settlers is minimal. They may be given a talk as soon as they arrive, but they are thereafter left to their own devices. The fertilizer that they may possibly buy to help them make a start is very quickly washed through the soils which have little or no holding capacity. They are not told that the best way to get a return on a sustainable basis is not to clear the land, as they were used to doing in the south, but to mimic the forest by growing a mixture of tree crops or by using some of the trees that are already on their land, such as breadfruit or Brazil nut. They are simply left to make their own mistakes. There are in the forest people who know how to use it. Apart from the Amerindian hunter-gatherers, there are rubber tappers and the caboclos (of mixed Indian and white stock) who live by shifting agriculture. Nobody has bothered to use their expertise or to ensure that the new settlements do not interfere with their traditional lands.

The Project is making three big mistakes. It is using a false justification for its existence, in that it is substituting a desire for giving land to the landless in Rondônia instead of addressing the reason for this, which is that land has been taken from these very people in both the north-east and the south. It is promoting totally unsuitable forms of agriculture in Rondônia, and it is driving out of the forest the very people who know how to use it in a sustainable way. The Project, as I said before, is a disaster.

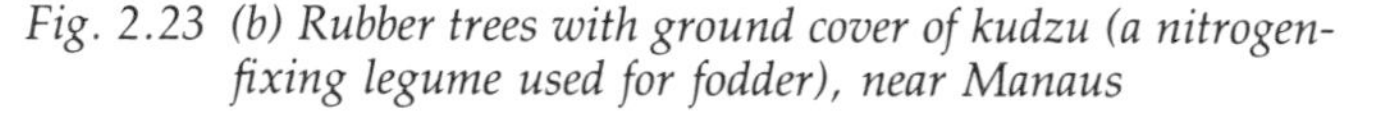
Fig. 2.23 (b) Rubber trees with ground cover of kudzu (a nitrogen-fixing legume used for fodder), near Manaus

Fig. 2.23 (c) A farmstead in Rondônia

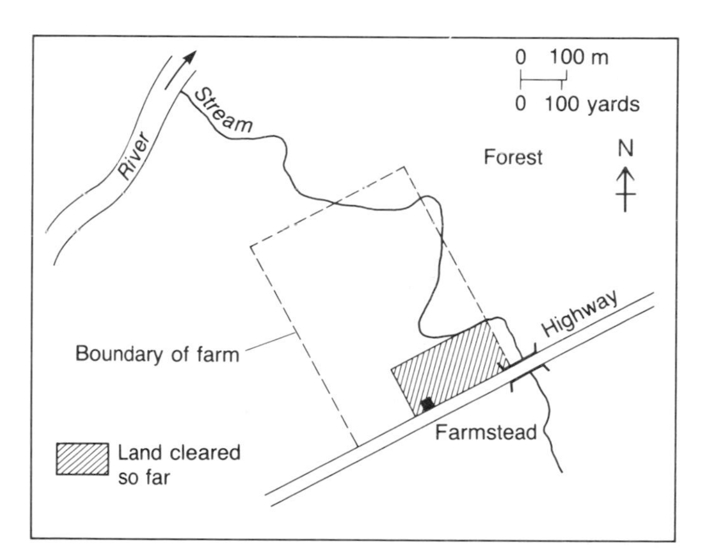

F

F

A statement from a climatologist
There are two big worries about deforestation such as that proceeding at this moment in Rondônia. One is that it will upset local weather patterns, and the other is that it will disturb world climates.

Not very much is at present known about the workings of the Amazonian climate. In order to find out more, an Anglo-Brazilian group has been set up to investigate the inter-relationships between climate and forest at a site 25 km north-east of Manaus. There they have constructed a tower which thrusts up through the canopy to measure incoming radiation, outgoing radiation and humidity levels. It has been established that up to 70% of the sun's energy is used up evaporating water. Thus the vegetation has an important cooling effect on the area. Without it, the area would quickly turn into a baking desert. The rainfall figures established by the tower show that 17% of precipitation returns direct to the atmosphere after being intercepted by the canopy and evaporated; 30% of the precipitation is also returned through plant transpiration, which leaves just over 50% to run off as surface flow. Previous experiments by Eneas Salati, director of the Brazilian Centre of Nuclear Energy for Agriculture, suggest even greater return of water via the plants. His figure is 75%. His overall figures are staggering. He estimates that the Amazon Basin, covering 5 million km^2, receives 12×10^{12} tonnes of rain per year and loses 6.43×10^{12} tonnes through evapotranspiration and 5.45×10^{12} tonnes through direct run-off. He has also shown that much of the rainfall occurring over the area comes about not from rain brought in from the Atlantic by the north-east Trade Winds, but from evapotranspiration from the forest itself. What would happen if the forest were cleared? Salati and others foresee not only a change in the climate of the immediate area, but also changes in other neighbouring farming areas of Brazil. Some climate models have shown that a change in vegetation in 5 million km^2 of the Amazon from trees to grassland will result in lower average rainfall (by some 0.5 to 0.7 mm per day) and a lower total rainfall by 250 mm. The temperature did not change significantly in this model since the lower cooling effect from vegetation would be offset by increased radiation due to the higher albedo of the new vegetation. The latent heat stored in water vapour is released when it turns to water droplets and falls as rain. This is a very important heating mechanism in the tropics. For instance, the latent heat released in 2 cm of rainfall is sufficient to raise the temperature in the upper air by 5°C. If there is less water vapour in the atmosphere, the radiation from the sun is not able to make up for the deficit. The circulation of air from the tropics polewards is an important factor in the determination of climates of these other areas. Without as much water vapour in the air over the Amazon, the climates of other areas of the world could be seriously disrupted. It may lead to a spread of desert conditions, for instance.

Another very important consideration to be taken into account is that the loss of forest also leads to an increase in the carbon dioxide content of the atmosphere. Plants absorb carbon dioxide and store it in their tissues. When forest trees are felled, they are often burnt, which increases the carbon dioxide content of the atmosphere immediately. The gas is then retained in the atmosphere since it is not used in such quantities by the replacement vegetation. The effect of this may well be to increase the average temperature of the earth, since carbon dioxide lets through the short wave radiation from the sun, but, when this is translated into long wave radiation from the earth, it cannot pass through. Heat is therefore retained in the atmosphere. The long term effect of this might be that the polar ice caps could melt and thus disrupt not only climates but also a good proportion of the world's commerce, since most of the world's population lives at present beneath the height to which it would be flooded! Is this the price we are prepared to pay for the reckless development of the rain forest?

Fig. 2.24 (a) Results of a daily forest downpour

Fig. 2.24 (c) Water balance and storage of water vapour in the Amazon Basin

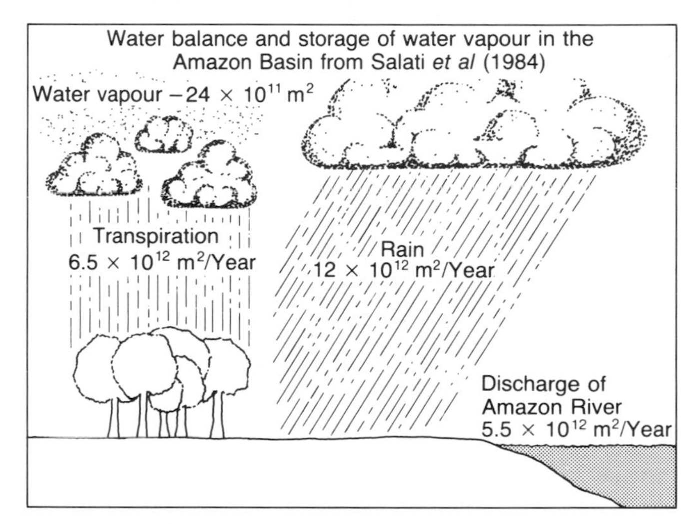

Fig. 2.24 (b) Water balance of the Model Basin

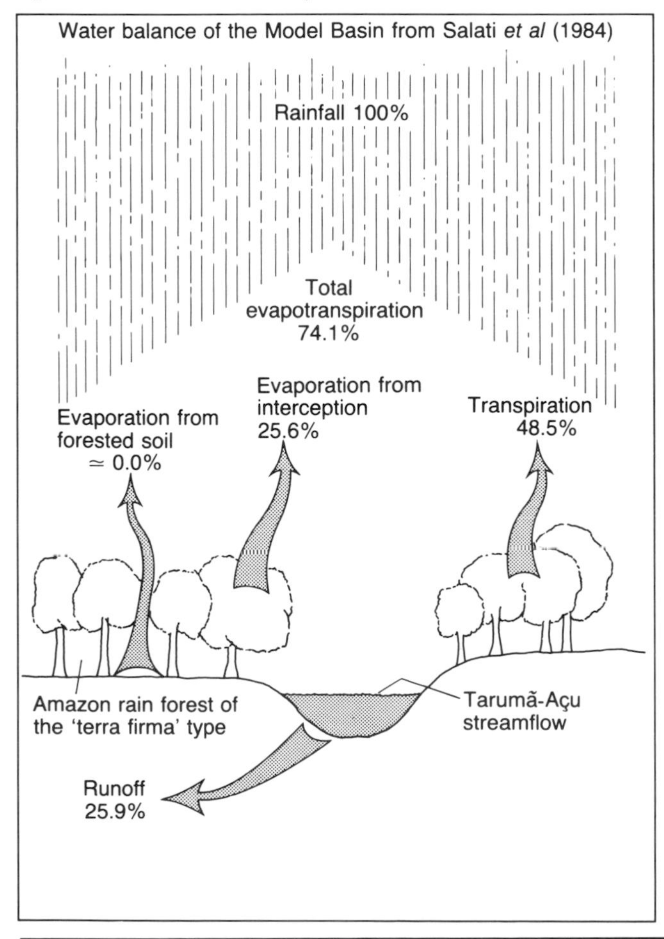

Fig. 2.24 (d) Patterns of atmospheric circulation

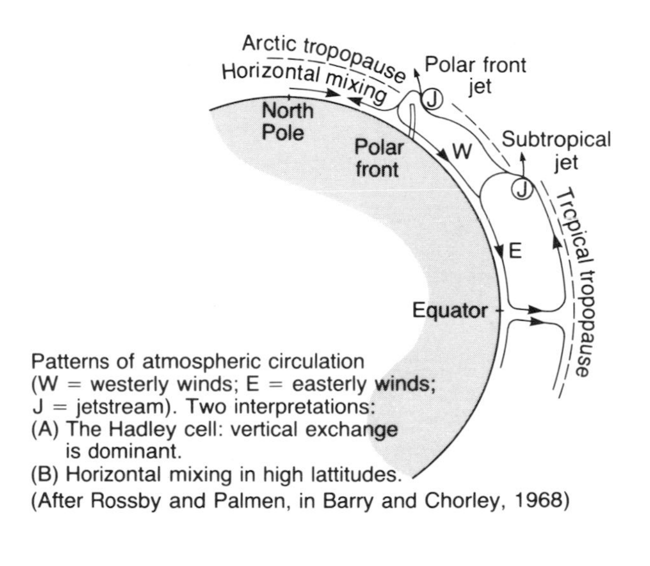

Case-study: game reserves

We have seen how many ecosystems can be abused if the 'development' idea does not fit in harmoniously with the ecosystem. There are other ways, however, in which we can make the ecosystem work with us and for us. One such way is to develop the existing ecosystem as a tourist resource.

Africa has very large expanses of tropical grasslands, known as **savannas**. The extent is shown in Fig. 2.25. These consist mainly of grasses, although some deep rooted bushes and trees such as acacia and baobab are also found there, especially where water is more abundant. The savanna in fact encompasses a range of habitats, from mainly trees and grass near the tropical rain forest to low, sparse grass near the desert border. Such a gradation, a response in this case to changing rainfall figures and variations in rainfall frequency, is known as an **ecocline** (Fig. 2.26). The ecosystem is extremely rich in herbivores, as would be expected from the dominant vegetation, which migrate from one part of the savanna to another in response to the rainfall patterns. One of the world's great sights is to see a massive herd of gnu moving across the vast, flat, dry plains of the Ngorongoro crater in Tanzania. It is for this sort of experience that more and more people are travelling to savanna areas 'on safari'. Countries such as Kenya, Tanzania and Botswana are opening up their savanna lands for this new trade, which brings the problem of how to manage the areas so that people are allowed access but the ecosystem is not destroyed.

Fig. 2.27 shows a model for planning a biosphere reserve as suggested by UNESCO after a conference in Paris in 1974. There are now some 214 such reserves in fifty-eight countries. The model suggests that a **core zone**, possibly the most important area in terms of vegetation type and animal species, should be maintained in as near to its natural state as possible. Around this there should be a buffer zone of only marginal human use, with more allowed in buffer zone 2. Tourism would be encouraged in the outer buffer zone where there would be much of interest to see as long as the management of the two inner zones was working well.

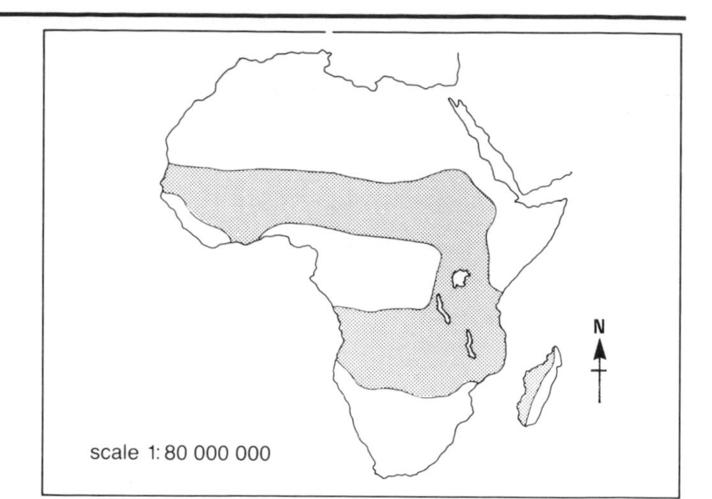

Fig. 2.25 The extent of the African savanna

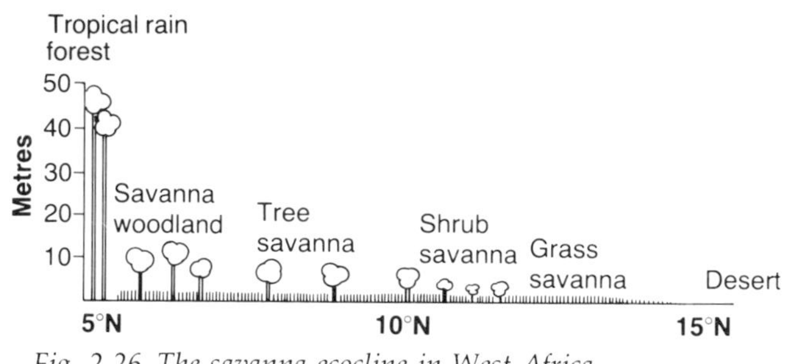

Fig. 2.26 The savanna ecocline in West Africa

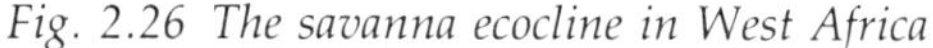

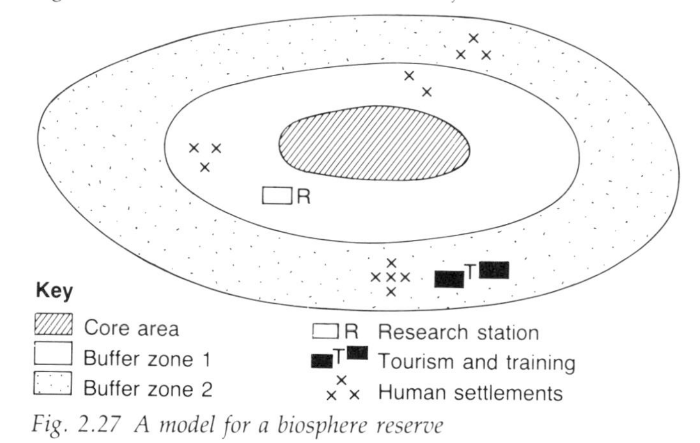

Fig. 2.27 A model for a biosphere reserve

Developing and managing a game reserve is one of the most significant applications of biogeography. The manager is trying to satisfy a wide variety of often conflicting demands on the area. He or she is concerned about the indigenous people, the visitors providing revenue, and maintaining the environment for the flora and fauna. Game parks provide substantial income to the government of the country and provide a spring-board for local development and wealth generation. Many game parks are situated in developing countries where resources are scarce and development options are restricted, but the expectations of the indigenous people for the benefits of development are high.

However, they do bring a series of concerns for you to investigate.

19 Draw a modified version of the biosphere reserve model for a savanna ecosystem where most animals migrate from one area to another, some distance away.

20 Share the following questions out around the group and investigate them. You should all report back later on the issues.

(a) Should more of the wealth of the country go into providing international resources to encourage tourists to the game parks?

(b) Should indigenous people be allowed to farm or hunt wild animals within the game parks?

(c) Should stricter measures be taken against ivory poachers in the game parks?

Take the position of someone concerned with the global situation and try to evaluate what would be best for the local people and the country against what would be best for the world.

Soils – the foundation of the ecosystem

The importance of soil in the ecosystem is suggested in Figs. 2.1 and 2.2 (p. 52). Without a fertile soil, the whole ecosystem is impoverished.

Soils are in a very delicate state of balance with other parts of the ecosystem, especially climate, underlying parent material and vegetation. The vegetation-soil relationship is illustrated in Fig. 2.11 (p. 59) for a system where nutrient transfers are very rapid.

One of the difficulties faced by geographers is that, since there are so many factors influencing soil formation, there is a huge variety of resultant soil types. In fact, it could be argued that a soil at any one location is unique. In their search for order, however, geographers have identified three basic processes at work in soil formation. These are weathering, water movement and nutrient transfer. This knowledge has allowed them to group soils together, classifying them to enable the construction of distribution maps.

The types and extent of weathering are shown in Fig. 2.13. Physical weathering is the breakdown of parent material by agents such as heating and freezing. Chemical weathering is the breakdown by oxidation, hydrolysis or solution, for example. Refer to a dictionary of geography or a suitable physical geography text for more detailed information on these types. Sometimes the two act together and sometimes they influence other factors. For instance, if a tree grows in an area of relatively shallow soil, its roots may well penetrate through that soil into the bedrock below, breaking it apart (physical weathering). The debris from fallen leaves may rot, leading to the production of acidified water which may attack the underlying rock by chemical action.

21 An experiment to assess the effect of particle size and shape on water movement within the soil.

Take three glass tubes, open at both ends. Pack with cotton wool one end of each tube and support the tubes, blocked send down, in clamps. Place them next to each other on a bench with beakers beneath each tube. Place a similar amount of three different soil materials (e.g. coarse sand, fine sand and clay) in the tubes. Add the same amount of water to each with a pipette and observe how long it takes for the water to pass through. Try the experiment with mixtures of material to see what happens.

22 Comment on the following statements:

(a) The rate at which water percolates through a soil depends on the size of the soil particles.

(b) The shape of the soil particles defines the size of the air spaces within the soil and therefore the rate of percolation.

(c) The type and degree of weathering influences the rate at which water percolates through the soil.

The size and shape of particles created by weathering greatly affect the movement of water through the soil and hence nutrient transfer and the development of that soil.

Water movement through a soil moves soil particles, humus material (decomposed biomass) and nutrients within it. This leads to the development of layers within the soil which are known as horizons. The relationship of water movement to horizon development is shown in Fig. 2.28. Factors which affect water movement include the actual amount of precipitation, its duration and intensity, temperature, vegetation, porosity of the soil and parent material. Human influences can greatly alter the balance of the soil profile.

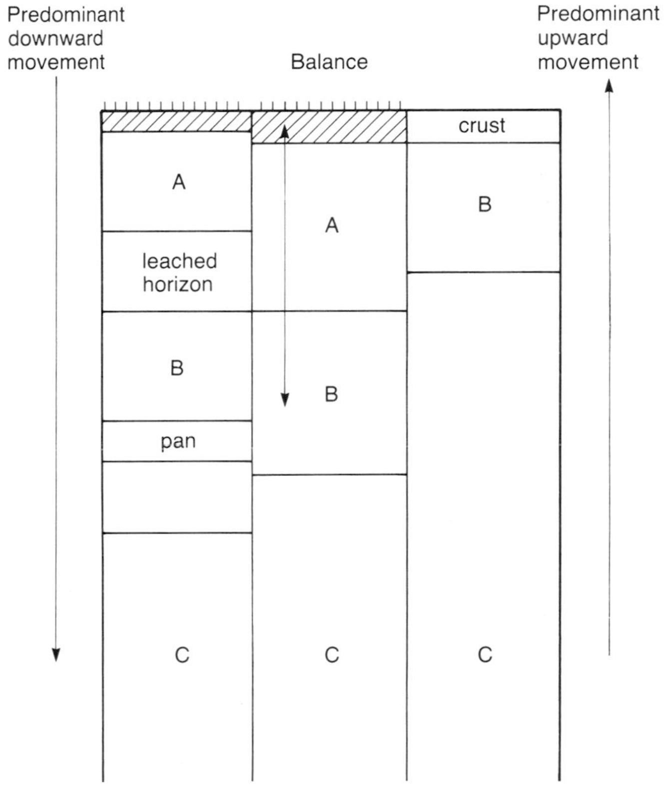

Predominant downward movement
Balance
Predominant upward movement
A
leached horizon
B
pan
C
A
B
C
crust
B
C

Definitions

A horizon – surface layers containing most finely weathered parent material and most humus material

B horizon – less well weathered parent material and less humus

C horizon – parent material

Leached horizon – a layer from which humus material and nutrients have been removed by percolating water. It is usually grey or white as a result.

Pan – a layer within the soil where minerals and humic material are redeposited to form a solid layer e.g. iron pan

Crust – a surface layer formed by the evaporation of water to leave behind the previously dissolved material as a solid layer e.g. salt crust

Fig. 2.28 Three soils developed on similar parent materials under differing conditions of water movement

Nutrient transfer is closely allied to climate, which helps to define the extent of weathering, the amount of water movement and the type of vegetation in the area. When leaves fall or animals die, they are decomposed on the surface of the soil. Temperature and humidity influence the types and numbers of bacteria and fungi which perform this function. Soil fauna then help to incorporate the resulting material into the lower layers of the soil. The downward movement of water also helps to take the decomposed material into the lower layers.

23 Redraw Fig. 2.11(b) to show the relative amount of nutrients in the three compartments and the rate of transfer between them in an area with the following characteristics:–
(a) Mean annual temperature of 5°C.
(b) Annual precipitation total of 800 mm.
(c) Vegetation of sparse coniferous trees and acidic lower storey (bilberry, heathers, sphagnum moss, lichens). Refer to Fig. 2.13 and Table 2.1 and decide how quickly litter would break down under these circumstances.

24 Name an actual ecosystem which would have the conditions referred to in question 22.

25 Attempt to draw another version of 2.11(b) for one other ecosystem shown in Table 2.1.

Using the chi² test

This statistical technique is used to see if there is any variation in a distribution and hence if there is any linkage between the factors under consideration, e.g. the angle of slope and the percentage of bare cover in a sand dune system.

Data have been collected from 40 sites in a sand dune system and this is set out in Table T2.1

The test will try to reveal whether steeper slopes have less vegetation on them or not. Follow the procedure through this example.

1. Formulate the **null hypothesis**. You start by assuming that there is no relationship between the variables unless the information satisfies a fairly high level of significance. In this case the null hypothesis would be that vegetation is found on any type of slope.

2. Formulate the **alternative hypothesis**. This is in case you can reject the null hypothesis and it would be that vegetation colonizes and is found on gentle slopes rather than steeper ones.

3. Agree on the level of significance. You have to state the level at which you have enough evidence to accept the alternative hypothesis and reject the null hypothesis. This is known as the **level of significance**, and is given as 95% or 99%. At 95% there is a 1 in 20 probability that the pattern in your observations occurred by chance, and at 99% there is a 1 in 100 probability that the pattern is a chance one. In geography, 95% is normally acceptable, although 99% is better.

4. Calculate the **expected frequencies** (E). This is assuming that there is an even distribution of cover over the various slopes. This is done by multiplying the sum for each column by the sum of the row and dividing by the total number of observations (40 in this case). For example, in the top left hand corner of the table of observations, i.e. less than 1 degree with between 0 and 20% bare cover, if the distribution had been even there would be 18/40ths with less than 20% bare cover on 11 slopes of less than 1 degree. This is simplified to:

$\frac{18}{40} \times 11 = 4.95$. The expected value is given in the table below

For the square below, i.e. slope angle of 2–5 degrees with less than 20% bare cover, it would be:

$$\frac{18}{40} \times 8 = 3.6.$$

The complete table of expected values is given below with the values rounded up to 1 decimal place.

5.0	1.4	1.1	1.4	1.9
3.6	1.2	0.8	1	1.4
5.0	1.7	1.1	1.4	1.9
4.5	1.5	1	1.3	1.8

5. Calculation of chi². This is given by the symbol X^2 and the following expression is used:

$$X^2 = \Sigma \frac{(O-E)^2}{E}$$

where O is the observed value and E is the expected value. Σ means sum of.

Therefore the calculation will go as follows:

$$\frac{(1-5)^2}{5} + \frac{(2-3.6)^2}{3.6} + \frac{(8-5)^2}{5}$$

and so on.
A calculator with a memory is very useful here. The final value is 26.7, and this is the value of chi².

6. **Degrees of freedom**. This is important to find out whether this value is high enough to reject or accept the null hypothesis. Degrees of freedom (df) are calculated as follows:

(number of rows – 1) × (number of columns – 1)

In this case the value is 12.

Slope Angle	Bare Cover 0–20%	21–40%	41–60%	61–80%	81–100%	Σ
Less than 1°	1	0	2	3	5	11
2–5°	2	1	1	2	2	8
6–10°	8	2	1	0	0	11
More than 11°	7	3	0	0	0	10
Σ	18	6	4	5	7	40

Expected value of $\frac{18}{40} \times 11 = 4.95$

7. Acceptance or rejection? This is shown by looking up the critical values of chi² in a book of statistical tables. The critical values are shown in Table T2.2. Find the degrees of freedom value (12) and look across at the values beside this. The first column (.05) is for 95% significance, the second (.01) is 99%. The calculated value of chi² must be the same or greater than the values given in the table. In this case the value is greater than the value for 99%. Therefore there is enough evidence to reject the null hypothesis and say that there is a strong relationship between the variables. Vegetation will not tend to colonize steep slopes in the sand dune ecosystem.

Table T2.2 Critical values for chi²

The chi-squared value for any given number of degrees of freedom must be equal to or *greater* than that shown for the level of significance required.

df	Levels of significance	
	.05	.01
1	3.84	6.64
2	5.99	9.21
3	7.82	11.34
4	9.49	13.28
5	11.07	15.09
6	12.59	16.81
7	14.07	18.48
8	15.51	20.09
9	16.92	21.67
10	18.31	23.21
11	19.68	24.72
12	21.03	26.22
13	22.36	27.69
14	23.68	29.14
15	25.00	30.58
16	26.30	32.00
17	27.59	33.41
18	28.87	34.80
19	30.14	36.19
20	31.41	37.57
21	32.67	38.93
22	33.92	40.29
23	35.17	41.64
24	36.42	42.98
25	37.65	44.31
26	38.88	45.64
27	40.11	46.96
28	41.34	48.28
29	42.56	49.59
30	43.77	50.89

(Source: Siegel S., *Non-Parametric Statistics for the Behavioral Sciences* (McGraw-Hill, 1956). This table is taken from Table 4 of Fisher, R. A., and Yates, F., *Statistical Tables for Biological Agricultural and Medical Research*, published by Longman Group Ltd., London, 1963 (previously published by Oliver & Boyd, Edinburgh), by permission of the authors and publishers)

Fig. T2.1 Stations taken along the transect of a dune system

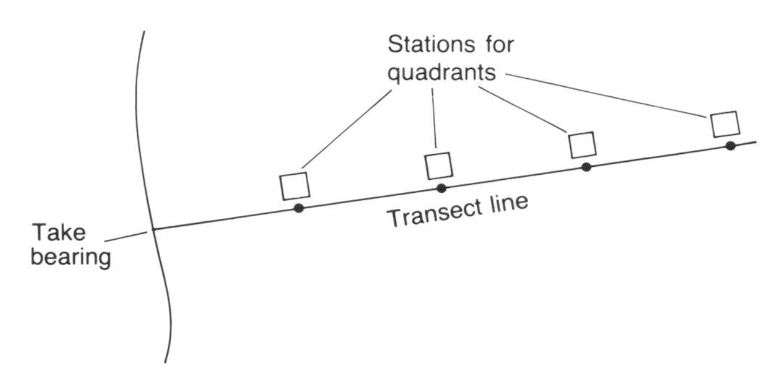

Sampling, using transects and quadrats

Collecting data in the field always requires some form of sampling. The sampling technique must always produce representative results, but must be easy to administer in the field. **Transects** and **quadrats** are good methods for collecting data from an area such as a sand dune ecosystem.

The transect is a line along which data is collected. As shown in Fig. T2.1, **stations** are then taken along the transect, or information can be recorded continuously. For the purpose of data collection in the sand dunes, stations at 15 m were used and information on slope angles and bare cover, number of species of plant etc. were gathered from the quadrats laid down at each of the stations. The quadrat itself is simply a square frame laid on the ground and the information taken from within the square is taken to be representative of that particular location.

Information is then filled in on a data collection sheet as shown in Fig. T2.2.

The procedure used in data collection in the sand dunes was as follows:

1. Agree a bearing for the line of transect.
2. Collect data at regular agreed intervals. In the sand dune ecosystem the surface was so irregular that a regular interval for measuring slope may have been inappropriate. In this case measurement was made to the nearest major break of slope. Distances were measured by tape and slopes with **clinometers**. Soil samples were taken using an **auger** and wind speeds were measured using a hand held **anemometer**. Vegetation information was collected using a quadrat.

Equipment checklist: clipboard, compass, ranging poles, 20-m tape, clinometer, auger, plastic bags for soil samples (don't make them too big – you will only have to carry more), plant reference book for identification, anemometer, quadrat, hygrometer (if you are collecting temperature and relative humidity data)

Fig. T2.2 A data collection sheet

SAND DUNE ECOSYSTEM				Group No —
Stations	Slope angle	% Bare cover	No. of species	Animal and Human Evidence
15m	3°	100	–	Litter-crisp bag
30m	2°	100	–	–
45m	5°	80	1	Remains of camp fire

Theme 3 *Agriculture and Rural Development*

Case-study: agriculture in the Soviet Union

Agriculture is an important part of the economy of the Soviet Union. It provides 16% of the country's Gross National Product and employs 14% of its population. Both of these figures are for 1982 and show a decline from the 1960 figures of 21% and 42% respectively. By comparison, the figures for the USA are 3% and 2%.

However, agriculture does not satisfy the demands of the Soviet peoples. In 1981 they had to import, for example, 40 million tonnes of grain, 4.5 million tonnes of sugar, 750 000 tonnes of rice and 600 000 tonnes of vegetable oils. The Soviet Union had long term agreements with Argentina, Brazil, Canada, India, New Zealand, Sweden, Thailand and the USA to supply wheat, coarse grains (for animal feed), soya beans and soya products, meat and butter. The agreement with the USA is for a minimum of 6 million tonnes of wheat and maize with an option on an extra 2 million tonnes.

How is it that the Soviet Union, with an area of 22.4 million km^2, a density of 11.9 people per km^2 and land which ranges from 37°N to over 70°N does not feed its population? The USA is only 41% of the size of the Soviet Union with a population density of 24.5 per km^2 and a latitudinal range of 25°N to 49°N (excluding Hawaii and Alaska).

1 Use the sieve analysis (see Techniques page 104) to help produce a map of the agricultural regions of the USSR.

(a) On a traced outline of the USSR shade out those areas with

(i) a growing season (above 5°C) of less than 120 days (Fig. 3.1)

(ii) A summer temperature below 18°C (Fig. 3.2)

(iii) precipitation less than 200 mm and over 800 mm (Fig. 3.3)

(b) Compare your map with Fig. 3.4 showing the actual pattern of arable land. Add areas not included from your sieve analysis.

(c) Place your completed tracing over the maps showing natural vegetation (Fig. 3.5) and permafrost and moisture deficiency (Fig. 3.6). Comment on any relationships between these factors and the pattern of arable land.

(d) Try to construct a model to show the interaction of these physical constraints and their influence on agriculture. Explain how these factors restrict agricultural patterns.

Part of the answer can be seen from looking at the physical environment of the USSR. Figs. 3.1, 3.2 and 3.3 give some of the climatic conditions.

Fig. 3.1 Snow cover and growing season of the USSR

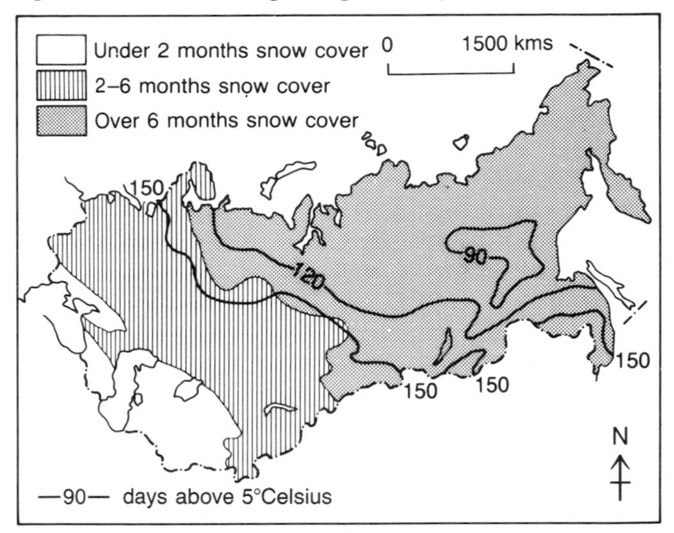

Fig. 3.2 Temperatures of the USSR 1.7 cms=1500 kms.

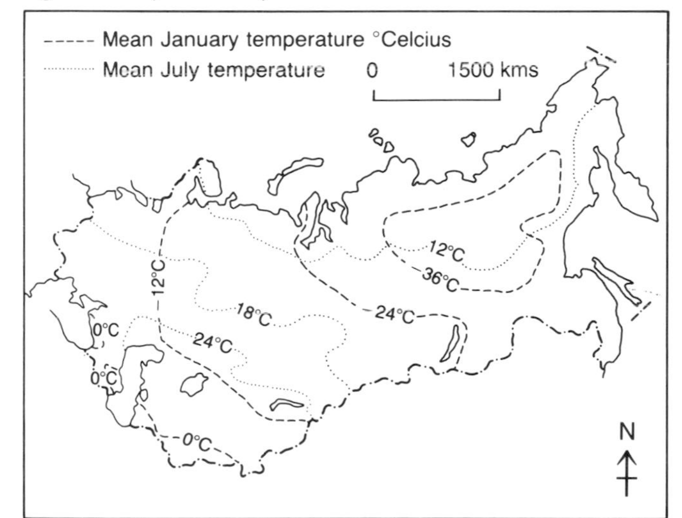

Fig. 3.3 Precipitation of the USSR

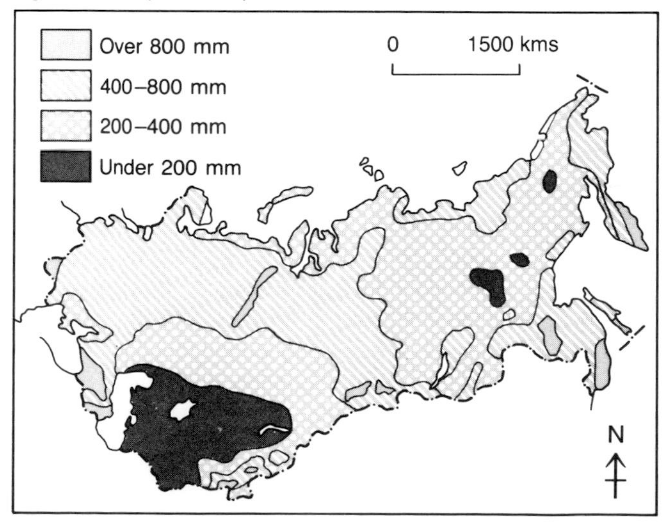

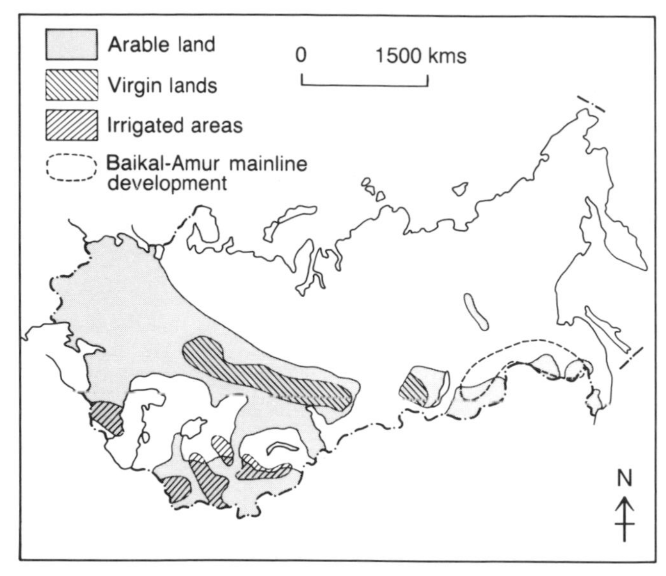

Fig. 3.4 Arable land, irrigation and recent developments

Fig. 3.5 Natural vegetation of the USSR

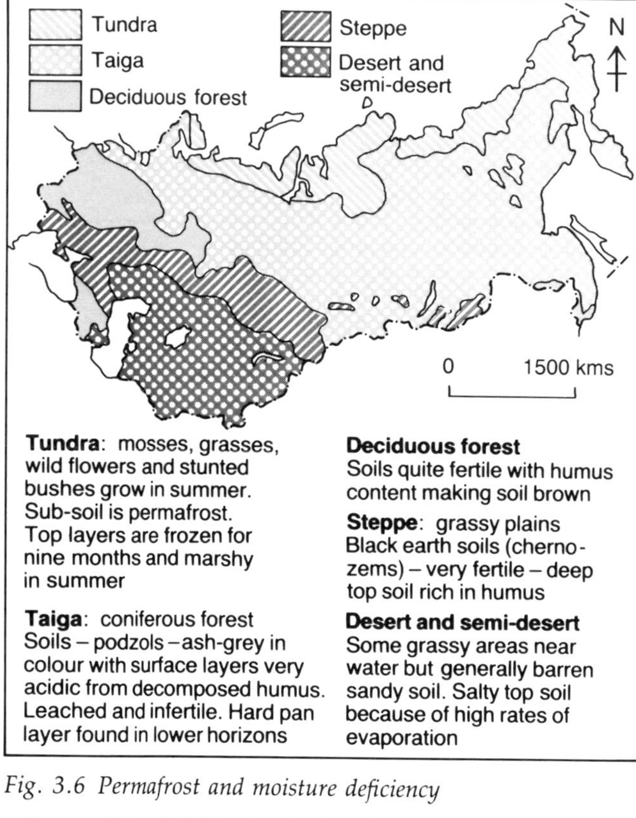

Fig. 3.6 Permafrost and moisture deficiency

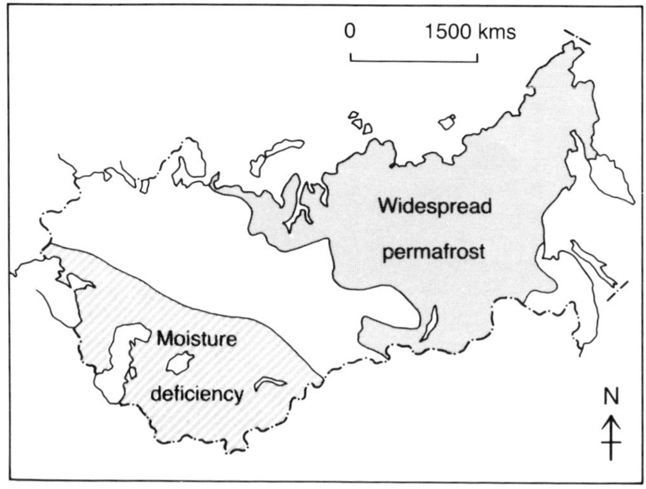

Table 3.1 Agricultural products of the USSR

Crop	Percentage of town area
Cereals	
Wheat	29
Barley	8
Rye	7
Oats	4
Maize	3
Others	8
Industrial Crops	
Cotton, flax, hemp, sugar beet, sunflower	7
Fruit and Vegetables	
Potatoes, market gardens, deciduous fruits, vines	5
Fodder Crops	
Sown grasses	14
Maize	12
Others	3

2 Refer to Table 3.1. Represent the percentages for the various crop areas in the Soviet Union in the form of a graph.

Wheat is the most widespread crop being grown throughout the main arable belt. Due to the harshness of the winter temperatures, only a quarter of the wheat is autumn sown, the rest being spring sown. Barley is also grown over a wide area, although the hardier crops of rye and oats are grown in the northern part of the arable belt. Maize occupies the southern part of the arable belt but cotton is grown in the irrigated parts of the south, while sunflowers are grown in the Ukraine. Sugar beet is found throughout the main arable belt and flax and hemp are produced in its northern range. Potatoes form a staple part of the diet and are grown over a wide area, many coming from private plots. Market gardening is found close to the bigger urban areas while fruits and vines are produced in the southern part of the Ukraine.

3 Add annotations to your map of arable areas to show the approximate locations of the main crops of the Soviet Union.

Recent progress

Since 1945, the aim of the Soviet farm policy has been to improve the diet of the people. Large areas of the Steppe were ploughed up in 1953 as part of the Virgin Land Scheme (Fig. 3.4). However, much of the improvement in Soviet agriculture has come through intensifying production in the existing areas, by modernizing grazing practices to increase feed conversion ratios and increased milk yield per cow, as well as by improving the quality of the crop land.

Table 3.2 Crop land improvements in the USSR

Republic	Area of irrigated cropland to be brought into use (000 ha)	Area of waterlogged and marshland to be drained (000 ha)	Area of irrigated pastureland (000 ha)	Total
Estonian SSR		170		170
Latvian SSR		520		520
Luthanian SSR		900		900
Belorussian SSR	110	960		1070
Russian SSR	3330	3700		7000
Ukrainian SSR	1000	1300		2300
Moldavian SSR	240			240
Georgian SSR	120	70	150	340
Armenian SSR	60		120	180
Azerbaydzhan SSR	160		285	445
Kazakh SSR	820		22 000	22 820
Kirgiz SSR	150		750	900
Uzbeik SSR	900		2600	3500
Tadzhik SSR	105			105
Turkmen SSR	185		8400	8585
Total	7150	7620	34 305	49 075
Total as percentage of USSR total area	0.32%	0.34%	1.53%	2.19%

Crop land improvements include irrigation of 2.2 million ha of pastures and hayfields to create new farmland, improvements to between 27 and 29 million ha of fodder land, drainage and the irrigation of 34 million ha of existing pastures. Table 3.2 gives details of this new irrigated crop land and areas to be drained for each republic.

4 Use the data in Table 3.2 to construct a multiple choropleth of farm land improvements in the Soviet Union. A multiple choropleth will show more than one piece of information and by careful selection of line shading, (e.g. vertical lines for one improvement, horizontal lines for another) you will be able to display all this information on one map. Make a tracing of Fig. 3.7 as a base map for your choropleth.

Two areas in particular are worth closer study.

(i) The Baikal-Amur Mainline Development

The Baikal-Amur Mainline Development (BAM) is an area of planned industrial development along the railway (Fig. 3.4) based on mining and communications. Work began in 1974, and by the year 2000 one million people are expected to live in this zone. Agricultural development is part of the scheme due to the high cost of transporting foodstuffs from elsewhere. The physical environment is harsh, with permafrost soils, low air temperatures and a growing season ranging from 40 to 100 days. Despite this, potatoes, barley, oats and flax are grown and dairying takes place in the more sheltered valleys. To overcome the short growing season, plants are raised under vast areas of glass and then transplanted. Cucumbers, tomatoes and cabbages have been grown in this way and have given reasonably high yields. Reservoirs on the Angara and Yenesei rivers have been constructed to control flooding and provide irrigation water. They have also been found to moderate the microclimate, giving milder winters and wetter summers.

Fig. 3.7 The republics of the USSR

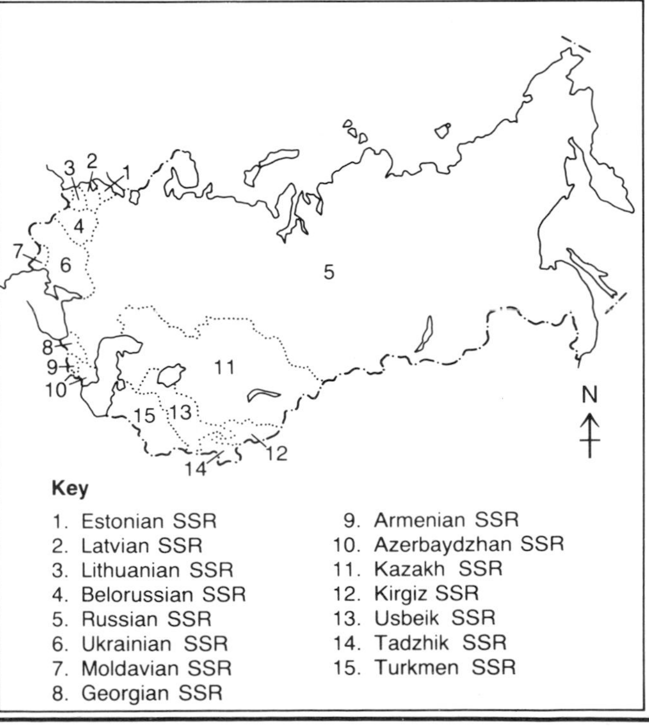

(ii) The Virgin Lands Scheme

The development of the virgin lands of Kazakhstan is very important to Soviet agriculture. Already over the last 25 years 1½ million people have moved to the area where there are over 36 million ha in locations that have less than 300 mm of precipitation. Hence, there is a need to continue to develop the irrigation systems of the area to keep pace with the required increases in agricultural production. At present, the demands on the waters of the Amu Barya and the Syr Darya are reported to be so great that the Aral Sea is shrinking, with a fall in water level of 1.5 m. To provide this irrigation water, some of the boldest irrigation schemes have been proposed, and some are being implemented at present. The map (Fig. 3.8) shows the movement of water from the Volga River to the River Don for irrigation in the southern Ukraine and from the Ural River to the Volga. Water is being transferred from the Syr Darya to the Bet-Pak-Dala desert and from the Irtysk River south-west towards Dzhezkazgan.

Future plans, however, may well have far reaching side-effects. Water from the Pechora River is to be transferred to the Volga to increase the supply flowing into the Caspian Sea. Fisheries have been adversely affected there by reduced flow and increasing salinity. It is also planned to pump water from the Pechora to the River Don, eventually to find its way into the Sea of Azov. Even larger plans have been formulated for the Ob and the Yenisei Rivers. Water is to be transferred via the Irtysh River to add to the flow into Kazakhstan, eventually flowing into the irrigation systems fed by the Syr Darya. These schemes would involve monumental engineering in construction of canals, dams and pumping equipment, and while they are within the compass of the USSR, it is their possible environmental impact which is worrying.

The Yenisei and Ob rivers have a combined flow 400 times greater than that of the Thames and about one seventh of the annual flow of the Amazon. These water transfer schemes, however, would involve a reduction of the flow of water into the Arctic basin of between 5 and 20% by the middle of the 21st century. The freshwater from rivers such as the Yenisei and Ob is far less dense than the saline water of the Arctic Ocean. The freshwater thus acts as a 'lid' keeping the warmer, denser water down and thus maintains the level of the ice cover. If the flow of freshwater diminishes, then it is feared that the saline water will flow to the surface which will cause the melting of much of the polar ice cap. The long term repurcussions of this would therefore be enormous.

One of the predicted events concerns the climate. The links between climate, river flow and ice cover are as yet poorly understood, but it has been observed that when ice cover is most strongly developed in the Kara Sea, this is reinforced by strong easterly winds which stop the ice drifting out into the Arctic Ocean. At the same time, there is the stronger development of low pressure systems which bring increased precipitation to the catchment areas of the Yenisei and Ob rivers and the areas of the south. This in turn increases discharge, which reinforces the development of the ice cover in the Kara Sea. Hence there would appear to be a relationship between ice cover, atmospheric disturbance, precipitation and river flow. If the flow of freshwater into the Arctic basin is reduced, what will happen then?

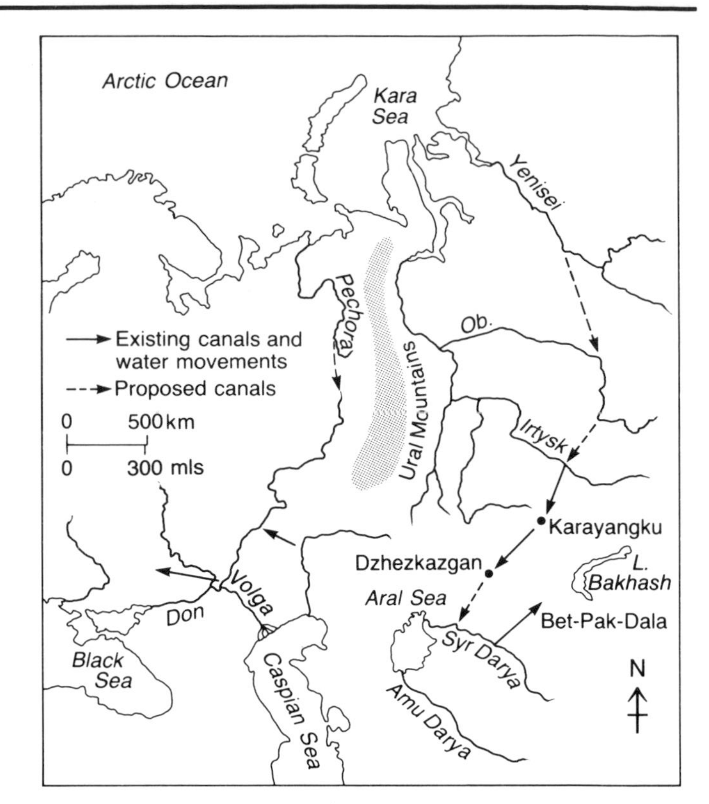

Fig. 3.8 Irrigating Central Asia

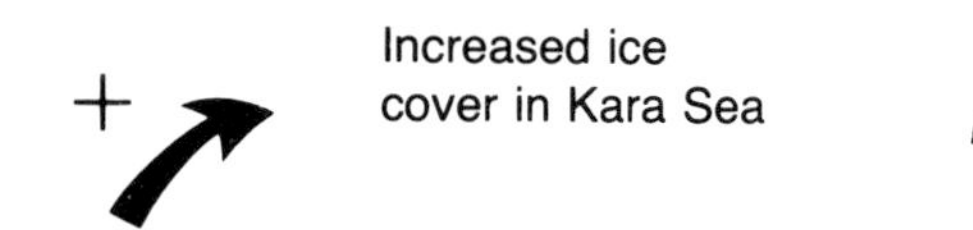

Increased discharge of freshwater

Increased incidence of low pressure systems

Increased rainfall to Siberia and south central Asia

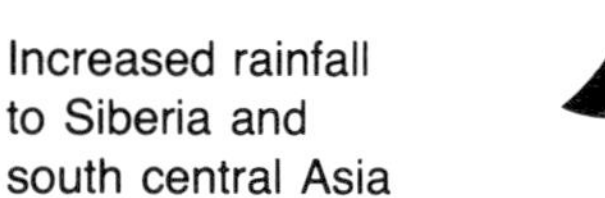

Fig. 3.9 A feedback diagram to show inter-relationships in northern USSR

5 Redraw the feedback diagram (Fig. 3.9) starting with the predicted series of events following a decrease in the discharge of freshwater. Give more detail of the processes if you can.

What might happen to the climate of the areas being irrigated if our assumptions are correct?

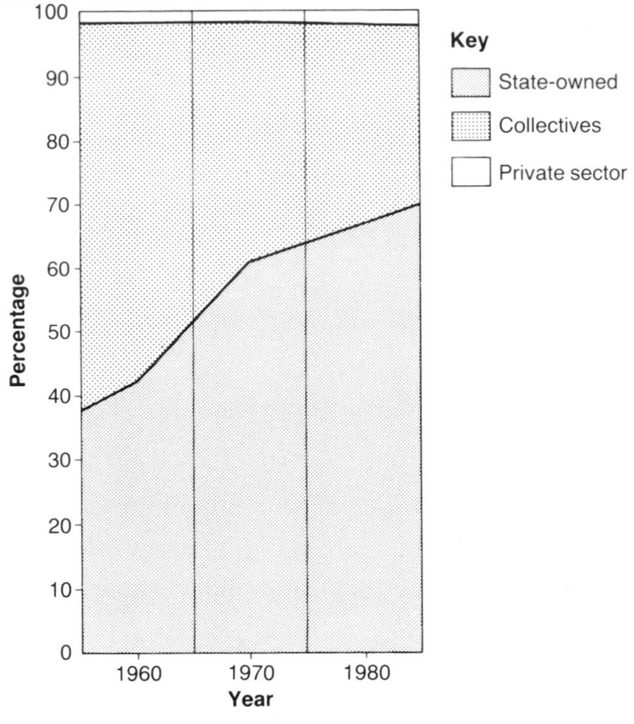

Fig. 3.10 Percentage share of the main farm categories in the total agricultural area

> **6** Complete the annotation of your arable land map with information about BAM and the irrigation schemes of the southern Soviet republics.

What has been the progress to date?

Farming in the Soviet Union is based upon the **collective farm (kolkhoz)**, the **state farm (sovkhoz)** and small private plots. The sovkhoz lease the land from the state and it is worked collectively with decisions being taken by an elected committee. The kolkhoz earns money by selling produce to the state and by selling to the local market. Incomes are then allocated by the committee. The sovkhoz is a farm totally owned and run by the state and the labourers are employed as in a factory, earning regular wages.

Fig. 3.10 shows the relative percentages by type of farm of the total agricultural area. Table 3.3 gives the percentages for each of the republics.

Soviet agriculture before 1985

> **7** Study the data in Table 3.3. Do all the republics follow the main trend revealed in Fig. 3.10? List any odd ones out (residuals) and compare your findings with others in your class if possible. Try to account for these residuals. Remember to consider their location in the Soviet Union.
>
> **8** Table 3.4 gives the average size of the sovkhoz and kolkhoz. Comment on the trends shown in these figures. Is it possible to identify any groups of republics with similar characteristics?

The private plots, although small, do provide a significant amount of food for the local market. The Food Programme is also involved in promoting improvements to the social and cultural amenities of the rural areas. This has been done in some areas by amalgamating villages and providing a wide range of amenities including housing, a sports stadium, a department store, a machine shop and a garage.

Recent changes and issues

> **9** Using Fig. 3.10 as a basis, what will the percentages of agricultural land be for the state owned farms, the collectives and the private sector for the years 1990 and 2000?

Table 3.3 Percentage share of the main farm categories in the total agricultural area – for the USSR by individual republic

	State-owned agriculture State and institutional farms			Collective farms (excluding private plots)			Private sector (private farms and private plots)		
	1960	1970	1980	1960	1970	1980	1960	1970	1980
USSR	42.2	61.0	67.3	56.4	37.5	30.9	1.4	1.5	1.8
Estonian SSR	30.6	43.7	46.7	66.2	50.0	46.7	3.2	6.3	6.6
Latvian SSR	25.0	37.0	44.0	67.9	59.3	48.0	7.1	3.7	8.0
Lithuanian SSR	21.0	31.4	30.6	73.7	60.0	61.1	5.3	8.6	8.3
Belorussian SSR	29.6	34.7	35.4	64.3	59.2	59.4	6.1	6.1	5.2
Russian SSR	38.3	53.5	58.4	60.2	44.9	39.9	1.5	1.6	1.7
Ukrainian SSR	16.7	21.7	24.0	77.9	72.4	70.1	5.4	5.9	5.9
Moldavian SSR	7.4	18.5	30.8	88.9	74.1	50.0	3.7	7.4	19.2
Georgian SSR	11.1	33.3	45.1	81.5	63.4	45.2	7.4	3.3	9.7
Armenian SSR	15.0	46.2	66.9	82.7	46.2	30.0	2.3	7.6	3.1
Azerbaydzhan SSR	10.2	37.5	53.6	88.5	60.0	41.5	1.3	2.5	4.9
Kazakh SSR	60.8	90.7	92.2	39.1	9.2	7.6	0.1	0.1	0.2
Kirgiz SSR	29.0	31.8	48.5	70.5	67.6	49.5	0.5	0.6	2.0
Uzbeik SSR	44.7	56.4	88.3	54.4	42.8	10.6	0.9	0.8	1.1
Tadzhik SSR	20.5	29.0	59.5	76.9	70.0	28.6	2.6	1.0	11.9
Turkmen SSR	33.4	34.4	46.5	66.5	65.5	52.2	0.1	0.1	1.3

Source: Calculated on the basis of data in TsSU SSSR (1961., p.386), TsSU SSSR (1971, p.291), TsSSU SSSR (1981, pp. 222).

Table 3.4 Average sizes of USSR state farms (excluding institutional farms) and of collective farms (excluding private plots) in 100 ha of agricultural land by individual republic

	State farms			Collective farms		
	1960	1970	1980	1960	1970	1980
USSR	261	205	173	65	61	65
Estonian SSR	42	41	44	20	27	49
Latvian SSR	43	44	45	17	24	38
Lithuanian SSR	35	37	35	15	15	29
Belorussian SSR	97	41	38	27	26	32
Russian SSR	210	138	108	64	73	73
Ukrainian SSR	79	57	48	34	34	42
Moldavian SSR	33	35	23	43	36	33
Georgian SSR	20	43	28	12	15	20
Armenian SSR	24	23	20	14	13	13
Azerbaydzhan SSR	47	37	32	31	24	28
Kazakh SSR	1098	1034	858	482	397	373
Kirgiz SSR	373	301	212	201	269	278
Uzbeik SSR	542	374	261	124	102	33
Tadzhik SSR	205	135	107	85	103	73
Turkmen SSR	2020	1981	1306	590	612	505

(Source: Calculated on the basis of data in TsSU SSSR 1961, pp. 386, 500, 509, TsSU SSSR 1971, pp. 291, 388, 396, TsSU SSSR 1981, pp. 222, 260, 272)

Table 3.5 Sown areas of the Soviet Union (in million ha)

Crop	1960	1965	1970	1975	1980	1988
All grains and legumes including:	115.6	128.0	119.8	127.9	126.6	108.5
winter wheat	12.1	19.8	18.5	19.6	22.6	48.0
spring wheat	48.3	50.4	46.7	42.4	38.9	
barley	11.0	19.7	21.3	30.9	30.1	30.0
industrial crops	13.1	15.3	14.5	14.1	14.6	19.8
potatoes and vegetables	11.2	10.6	10.1	10.1	9.2	8.5
fodder crops	63.1	55.2	62.8	65.6	66.9	N/A

Table 3.6 Growth in agricultural production in the USSR. Figures are annual averages for the periods shown. The first three correspond to three successive Five-Year Plan periods.

Product	1961–65	1966–70	1971–75	1980–88[1]
Grain	130.3	167.6	181.6	177.4
Raw cotton	4.99	6.1	7.67	2.6[2]
Sugar beet	59.2	81.1	76.0	82.6
Vegetables	16.9	19.5	22.8	33.1
Meat	9.3	11.6	14.0	17.8
Milk	64.7	80.6	87.4	97.88
Eggs	28.7	35.8	51.5	42.03

(Source: USSR over 60 years, Anniversary Statistical Yearbook, quoted in *Agrarian Relations in the USSR*.)

1 Average Figs for 1980, 1984, 1988 from FAO Yearbooks
2 Average for 1980 and 1984 only

Your answer to question 9 assumed that the economic situation was static. However, there have been a number of significant changes in the Soviet Union. The country is going through a period of restructuring, politically and economically. Perestroika is now a popular expression of political commentators.

The people of the Soviet Union now have greater political freedom, and issues which were not discussed are now coming into the open. The Soviet Union has its 'Green' groups including the All-Union Movement of Greens and the Soviet equivalent of Greenpeace, Zelenyi Mir. President Gorbachev has said that the problems of the environment must be sorted out quickly because they are 'a question of the very life of our people, their present and future.'

The problem of Chernobyl is fairly well known and is dealt with on p. 191. The area affected by the radioactive fallout is now recognised as being much larger than was first realised. Immediately after the accident a zone 30 km around the station was evacuated. Now areas of up to 400 km away are affected. Agricultural land has been damaged as the radioactive fallout has affected soils and pasture. Livestock is affected and has to be moved to clean pasture for two months before slaughter. Other restrictions are still in force on dairy products. Fishing in the Pripyat river and its tributaries is prohibited.

The intensification of agriculture is also beginning to show negative results. The use of water from the Aral Sea to irrigate regions in the south of Kazakhstan is dramatically reducing the size of the Aral Sea, as shown in Fig. 3.11. Between 1960 and 1989 the level in the lake dropped by 13 metres and its area decreased by 40%, from 69 000 square km to 39 000 square km. Much of the water taken from the Aral

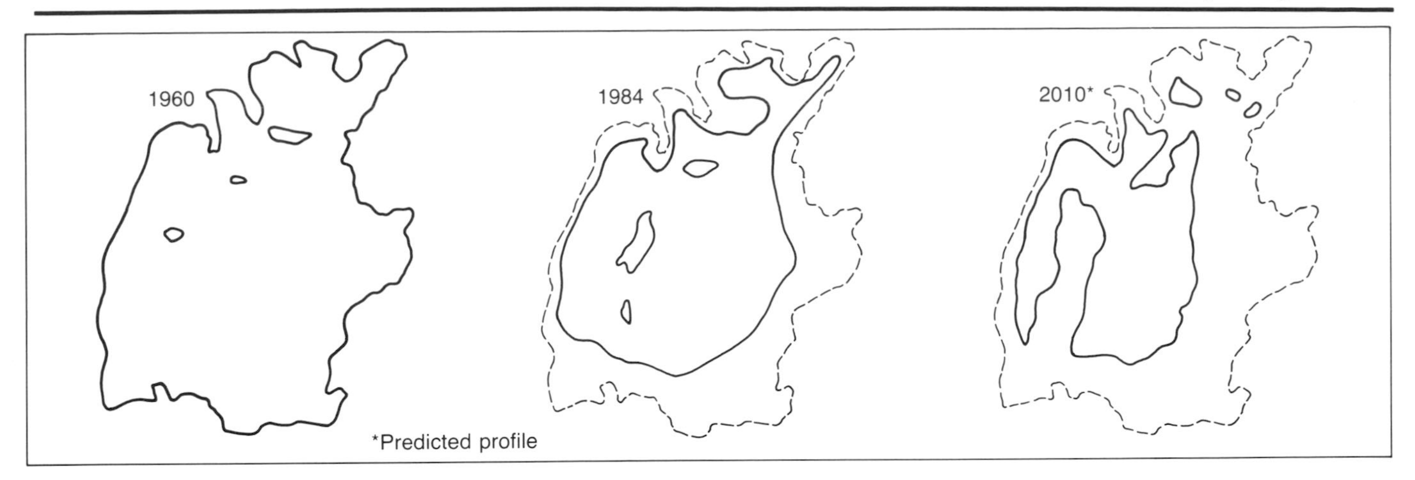

Fig. 3.11 The changing profile of the Aral Sea

Sea was carried by poorly constructed canals and a lot of the water was lost through seepage. The high rates of evaporation in the area have led to the salination of the soils reducing their productivity. Further water is then needed to flush the salts out of the surface layers.

The exposed lake bottom of the Aral Sea has uncovered vast amounts of salts which have been blown many hundreds of kilometres away from the area. The commercial fishing of the lake has been affected and many areas around inflowing rivers have been negatively affected. Pastures and livestock farming has declined. The natural ecosystem has been damaged with a reduction of 173 animal species down to 38 species.

It is also thought that the climate of the area has been affected. The summers have become warmer and the winters cooler. The frost free period has decreased with spring frost becoming common. Cold winds bring snow at the beginning of May when the farmers would be expecting very warm conditions. This has affected the area under cotton with over 500 000 hectares affected, while 70% of all grain fields were lost and crops such as grapes, melons and fruit trees badly damaged.

The intensification of agriculture in the region also encouraged the authorities to use pesticides including DDT, banned elsewhere, in an indiscriminate manner. These have poisoned the water supply and have increased the incidence of cancers and mental diseases. The monoculture of cotton, known as beloye zoloto or white gold, and the use of child labour has added to the problems of fertility and health. The area has a very high infant and child mortality rate of over 50 per 1000 live births.

Despite these massive problems of environmental deprivation, changes are taking place in the structure of farming in the Soviet Union. In 1985 the six agricultural ministries were merged into one known as Gosagroprom, and made directly responsible to the President. The Agricultural Plan formulated in 1986 aims at increasing grain production by 6%. This plan introduced a series of reforms to provide incentives by paying a 50% bonus on production above the 1981–85 average for grain, meat, milk, cotton, soya, sugar beet, tea and wool. State and collective farms are authorised to sell 30% of the produce of fruit and vegetables to local shops and markets. These shops can set local prices rather than using the nationally set prices. This should provide greater flexibility and increase incentives to farmers.

The trend towards large farms is being reversed and a system of 'lease contracting' is being piloted. In this system a farmer will lease a certain number of animals from the collective, for example, and buy feed and sell the product on the open market. He would contract to sell a proportion of his product to the farm. When the system is fully operational farmers will have security of tenure which will encourage them to invest in improvements. Loans would be made available to finance this investment.

The recent history of the rural areas of the Soviet Union have not been very encouraging. The people of the countryside have been encouraged to move to the towns while farming was dealt with by highly mechanised farms. Now they are being encouraged to return to fill up the empty shelves in the shops and keep the people in the towns happy. A major problem to be addressed is the lack of infrastructure to provide the transport to get the products to market. There is still considerable wastage of crops. Agriculture is very important to perestroika and the future of the Soviet Union.

10 How far do you agree that effective decision making on agricultural matters should be made at the level of the farmer? Use the information above to justify your case.

11 How much does politics influence agricultural decision-making? Use local libraries to investigate the influence exerted by the British government and the European commission over the British farmer.

Case-study: change in India

The present situation

Agriculture is very important to India; 69% of the working population is engaged in farming. This percentage has fallen from 74% in 1960. Between 48% and 56% of the area of India is under cereal crops, mainly wheat and rice. India's population of 690.2 million in 1981 puts a considerable pressure on agriculture. Plots therefore tend to be very small and fertile areas very attractive.

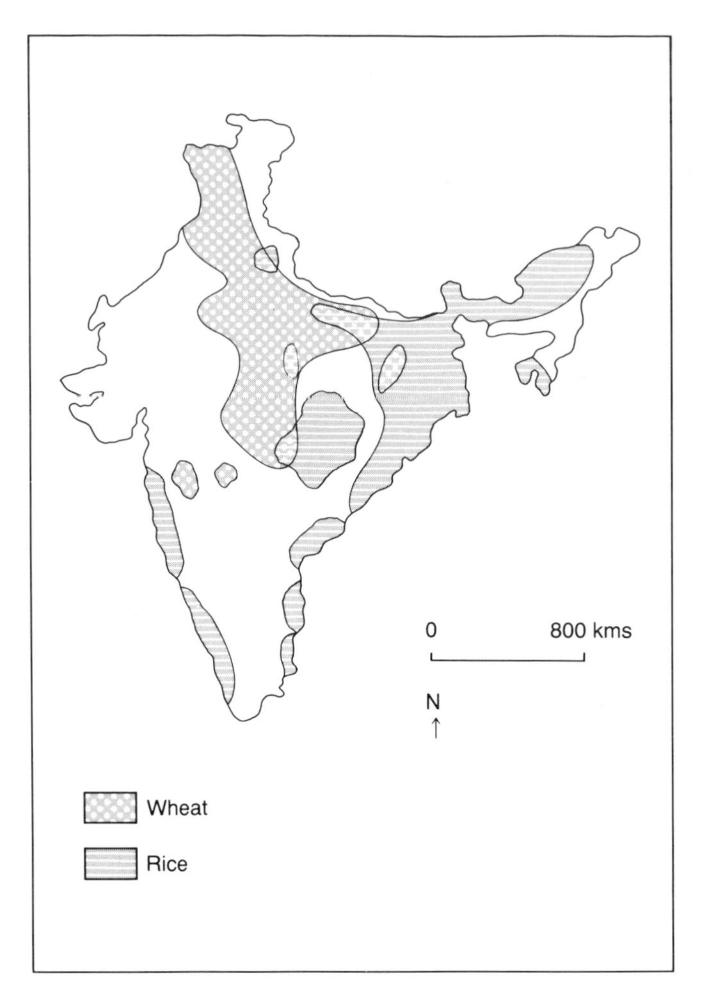

Fig. 3.12 Principal wheat and rice-growing regions in India

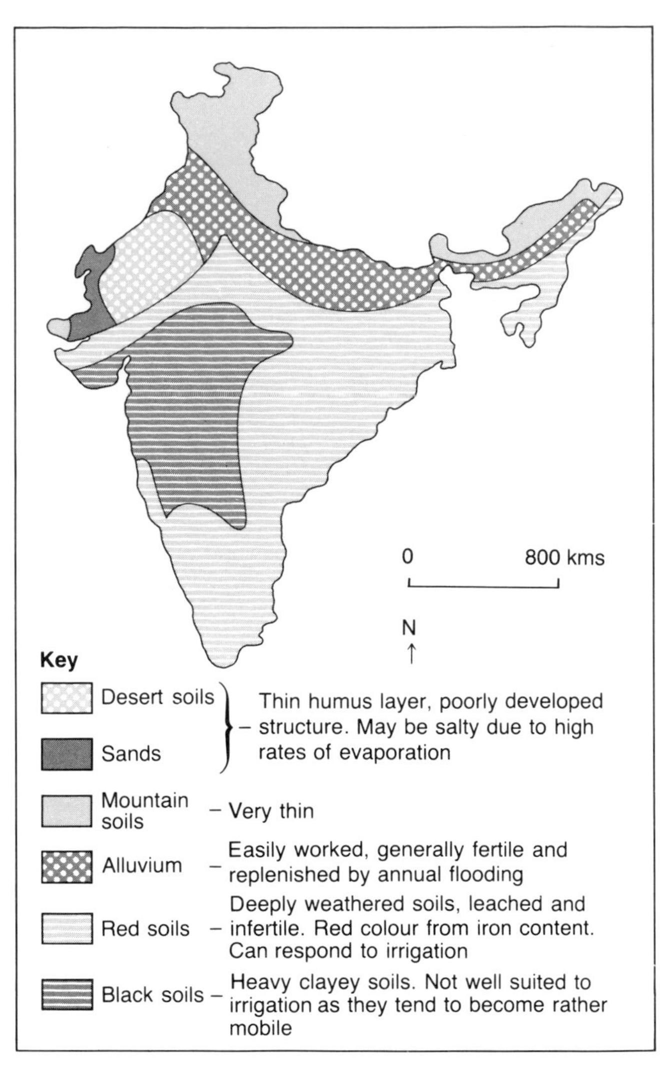

Fig. 3.13 Soils of India

Fig. 3.14 Irrigation in India

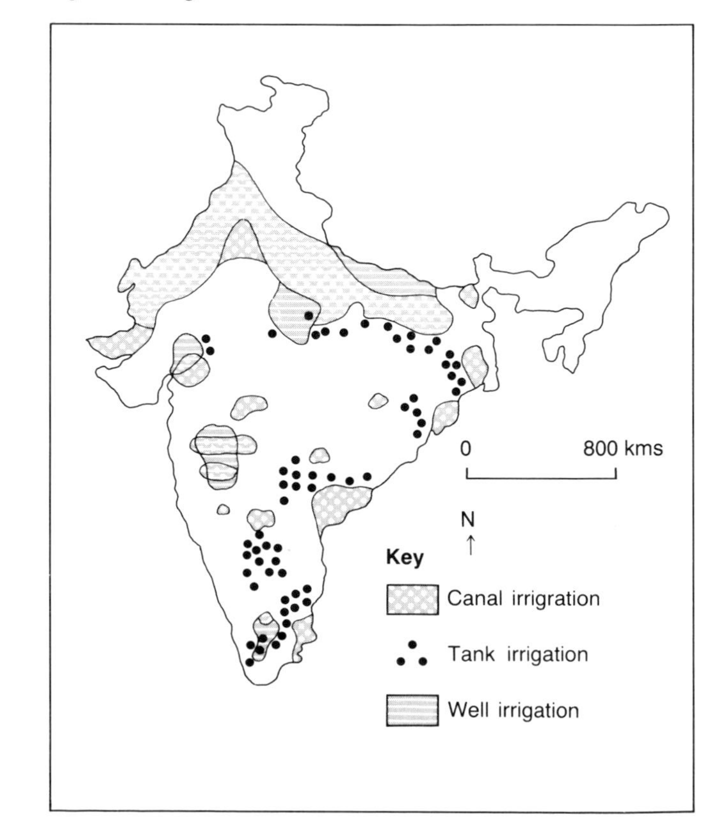

12 Make a tracing of the outline of India and use your atlas to mark on the principal physical features including mountains and rivers.

13 Summarize the main features of the Indian climate shown in your atlas by adding notes to the map. Ensure that the word **monsoon** occurs in your notes and is fully explained.

14 Make a second tracing of the principal wheat and rice growing regions (Fig. 3.12).

15 Place this tracing over Fig. 3.13 (soils) and Fig. 3.14 (irrigated areas) and using the information available to you, try to produce a series of generalizations regarding the growth of rice and wheat in India. You may be able to draw a model to show how these factors operate to restrict and limit the growth of these two cereal crops.

16 Use the data in Table 3.7 to construct three choropleths of the percentage of the total area for each state under the three main types of cereal crops. Comment on the effectiveness of this method of presentation and compare it with the tracing of the principal wheat and rice growing regions (Fig. 3.12).

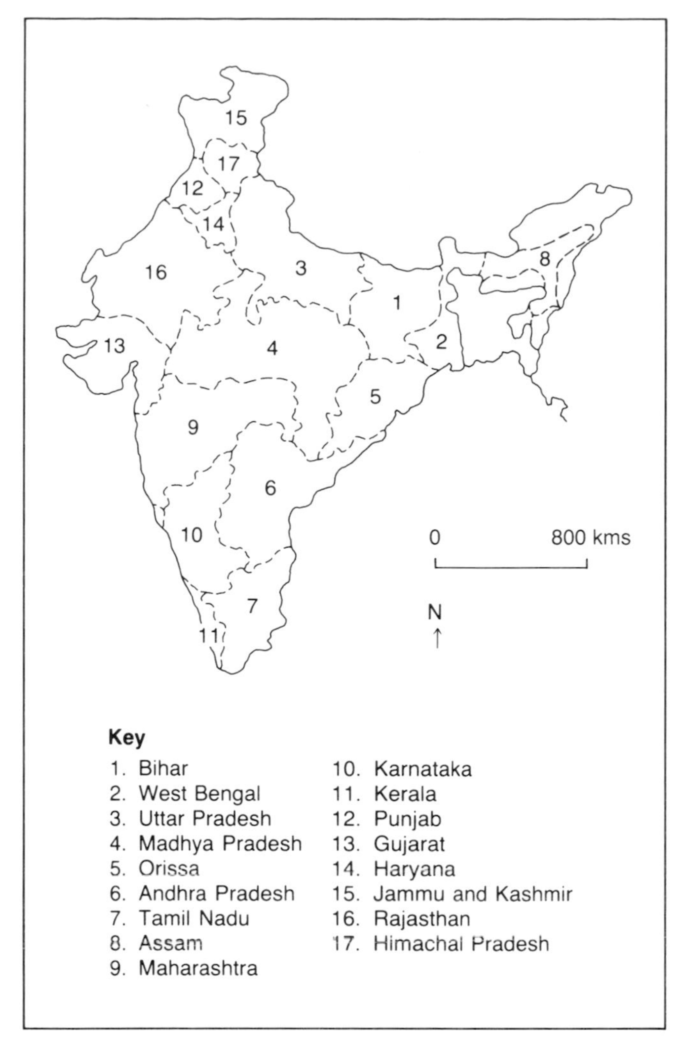

Fig. 3.15 The states of India

Table 3.7 gives more information regarding the water requirement of the crops. Most strains of rice require mean temperatures in the growing season of over 24°C and need daylight to flower and mature. **Kharif** is the name given to the south-west monsoon (summer) and **Rabi** is the north-west monsoon (winter). These names are also used for the rice crops produced in these monsoons. Wheat, however, needs temperatures of only 16°C in the growing season. Higher temperatures actually lead to a decreasing yield. Jowar, Bajra and Ragi are varieties of sorghum or millet which, after harvesting, are ground down to produce flour. These are indigenous to the tropics.

17 Use the information in the paragraph above to add detail to your agricultural model.

Identifying farming regions

Table 3.8 gives information on the cash crops of India, or at least the main ones. These are crops which are grown to sell to middlemen and industries and form the raw materials of industries in India or are transported elsewhere to be processed. However, they form an important part of the agricultural system of the country.

Table 3.7 Water requirements of crops

Water requirement of crops

Crop	Duration (days)	Total water requirement (mm/ha)	Daily water requirement (mm/ha)
Rice			
Kharif	120	965	8.05
Rabi	130	1057	8.20
Wheat	125	356	4.32
Bajra		508	
Ragi	125	508	5.84
Jowar		635	
Groundnut	120	457	5.33

Table 3.8 Area under major cereal crops as a percentage of the total area under cereals

Area	Rice	Wheat	Millets
Bihar	62	22	–
West Bengal	90	8	–
Uttar Pradesh	30	42	10
Madhya Pradesh	38	26	17
Orissa	86	–	–
Andhra Pradesh	46	–	36
Tamil Nadu	57	–	26
Assam	96	1	–
Maharashtra	13	10	72
Karnataka	19	5	46
Kerala	100	–	–
Punjab	20	63	12
Gujarat	13	18	60
Haryana	13	47	37
Jammu and Kashmir	39	–	25
Rajasthan	2	22	59
Himachal Pradesh	13	38	–

The information in Table 3.9 is in the form of a **location quotient**. This is a measure of how important the commodity is for each particular area (see Techniques, page 161). A high value indicates that the area has a high concentration of the commodity. A value between 0.80 and 1.20 indicates that the region has an average amount of that commodity for its size. Values larger than 1.20 indicate that the commodity is concentrated in that area. It would appear from the table that tea is very concentrated in four states and is also significant in Tamil Nadu.

The location quotient is very useful in producing maps showing **regions**. A region is an area characterized by the factors under consideration. In this case, values larger than 1.20 are considered to have a larger than average amount of a particular

States	Oilseeds	Cotton	Sugar Cane	Tea
Andhra Pradesh	1.25	0.69	0.65	–
Assam	0.79	0.03	0.86	32.30
Bihar	1.10	0.01	3.03	–
Gujarat	0.95	1.51	0.12	–
Haryana	0.50	1.40	2.54	–
Himachal Pradesh	1.73	0.14	1.34	–
Jammu & Kashmir	1.61	0.08	0.39	–
Karnataka	0.87	1.49	0.55	0.06
Kerala	0.80	0.26	0.63	28.74
Madhya Pradesh	1.23	0.89	0.28	–
Maharashtra	0.59	2.05	0.53	–
Meghalaya	0.99	1.47	–	–
Orissa	1.55	0.44	0.78	–
Punjab	0.51	2.03	0.95	–
Rajasthan	1.23	0.86	0.35	–
Tamil Nadu	1.19	0.65	0.85	1.66
Tripura	0.65	0.43	1.05	27.84
Uttar Pradesh	1.19	0.02	2.60	0.03
West Bengal	0.96	–	0.94	24.66

Table 3.9 The location quotient for the most important commodities of India

commodity. When these areas have been identified on the map, they will be a region. Some of them may be next to each other and will form a continuous region.

18 Comment on the effectiveness of the two methods of showing agricultural patterns, i.e. the location quotient map and the map drawn from percentages. Which is the more informative?

19 Describe and explain the pattern of agriculture in India using the information you have so far.

Drawing regions on the map is another way in which the geographer is attempting to simplify and classify complex information.

20 Use the information in Table 3.9 to construct a regional map of the main cash crops of India. Follow these instructions:

(a) For each commodity identify those states with location quotients over 1.20. (You may wish to identify an even higher level of super-concentration for tea in particular.)

(b) Devise a means of showing this information on ONE traced outline of India, Fig. 3.15. You may consider various types of shading.

(c) Draw a heavy black line around those states which have high concentrations. If the states are side by side draw the line around the outside of the two states.

The vicious cycle
Fig. 3.16 summarizes some of the problems facing the Indian farmer in the traditional system. The farmer may not own land and will be forced to rent from a landlord who may exact payment on a share cropping basis (where part of the crop produced goes as payment for the land). Human labour is the most important form of power used in such a system and its quality may be reduced by the subsistence diet. To improve the yield, the farmer must try to manage the physical environment more and this may require irrigation, which in turn may require machinery and power. Buying new seed, fertilizer and pesticide may also improve the yield. For these, money is needed and the money lender may be on hand to help. Money loaned will earn high rates of interest for the lender.

Fig. 3.16 The vicious cycle of poverty

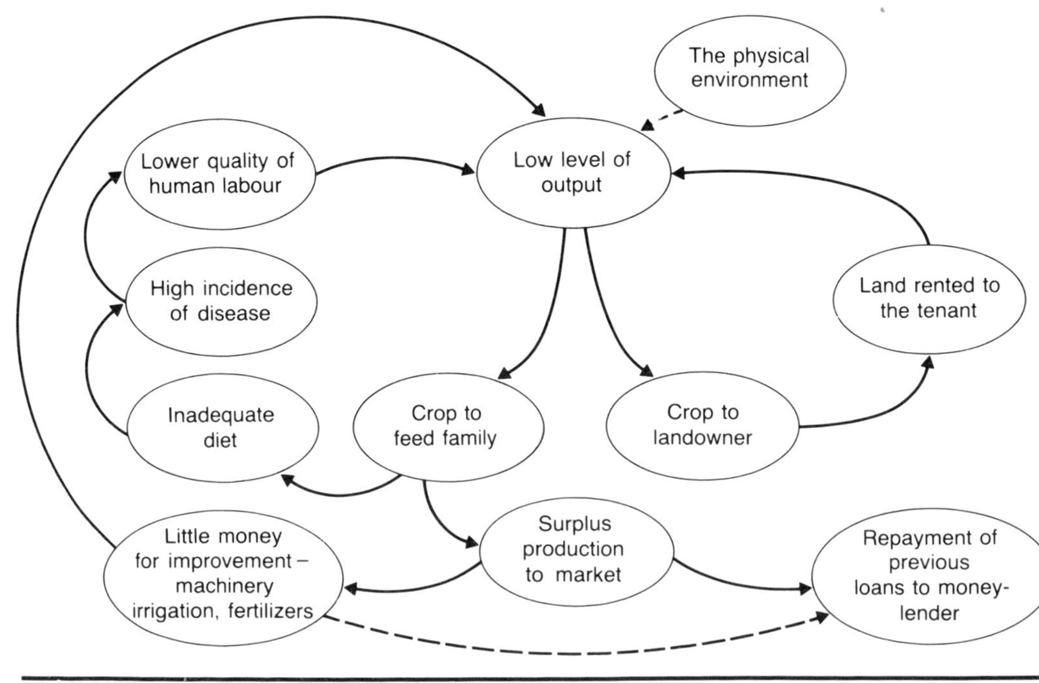

21 Comment on the ways that the working of the cycle may hinder agricultural change.

22 At what point in the cycle would you try to alter it to enable more food to be produced?

The Green Revolution

The **Green Revolution** is the name given to a number of measures designed to increase agricultural yields in the Third World. These include the development of **high yielding varieties** (HYVs) of wheat and rice, which demand in turn irrigation, fertilizers and pesticides. India was one of the countries to benefit from these developments in the mid-1960s. Yield per hectare rose by 18% from 1969 to 1977, while the area under crops rose by 5% and production by 25% over the same period. Rice yields have increased by over 51% since the mid-1960s and wheat by 50%. The Punjab has shown considerable change, which is shown in Table 3.10.

23 Use the information in Table 3.10 and calculate the percentage changes to 1969–70 and 1971–72 using 1965–66 as a base year.

24 Compare your results with the figures given above for the whole of India.

Punjabi farmers were quick to adopt high yielding varieties in 1966, and by 1972 farm incomes had doubled. The number of tube wells had increased fourfold and the number of tractors sixfold. New areas of land were brought under cultivation using the available irrigation water, so that the area under wheat increased by 50%. Fertilizer application increased sixfold. While the farmers owning large areas benefitted most at first, those owning smaller farms also gained. Share cropping agreements were replaced by fixed rent tenancies. With the development of wealth came the demand for industrial products and services, which provided non-farm employment.

Punjabi success can also be attributed to the existence of good groundwater supplies and an existing canal system, government investment in roads and rural electrification projects. The proximity to research stations producing innovation for farmers was also significant for the speed of take-up of new ideas.

25 Draw line graphs to represent the data in Table 3.11, showing the adoption of HYVs in the Punjab.

26 Re-draw the cycle of poverty (Fig. 3.16) to show the changes that have occurred in the Punjab.

Table 3.10 Early impact of the Green Revolution in the Punjab

PRINCIPAL CROPS: increases in yield per hectacre (kg)				
Crop	1960–1	1965–6	1969–70	1971–2
Wheat	1244	1236	2245	2406
Rice	1009	1000	1490	2045
Maize	1135	1653	1469	1564
Gram	813	615	1033	841

(Source: *Statistical Abstract of Punjab 1972*)

Table 3.11 Changes in cereal varieties during the Green Revolution

ADOPTION OF HIGH YIELD VARIETIES Area under HYV as a % of cropped area				
Year	Wheat	Rice	Millet	Maize
1967	34.6	5.4	24.4	6.1
1968	57.9	0.5	42.5	10
1969	69.3	20.0	47.0	9.7
1970	69.1	33.3	60.8	8.8
1971	72.5	69.1	47.5	7.3
1972	78.6	80.5	7.0	5.2
1973	84.2	86.8	4.0	4.6
1974	88.2	84.5	24.6	16.8
1975	89.6	91.0	22.6	17.2

(Source: *Statistical Abstract of Punjab (1972)*)

Impact of the Green Revolution

Tables 3.12 and 3.13 show changes in the production, yield and area under rice and wheat for the principal recipients of the ideas of the Green Revolution.

27 Use the figures and the information above to write two speeches for the promoters of the Green Revolution, one to be given to a group of farmers considering changing to HYVs in a developing country and one to be given to the government of that country. Discuss the different styles needed in the speeches and the type of information they would need to carry.

28 Write a comment on the effectiveness of the Green Revolution.

The success of the Green Revolution depends upon the support of infrastructures such as transport links and electricity to provide power for pumps. It also requires a high degree of education and a willingness to change. Most HYVs are short stemmed plants which in the case of rice can be drowned by high water levels in paddy fields. Fertilizer application must be carefully controlled or plants become 'leggy' and yield diminishes. Yield increases can be spectacular, although fertilizers must be available and this may add a further burden to the developing country. In parts of the world, fertilizers may be controlled by foreign multinational companies

Table 3.12 Percentage change in area, population and yield of rice during the key years of the Green Revolution

Country	Change in rice area million ha	Change in rice area percentage	Change in rice production million tonnes	Change in rice production percentage	Change in rice yield metric tonnes/ha	Change in rice yield percentage
China	3.1	10	67.9	73	1.76	57
India	4.0	11	31.5	69	9.66	51
Indonesia	1.6	21	18.8	122	1.72	83
Bangladesh	1.4	15	7.0	46	0.44	27
Thailand	2.8	43	5.2	42	−0.02	−1
Burma	0.1	2	7.0	97	1.45	94
Vietnam	0.9	19	5.3	60	0.66	75
Philippines	0.3	10	3.8	93	0.99	75
S. Korea	<.1	0	2.3	45	1.96	47
Pakistan	0.6	43	3.2	160	1.20	83
N. Korea	0.2	33	2.7	117	2.47	66
Total	+ 15.0	+ 14	+ 154.7	+ 74	+ 1.05	+ 52
World less 11 above	+ 3	+ 15	+ 5.3	+ 11	− 0.10	− 4

(Source: Palacpac 1982 and US Department of Agriculture 1984)

Table 3.13 Percentage change in area, population and yield of wheat during the key years of the Green Revolution

Country	Change in wheat area thousands of hectares	Change in wheat area percentage	Change in wheat production thousands of tonnes	Change in wheat production percentage	Change in wheat yield tonnes/ha	Change in wheat yield percentage
Egypt	47	+9	728	+57	1.08	+45
Morocco	318	+19	874	+80	0.34	+52
Afghanistan	294	+13	800	+36	0.20	+21
Bangladesh	405	+405	960	+813	0.96	+81
China	4106	+17	51 671	+173	1.63	+135
India	9685	+72	25 900	+156	0.61	+50
Iran	2000	+50	2582	+66	0.10	+10
Pakistan	2052	+39	6284	+103	0.54	+47
Turkey	1574	+22	3264	+33	0.12	+9
Total	20 481	+39	93 063	+131	0.82	+66
World less 9 above		−4		+ 24		+29

(Source: US Department of Agriculture 1984 and FAO 1967)

and if they are synthetic will be based on natural gas. Price changes occur in line with supply from fertilizer plants. Fig. 3.17 shows the yield improvements that can be effected by controlled application of fertilizers in Indonesia. Table 3.14 gives information on yield and fertilizer input for South-East Asian countries. The levels of application are generally lower when compared with European countries.

Table 3.14 Rice area and nitrogen used during a recent year

Country	Rice area planted thousand hectares	Nitrogen kg/ha	Yield rough rice tonnes/ha
Burma	5013	11	2.19
Bangladesh	10 308	21	2.02
India	40 200	19	2.01
Pakistan	2026	52	2.37
Indonesia	8495	73	3.29
Philippines	3543	33	2.15
Thailand	8288	10	1.82
S. Korea	1314	131	5.90

29 Test the relationship of yield and fertilizer application by calculating the Spearman's R correlation coefficient for the eight countries (see Techniques page 192). What does your result tell you?

How can we increase food supply? Should we use the European Economic Community model of regional specialization and trade as the model for the world, or should we look to improving the chances of survival of each farmer? Should we work towards specialization or diversification? These are two of the broad options open, but Fig. 3.18 gives some others.

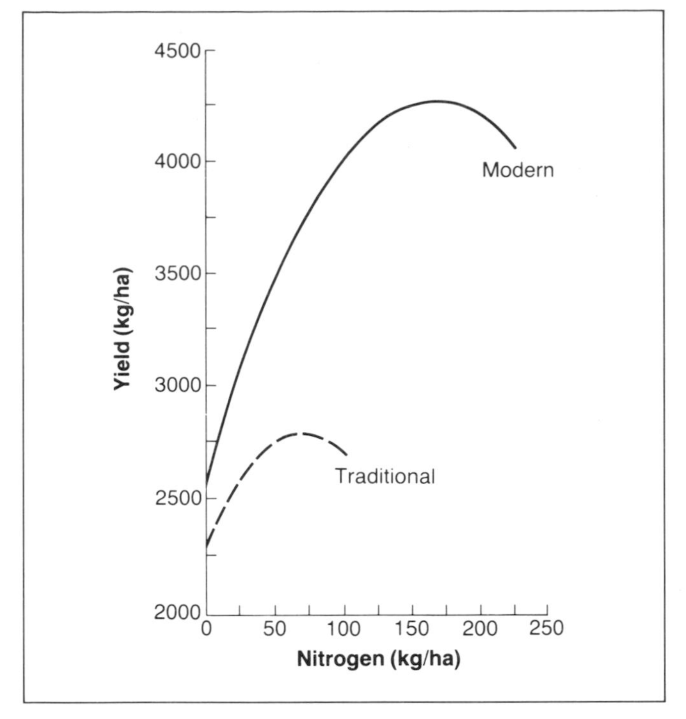

Fig. 3.17 Yield response of nitrogen on irrigated rice in Indonesia

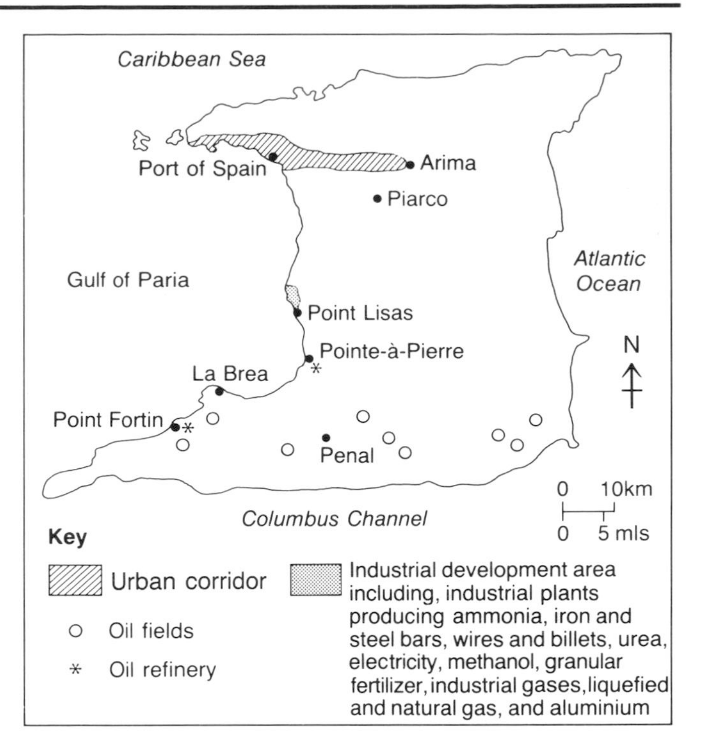

Fig. 3.19 Distribution of oil reserves in Trinidad

30 Make a copy of the table which forms the lower part of Fig. 3.18 and discuss the preconditions, duration and scale of the changes, and the environmental impact of as many as possible. (One example has been completed to give you a better idea of what is required.)

31 If possible, divide the solutions among the group and research them. Present your findings to each other, remembering to find actual examples wherever possible.

Case-study: the impact of industrialization on agriculture in Trinidad

As a country develops its economy, industry may develop quicker than agriculture. The benefits and issues of industrialisation are dealt with in Theme 5. However, this development may lead to land use conflicts as industrial priorities take precedence over agriculture. Trinidad's economy was based heavily on agricultural products until the recent discovery of oil. This has initiated industrial development. The distribution of oil reserves is shown in Fig. 3.19. This has led to the establishment of an industrial area at Port Lisas and refineries at Pointe-a-Pierre and Pointe

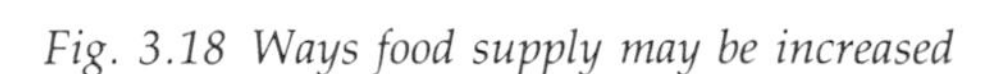
Fig. 3.18 Ways food supply may be increased

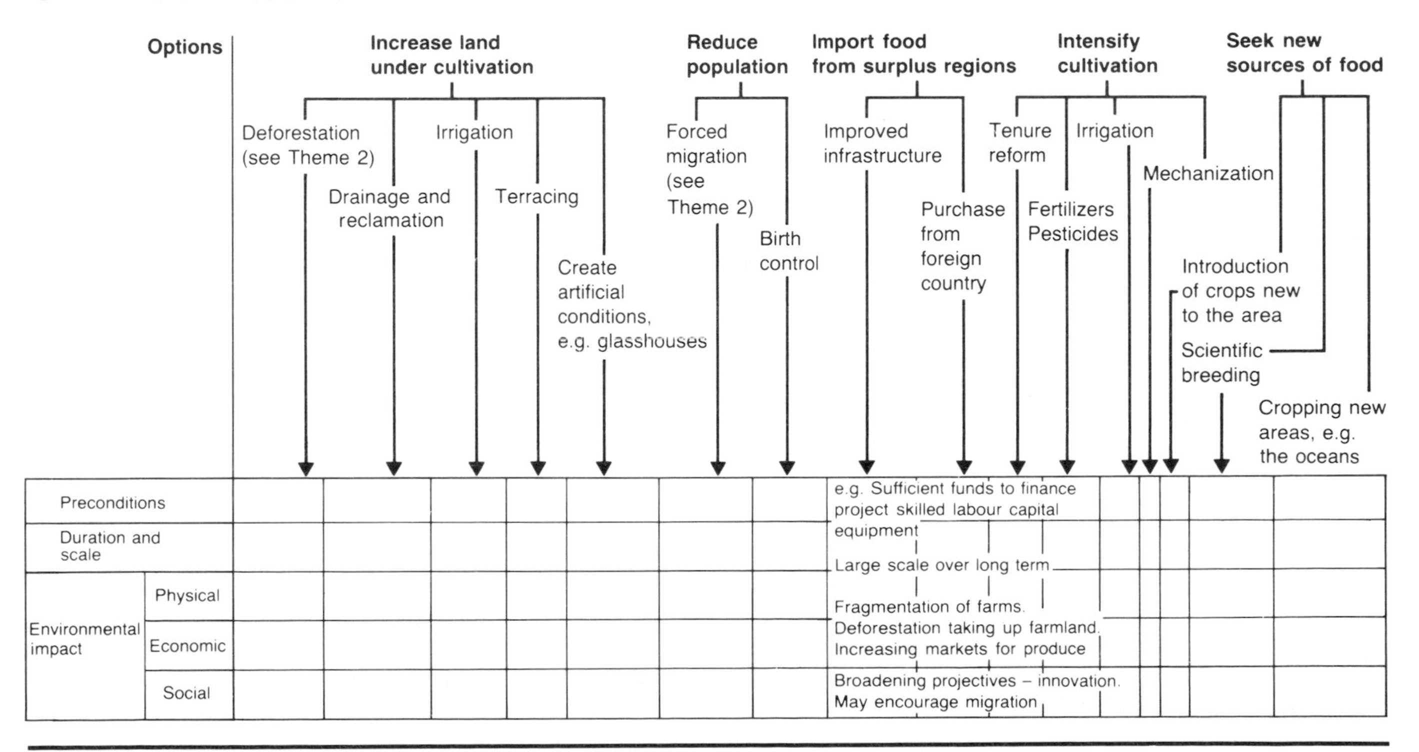

Table 3.15 Sectorial contributions to the Gross Domestic Product in the economy of Trinidad

Sector	1970 (Tr$m)	1980 (Tr$m)
Agriculture, forestry and fishing	95.7	68.4
Petroleum	361.8	1214.4
Manufacturing	149.3	172.5
Construction, quarrying	95.7	238.9
Public utilities	32.4	34.2
Services	887.5	1170.6
Gross Domestic Product	1622.4	2899.0

Fortin. Oil production peaked in 1978 and is now declining due to a depletion in reserves and due to the world glut in the mid-1980s, leading to a lowering of the price on the world market. There has only been one new find of oil since 1976. Exploitation of the known reserves has led, however, to a moderately high rate of growth in the economy (3.1% per annum in 1981), increased revenue for the government and a higher standard of living for the people of Trinidad. Revenue from oil provided 63% of government income in 1981.

32 Using the information in Table 3.15, draw pie diagrams to show the changes in the **Gross Domestic Product** and the **sectoral contributions** of the economy of Trinidad.

33 Describe these changes for the period 1970–1980 and give reasons for the changes.

Agriculture has been important to Trinidad in providing export revenue and employment as well as food. Production of crops has declined as the oil industry has developed which has led to a reduction in the volume of exports and a decline in the importance of agriculture as a source of employment (from 22% to 16% during the period 1960 to 1980). Industrial employment has increased from 34% to 36% during the period 1960 to 1980.

34 Table 3.16 gives the export crop production for selected crops from 1962 to 1982. Draw a series of line graphs to show the changes in production of these crops.

35 Calculate five-year moving means using the same figures and plot the results on the same graph. (See Techniques page 193.)

36 Describe the trends shown and comment on the moving mean technique.

37 Describe the likely environmental impacts of these changes in the economy as seen by the following people:
(a) the government representative of Trinidad,
(b) the managing director of a multinational oil company investing in the Trinidad economy, and
(c) a young school leaver from a rural area looking to take his or her first paid job.
What changes would they welcome? What sort of rural environment would be created for the future?

The attractiveness of employment in the oil or oil related industries has made agriculture more

Table 3.16 Trends in export crop production during the main development of the Trinidad oilfields.

Year	Cocoa 000 kg	Coffee 000 kg	Grapefruit 000 crates	Oranges 000 crates	Copra 000 tonnes	Sugar 000 tonnes
1962	6000	1700	762	274	12.3	206
1963	6532	3600	572	190	13.8	230
1964	4626	3700	598	254	12.3	230
1965	4354	3500	981	316	12.2	250
1966	4843	2652	797	342	12.4	213
1967	4292	2836	655	331	12.6	201
1968	6101	4517	502	211	13.7	243
1969	3977	2986	445	220	14.2	241
1970	6121	2667	470	262	11.2	220
1971	3766	3893	316	163	12.4	217
1972	4821	3300	514	240	12.5	235
1973	3812	2717	90	14	11.8	184
1974	3475	1940	468	248	6.7	186
1975	5239	4023	173	66	8.9	163
1976	3249	2671	242	153	9.1	204
1977	3345	2919	74	18	9.0	176
1978	3398	2500	154	54	7.4	147
1979	2628	2497	109	74	6.8	143
1980	2381	2240	112	64	4.9	113
1981	3145	2433	n.a.	n.a.	5.2	93
1982	2235	1795	n.a.	n.a	5.9	80

marginal. Wage costs are over 70% of total farm costs. Even though there is a 10% unemployment rate farmers have difficulty recruiting labour due to the prospect of high incomes and good benefits elsewhere, and also to the low esteem of agricultural work. Hence the unskilled manual worker and the qualified engineer are moving out of agriculture to seek work in the industrial sector. While welcoming the diversification of the economy, the Trinidadians are concerned that agriculture might decline to a critically low level and not provide support for the economy should industrialization prove short-lived.

Where will the food supply come from? Will the rural areas be managed at all? Have they achieved the right balance between industrialisation and agriculture? Time alone will tell.

Case-study: agriculture and rural change in Botswana

Botswana is a large country of 600 000 km^2 and with a population of 900 000 people. This is two and a half times the size of the United Kingdom with only 1.6% of the population. Much of the land of Botswana is at present unused, being an eastward extension of the Kalahari Desert. Hence large areas of land have been designated as national parks and game reserves (Fig. 3.20). The agricultural development strategy of the Botswana government and its experience to date may present some strategies for development elsewhere.

Much of the agricultural activity is centred on cattle rearing, although some areas have been ploughed and arable farms established (Fig. 3.20). Further areas have been fenced to enable a more sophisticated style of cattle farming to take place. Considerable areas are still open, however, and are managed by

Fig. 3.20 Present land-use in Botswana

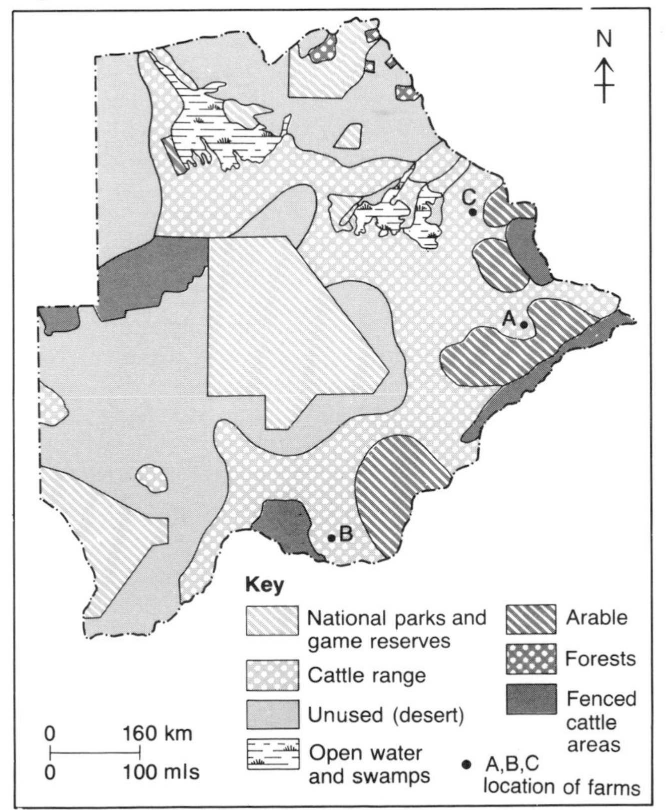

THE RANGE AND LIVESTOCK MANAGEMENT PROGRAMME IN BOTSWANA

The RLMP began in 1974 when six American technicians, funded by US AID (United States Agency for International Development), arrived in Botswana to support the Ministry of Agriculture. The technicians, all from Texas, were

- a Senior Livestock and Range Manager
- a Range Management Extension Training Officer
- a Livestock Production Specialist
- a Surface Water Land-use Engineer
- a Range Agronomist
- a Data Processing Technician

The programme was based on the belief that modern farming techniques should be used to obtain a higher productivity from Botswana's range; it aimed to achieve this by developing replicable groups of smallholders working livestock management systems.

Fig. 3.21 The Range and Livestock Management Programme in Botswana

Fig. 3.22 The Tribal Grazing Land Policy in Botswana

THE TRIBAL GRAZING LAND POLICY IN BOTSWANA

The TGLP was formulated in 1975, twelve years after independence, with the aim of helping groups of smaller livestock owners to own commercial ranches. That is, it was to initiate major land tenure reform.

The TGLP was presented as an outline policy for popular consumption via a well-organized radio learning campaign. Details were vague:

'A way will be worked out for groups of farmers to get leases for commercial land.'

'The Ministry of Agriculture is working out how best to form groups.'

The policy did include technical details, e.g. how to calculate the livestock-carrying capacity of the land, but these were later omitted as they were general and did not apply to individual cases.

The three national principles of Seretse Khama's government formed the explicit political basis of the policy:

1. *Democracy* Change must be based on consent and consultation. We must, by marrying the new institutions like District and Village Development Committees with the traditional methods of government, especially kgotla, achieve a system where the people themselves are involved in the choices.

2. *Self-reliancy* People have already built for themselves clinics, schools, and latrines. Now these efforts must include projects which help to increase production and raise rural incomes.

3. *Social justice* Government policy is directed towards reducing the large differentials in income and wealth which exist in Botswana. Social and economic change must not favour either the rich or the poor. Changes in land tenure can only be achieved by popular consent.

Fig. 3.23 Tsetsejwe: an RLMP ranch

TSETSEJWE: AN RLMP RANCH (. A)

Farmers
Subsistence farmers but dominantly cattle-owning; 63% families own 1–20 cattle, 14% own more than 60 cattle.

Ranch history
1. 1973: related neighbours told a local agricultural extension worker that they wanted to fence the grazing land of their cattle-posts and make one area of it. The ministry of agriculture said this was not possible in a communal area.

2. 1974: Extension staff interested the whole community in developing a communal ranch and brought the proposal to the RLMP.

3. November 1974 to June 1975: the RLMP team made periodic visits to the area and decided it would be a suitable pilot scheme.

4. 11 August 1975: A widely publicized kgotla meeting at Tsetsejwe. It was the official opening of the scheme and was attended by the minister of agriculture. No opposition to the plans was noticed.

5. 24 September 1975: Application forms to join the ranch project were distributed to the community and 300 were received back.

A selection board made up of RLMP staff and local government officials chose eighty-four applicants whose cattle-posts could be enclosed as a continuous area which measured 32 000 ha. (The board decided that the whole area owned by the community was too large to develop at once.)

The RLMP team calculated in detail the requirements for a communal cooperative ranch – carrying capacity, fencing and water costs, herd projects etc.

6. 16 October 1975: The proposals were presented to the group members but only thirty-eight out of eighty-four attended the meeting.

7. January to April 1976: A series of inconclusive meetings was held with representatives of cattle-post owners in the North-eastern corner of the project area. These people objected to the ranch as the area was part of their traditional grazing land, but they were not members of the Tsetsejwe community.

Other problems came to light at these meetings: some people in Tsetsejwe knew nothing of the proposed ranch; no account had been taken of the non-grazing uses of the land (tree-cutting, thatching grass, cultivation, driving cattle to water).

8. May 1976: The RLMP team decided that their grazing capacity calculations were wrong; no detailed land-use map was produced.

Fig. 3.24 Sebelalo: an RLMP ranch

SEBELALO: AN RLMP RANCH (. B)

Farmers
Sixteen cattle-owners, with 362 cattle altogether; two owners had only two cattle each.

Ranch history
1. Mid-1975: RLMP team proposed to kgotla at Sebelalo that local cattle-owners manage a 6400 ha ranch round a few boreholes. The question 'Who wants to join?' was put at kgotla. An Agricultural Demonstrator was assigned to the group until 1978.

2. 1975 to 1977: Physical development of the ranch; government drilled the borehole; 37 km of perimeter fencing plus fencing round bull and weaner camps were constructed by the group with RLMP help.

RLMP staff decided that members form a committee and meet monthly; prepare monthly financial report/herd register; appoint manager to look after cattle as one herd.

3. July 1977: The group's cattle moved onto the ranch.

4. November 1977: Regional agricultural staff noted that no monthly meetings were being held, no manager had been appointed and no-one looked after the water engine.

5. April 1978: RLMP staff noted that no firebreaks had been made and no controlled grazing practised. The RLMP team constructed the firebreaks.

6. 1978 to 1979: More advice was given at meetings with the group about cattle management (separating heifers, dehorning, tick control).

7. 1980: Ranch fence falling into disrepair; no improved methods being practised; regional staff lost hope . . . complete indifference of the group . . . 'gross negligence'.

38 Read the information given on these three schemes and analyse the reasons for success or failure. Make a list of the reasons and discuss these with your class.

39 What lessons could be learnt from these experiences?

tribal groups. It is the management strategy of these which might provide a model for development.

Two programmes were initiated in the mid-1970s. They were the Range and Livestock Management Programme and the Tribal Grazing Land Policy. A ranch at Tsetsejwe (Fig. 3.23) and a ranch at Sebelalo (Fig. 3.24) were part of the RLMP, and the Nkange-Maitengwe scheme is an example of the TGLP. Information on the programmes is given in Figs. 3.21–3.25. A **kgotla** is a meeting of all the local community.

The decision-making system of the Tribal Grazing Land Policy and its implementation is given in Fig.

40 Critically evaluate this decision-making model. What are its strengths and potential weaknesses? Are there any wanted or unwanted consequences to development using this type of method?

Fig. 3.25 Nkange-Maitengwe: land enclosure fence

NKANGE-MAITENGWE: LANDS ENCLOSURE FENCE (. C)

Farmers

Almost all the households in two villages participated; the majority own a handful of cattle and/or smallstock and work lands; a minority have more cattle on nearby grazing land.

The project

Both villages wished to fence off their land areas because

- no permanent water in the nearby grazing areas meant that the cattle congregated around the rivers Nkange and Maitengwe at the beginning of the winter and moved into the lands, damaging the crops
- many farmers wished to expand and improve their lands but this was impracticable while cattle were able to roam freely.

The project started from ideas expressed at kgotla meetings in the two villages in 1979; the Agricultural Demonstrator and Land Use Officer who served both villages called in regional staff to conduct a land-use survey, which showed soils with potentially high fertility, and informed the villagers about the availability of AE10 grants. AE10 grants are matching grants funded through the TGLP for group fencing projects; the group is required to provide 10% of the hardware costs and all the labour, the government provides 90% of costs.

Work started straight away on clearing the fence line; in some places a thornbush fence had been constructed in the early 1970s. Difficulties in getting to the extreme corners of the area were solved by the Land Use Officers providing daily transport.

Nkange formed a fence committee which reported back to the kgotla regularly. Maitengwe used an existing farmers' committee to deal with the fence business. Nkange was responsible for 43.5 km of the fence, Maitengwe for 56 km.

Both committees

- collected fees from group members; this was organized by village ward headmen who collected from 3 to 5 Pula (25p) from each household.
- drew up by-laws to solve problems and potential problems, e.g. grazing of smallstock was allowed during the growing season within the fence boundary provided that the herds were tended all the time (it was felt to be unreasonable to ask small owners to move far with their flocks). Urban workers were asked to contribute 60 Pula to the scheme to make up for the loss of their labour; fines for 'loafing' were put at 50 thebe (½ Pula) per day.

The government built a demonstration fence at Nkange. It was 1 km long and consisted of five types of fence of various costs and strengths; the communities decided on the fencing they wanted.

The final cost of the fence was

- Nkange: group raised 500 Pula (13%); government provided 3350 Pula
- Maitengwe: group raised 293 Pula (6%); government provided 4757 Pula.

3.26. The rôle of the Agricultural Demonstrator is very important and entails working alongside the farmers, advising, checking plans and eventually evaluating the results. Another feature is the importance placed on feedback to the kgotla to ensure that the people who will eventually carry out the scheme are fully aware of progress and play a key role in shaping the Activity Plan.

The improvement of the rural sector of the economy is very important to the development of the country. Not only does it provide food but also raw materials for industry. Low levels of development can lead to downward spirals in the periphery as shown in Fig. 3.28. At low levels of development the urban area may appear to be the land of opportunity with regular, higher paid factory or office work. The people tempted by the move tend to be the most able workers. This will denude the rural areas further. The urban areas, however, may not live up to expectations and low wages may result from a high supply of labour. Land prices will increase and accommodation for the people who migrate will probably be in the shanty town.

Traditionally, government has treated the symptoms of the problem rather than the cause, by allocating higher levels of investment to the urban areas. This further reduces the investment levels in the periphery as investment capital is limited and choices have to be made.

41 Redraw the lower part of Fig. 3.28, assuming that the government is investing in the agricultural areas.

42 What consequences would this reallocation of investment have on the urban and rural areas?

The government of Botswana has implemented a development strategy based on small scale industries and using appropriate technologies, as shown in Fig. 3.27. These industries, like the agricultural development, have been low cost and have depended on community development. This development has also attempted to break the dependency on South Africa by encouraging import substitution. Other important advances include higher levels of training and the provision of employment for women.

The main schemes are outlined in Table 3.17. Many of them suffer from a dependency on foreign expertise in management and training and from foreign funds. Some also produce products for western tastes and the weavers' cooperative at Odi uses imported raw materials. This is a start to providing a counter-attraction to the allure of the urban areas and will in time help stabilize the development of the rural areas.

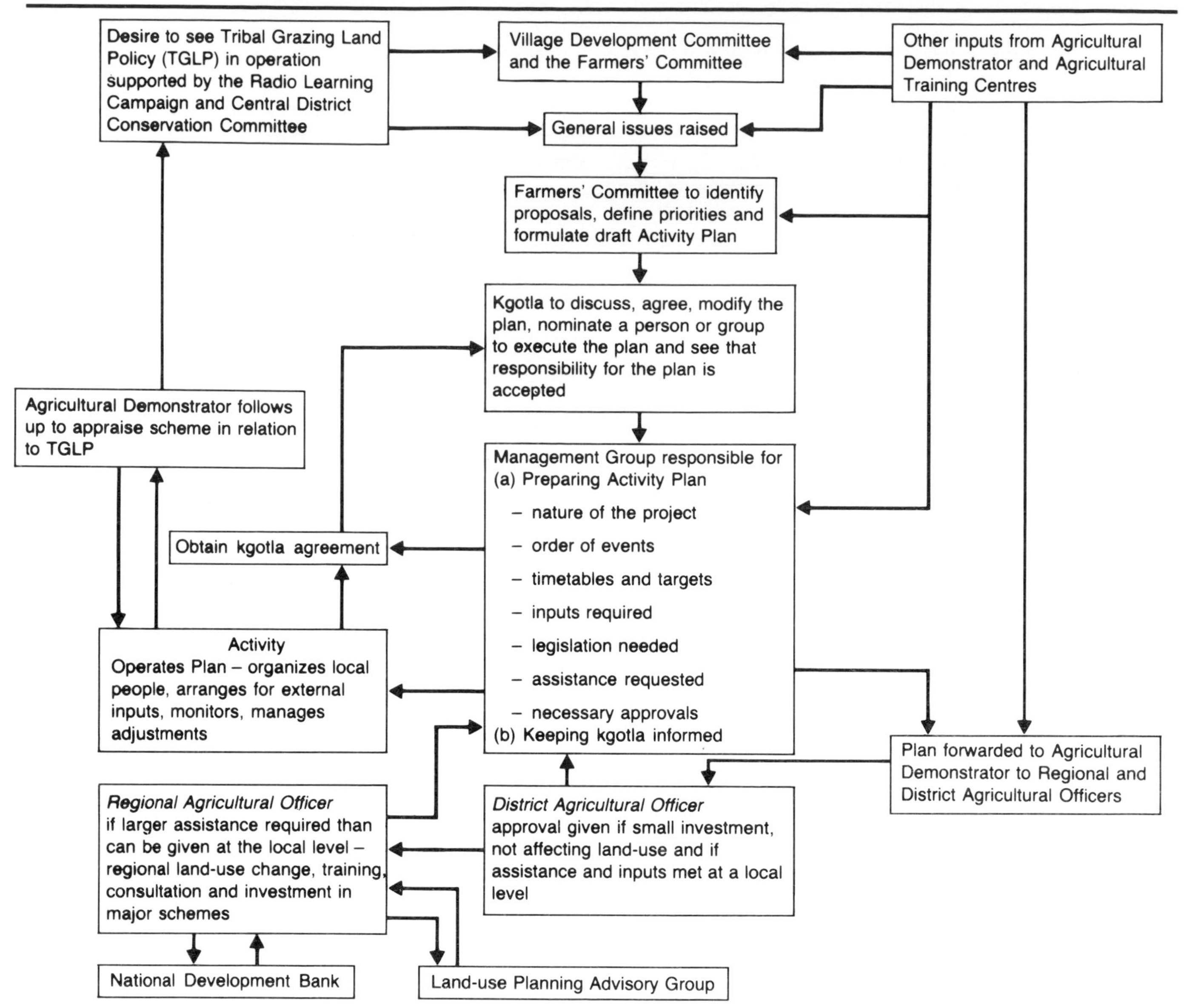

Fig. 3.26 *The extension approach to community-based rural development*

Fig. 3.27 *Small-scale rural industries in Botswana*

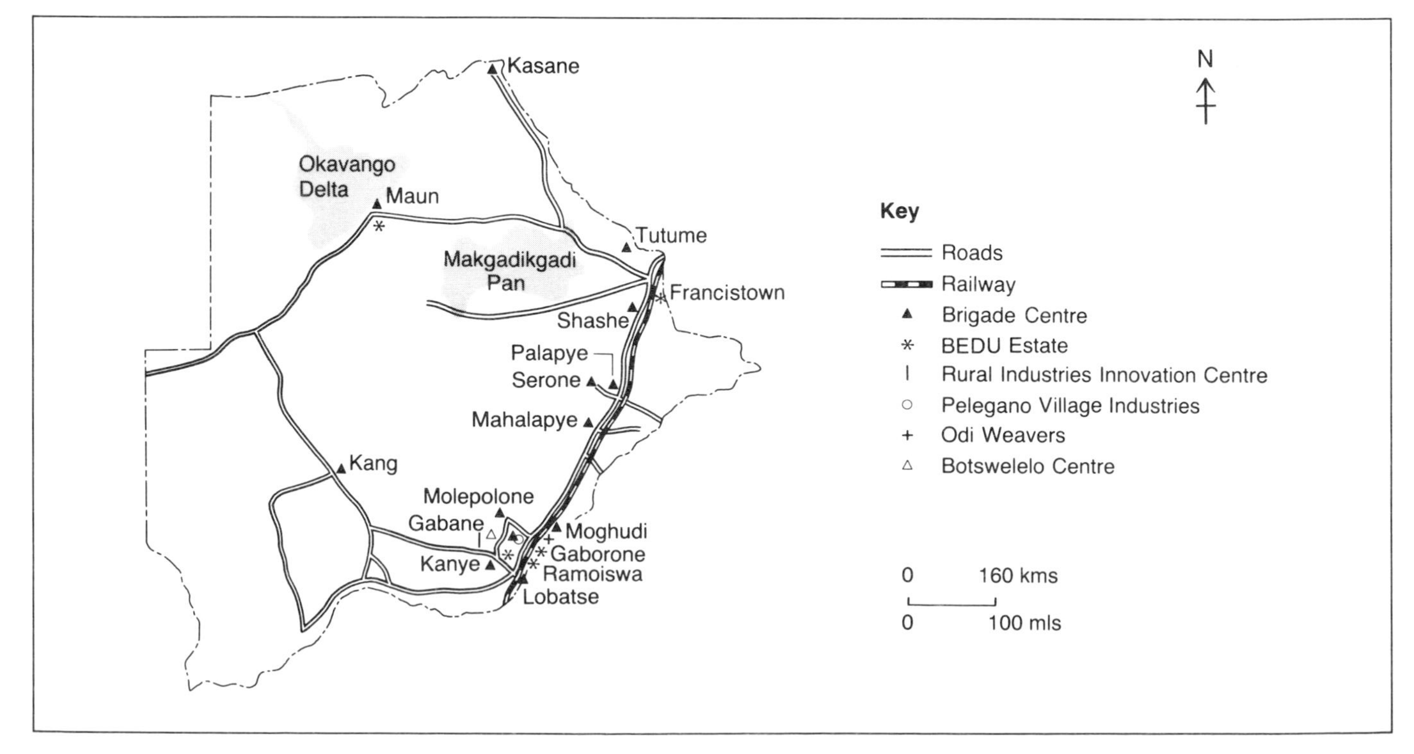

Fig. 3.28 Downward spiral of the agricultural periphery and upward spiral of the industrial core

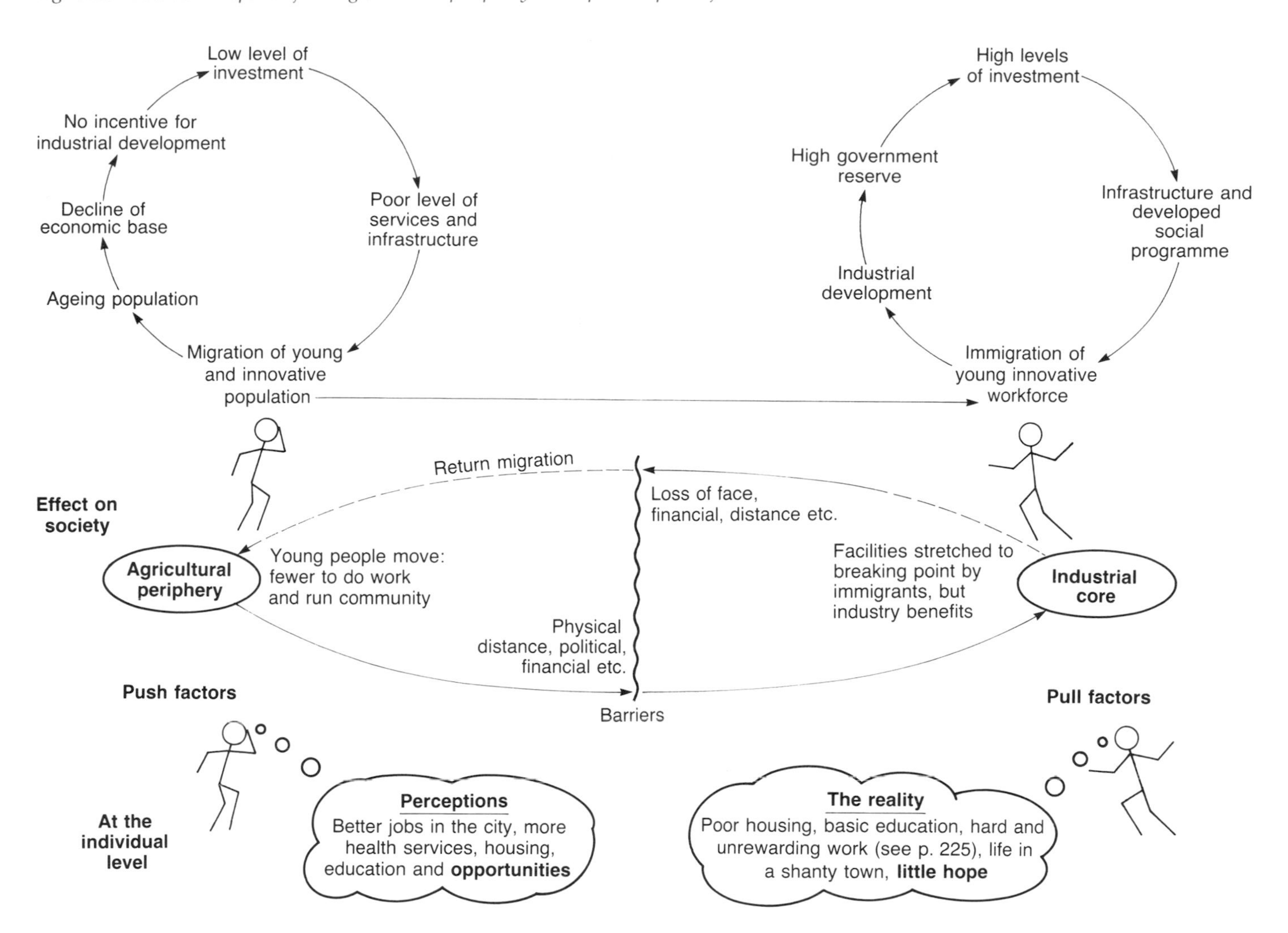

Table 3.17 Small-scale rural industries in Botswana

The Brigades	15 Centres providing vocational training for school leavers in building, carpentry, welding and dressmaking. Have provided training for 16 000 people.
'Nursery' Estates of the Botswana Enterprises Development Council (BEDC)	10–20 Workshops at each site, rented for about one third of the normal commercial cost. Each estate has a manager and advisory service centre.
Rural industries Innovation Centre	Established with West German assistance. Centre for development of rural industries in Botswana, including alternative energy from windmills, bridges and solar sources, building methods and materials, production of tools, mesh fencing, tanning equipment manufacture. Centre has library facilities and runs training courses for adults, with the idea of testing technologies before passing them on to the villages.
Pelegano Village Industries (PVI)	Encouragement of family-centred employment for the poorest families, e.g. chicken farming, rabbit raising, sorghum milling, metal and carpentry workshops, manufacture of corn husk dolls and developing citrus orchards. Provides 80 jobs.
Botswelelo Centre	Similar to PVI but funded by Canadian aid. Provides on-the-job training in textiles, pottery, cement block and brick manufacture. Provides 25 jobs. Runs a craft shop.
Odi Weavers Cooperative	High quality rugs and tapestries

Decision-making Exercise G: Developing Botswana

G

You have a good background to the development schemes which have been implemented in Botswana. Now your task is to evaluate them to try to produce a model of development which could be applied elsewhere. Use the following framework for your report:

(a) Briefly describe the schemes tried including the rural industrialization schemes

(b) Analyse the reasons for the success or failure of the schemes

(c) Make suggestions for future development in the form of a check-list which could be replicated elsewhere in the world based on the experience of Botswana. Suggest the sequencing of these schemes and ideas and consider any extra factors which might influence the model in other parts of the developing world.

Fig. 3.29 A farmscape in Botswana showing incursion of maize into grazing land

Table 3.18 Rural change model for the developing world

	Stage I **Traditional farming system**	**Stage II** **Commercialization of agriculture**	**Stage III** **Drive to maximising development**
Population	Population increase in equilibrium with farming productivity. Starvation and migration responses to natural increase higher than productivity	Increasing fluctuations of natural increase as goods are traded on world markets. Larger scale population crises as booms and slumps in agricultural products	Continue to increase naturally and suffer from fluctuations. Urban migration reducing number in countryside.
Physical environment	Natural ecosystem and man in equilibrium. Vegetation and soils having sufficiently long recovery time in areas of low density	Deforestations, soil erosion, climatic change, denudation of the ecosystem	Continuing trends with more peasant farmers pushed to the more marginal lands – steeper slopes, unstable conditions
Level of inputs	Low, primitive technologies and human labour	Increasing sophistication, still human labour. Maybe very high in plantations	Wide disparity of highly technical farming and peasant 'shambas'
Level of dependency of the economy	Independent local economy	Becoming more dependent on trade agreements, machinery expertise from developing world	Increasing dependency with aid programmes changing nature of dependency. Development attracting multinational companies.
Integration of the economy		Increasing integration	
Stress on the rural area		Increasing stress. Fragmentation due to transportation system development and mining operations	Industrial pollution, growth of urban areas. Development of non-farm activities
Agricultural systems. Examples	Bush fallowing	Cash cropping plantation systems	Wide range – small peasant farms, plantations – dependent on political ideology

Conclusions

The agricultural sector of the economy is highly significant in all but a minority of countries. Agriculture was the first type of industry and it demands a high degree of management. It is an activity where people are directly manipulating and attempting to control the natural ecosystem for their own benefit and survival. Hence development and change should be compatible with the ecosystem and should attempt to enhance the system rather than unleashing a series of effects which may set the system to destroy itself. The dynamic equilibrium of the natural ecosystems is finely balanced and small alterations can trigger a series of feedbacks which can destroy the change itself. An example of this is clearing land for agriculture by taking forest down on sloping ground. The result may be the loss of topsoil, the very component essential for agriculture.

Table 3.18 suggests changes in rural areas which may take place over time as these areas move from traditional farming systems towards maximizing development. Increasingly pressures develop in the rural areas as demands are made on it, some of them not being agricultural. These demands may force the pace of agricultural change.

43 Consider the model (Table 3.18) and discuss modifications and additions to it.

44 Attempt to outline a fourth stage to the model.

45 'Increasing world food supply depends on overcoming human obstacles rather than physical limitations.' Discuss, using a full range of illustrations and examples. (Use the information provided in Theme 3 and material from your own research in answering this essay question.)

Increasingly, the change in the rural area is being seen as, first, dependent on the change in the urban areas in both developed and developing situations. Secondly, agricultural concerns are just one of many considerations in the rural area. In the developed world, many rural areas are thought of as retirement areas because of their tranquillity. The labour force may be moving elsewhere and services may be in decline. The concluding exercise looks at just one such area in North Wales.

Decision-making Exercise H: Overcoming Rural Isolation

Rural areas are said to suffer from a series of problems. These include:

(a) Problems of rural employment identified in terms of the dominance of primary industries, low wage levels, low levels of skills, high unemployment and poor job opportunities

(b) Problems in the provision and/or maintenance of services (public and private) due to the high costs of servicing a dispersed and declining population. Such problems result in either the withdrawal of services or reductions in their availability

(c) Problems of accessibility to services and facilities due to their spatial distribution in rural areas and the inability of individuals and groups to overcome the friction of distance. Older and younger people have the most extreme problems and they make up most of the population

(d) Problems of rural people in the housing market due to their inability to compete with outside interests coupled with a low level of alternative housing opportunities

(e) Problems relating to the dominance of a prevailing political ideology in rural areas which emphasizes enterprise and the market at the expense of public sector responsibility and support.

(from B. P. McLaughlin.)

Fig. 3.30 The study area – a view north of Dolgellau

Your task is to investigate an aspect of these problems for the area of North Wales shown on the map (Fig. 3.31) and attempt to provide a solution. You should analyse the fieldwork (Table 3.19) and/or the census material provided, and represent this information with an appropriate spatial technique. You should present your answer in the form of a report using the following headings:

(a) *Identification of the problems* – justify your choice of data; state the problem

(b) *Analysis* – presentation of the data; justification of the technique; description of the pattern

(c) *Solution* – annotated sketch-map to present solution(s)

You may like to consider the five ideas given in Fig. 3.30 to assist this last section.

The written part of this report should be concise and to the point.

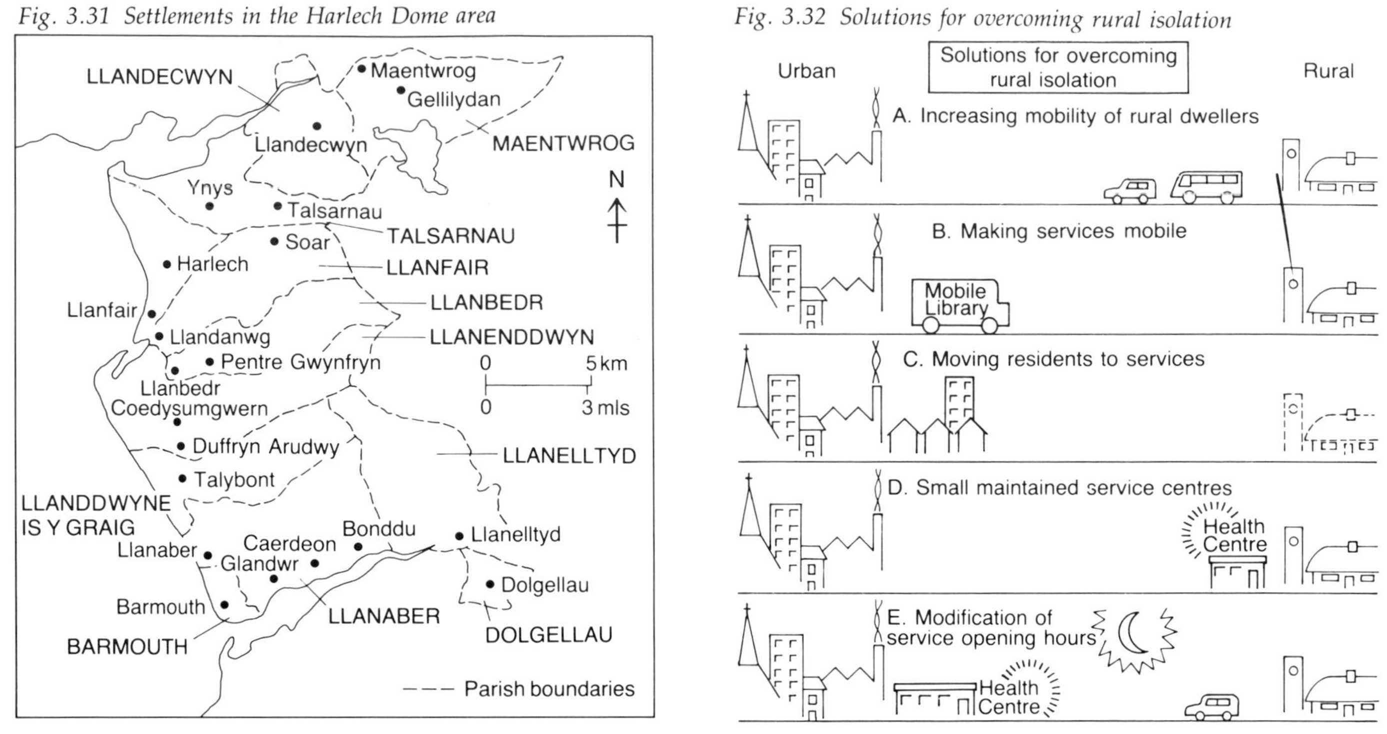

Fig. 3.31 Settlements in the Harlech Dome area

Fig. 3.32 Solutions for overcoming rural isolation

Table 3.19 Harlech Dome area: fieldwork data

	Tourist Services	Mobile Services	Off Licence 1985	Off Licence 1979	Post Office 1985	Post Office 1979	Hardware 1985	Hardware 1979	Bakers 1985	Bakers 1979	Dairy 1985	Dairy 1979	Butchers 1985	Butchers 1979	Public Telephones	Clothes Shops 1985	Clothes Shops 1979	Schools	Newsagents 1985	Newsagents 1979	Police	Hairdressers	G. Grocers 1985	G. Grocers 1979	G. Stores 1985	G. Stores 1979	Garages 1985	Garages 1979	Doctors
Maentwrog	2	1	√		√	√			√						√					√						√			
Gellilydan	1	1			√	√									√			√	√	√					√		√		√
Llandewyn	1	1													√								√						
Talsanau	1	1			√	√	√											√		√						√			
Soar	1	1													√														
Ynys	1	1	√																						√				
Harlech	3	1	√	√	√	√	√	√			√	√	√	√	√	√	√	√	√		√	√	√	√	√	√	√	√	√
Llanfair	1				√				√				√		√					√			√		√	√	√		
Llandanwg	4	1	√												√											√			
Pentre Gwynfwyn	1	1																						√					
Llanaber	2	1	√				√								√										√		√		
Llanbedr	2	1	√	√	√	√		√	√					√	√		√	√	√	√	√	√							
Boedystumgwern	2	1	√		√	√	√				√		√		√			√	√		√		√		√		√		√
Duffryn Ardudwy	1	1	√	√	√	√			√				√	√	√		√	√	√	√	√			√	√	√	√	√	
Talybont	3	1			√	√		√	√						√								√	√					
Barmouth	4	1	√	√	√	√	√	√		√	√			√	√	√	√		√	√	√		√	√	√	√	√	√	
Caerdeon	1	1																											
Bonddu	1	1			√	√																			√	√	√	√	
Glandwr	1	1																											
Llanelltyd	1	1										√			√			√								√			√
Dolgellau	4	1	√	√	√	√	√	√		√			√	√	√	√	√	√	√	√	√	√	√	√	√	√	√	√	√

Tourist Service: 1 = < 5 services; 2 = 6–10 services; 3 = 11–25 services; 4 = > 25 services

Mobile services include library, bread, butcher milk, doctor, grocers

Services include accommodation car parking, gift shops, field centres, information offices

Table 3.20 Harlech Dome area: census data (1981, unless stated)

Districts	1	2	3	4	5	6	7	8	9	10	11	12	13	14	15	16
Barmouth	2108	490	586	36	50	32	18	45.7	34.7	89	96	8	793	1	3	1
Dolgellau	2261	514	575	−240	49	36	15	36.7	25.6	85	94	11	852	2	4	3
Llanaber	257	84	48	−83	67	12	21	21.5	12.8	76	93	22	106	3	4	6
Llanbedr	448	81	128	54	21	21	22	27.9	15.6	89	96	8	170	3	4	3
Llandanwg	1313	295	355	119	70	19	11	23.9	16.3	90	97	8	485	0	2	1
Llanddwywe is Y Graig	424	128	103	47	83	4	13	21.6	11.3	93	99	6	169	1	1	1
Llandecwyn	227	36	71	45	58	26	16	16.7	8.8	74	95	28	76	4	4	4
Llanelltyd	361	79	102	43	52	30	18	21.6	11.1	85	95	12	126	2	5	2
Llanenddwyn	940	270	232	50	66	22	12	28.7	20.5	82	92	12	383	5	5	6
Llanfair	374	75	101	74	71	17	12	18.8	11.8	90	97	8	139	2	2	3
Maentwrog	507	84	153	−47	49	44	7	30.0	22.1	78	94	33	185	2	6	2
Talsarnau	419	102	112	68	64	12	23	27.6	18.7	76	94	13	163	5	6	5

	17	18	19	20	21	22	23	24	25	26	27	28	29	30	31	32	33	34
Barmouth	18	24	30	33	45	20	19	6	34	7	62	37	103	38	1122	153	43	358
Dolgellau	14	19	23	28	41	20	20	4	34	6	35	23	82	23	1047	30	58	77
Llanaber	11	19	28	42	54	34	20	4	27	2	5	3	5	7	179	44	4	0
Llanbedr	18	19	28	27	37	20	19	3	35	6	4	0	10	8	221	16	16	0
Llandanwg	13	15	26	28	43	24	19	5	35	7	34	13	47	23	798	84	157	117
Llanddwywe is Y Graig	10	20	28	40	47	32	19	2	26	8	4	6	7	6	252	12	37	0
Llandecwyn	6	13	16	20	29	18	28	4	43	7	3	3	8	5	105	4	18	0
Llanelltyd	18	17	33	32	48	27	21	2	34	10	4	4	10	5	182	4	25	33
Llanenddwyn	17	20	27	36	52	27	17	5	31	4	15	9	22	16	590	35	122	14
Llanfair	11	14	21	27	40	28	19	6	36	5	3	1	9	4	188	6	29	0
Maentwrog	16	16	24	25	37	21	18	6	40	6	11	1	22	10	250	3	35	39
Talsarnau	15	18	24	31	47	27	17	4	36	9	8	0	14	3	237	36	18	14

Census Data (1981 unless stated)

1. Resident population
2. Pensionable age numbers
3. Numbers of pre-school and school age children
4. Population change 1971–81 (positive unless indicated)
5. Percentage of houses owner-occupied
6. Percentage of council houses
7. Percentage of private rented houses
8. Percentage of households without cars
9. Percentage of people in private households without a car as a percentage of all private households
10. Percentage of households with all amenities 1971
11. Percentage of households with all amenities
12. Percentage change in all amenities, 1971–81
13. Number of households in 1981
14. Percentage of households lacking bath and toilet
15. Percentage of households lacking bath
16. Percentage of households lacking toilet inside
17. Lone pensioners households, 1971 (%)
18. Lone pensioners households, 1981 (%)
19. Households with persons of pensionable age only, 1971 (%)
20. Households with persons of pensionable age only, (%)
21. Households with one or more pensioner (%)
22. Households of married couple and child less than 16 (%)
23. Percentage of households of married couples and children less than 16
24. Percentage of households with single parent and children
25. Percentage of households with dependant children
26. Percentage of households with 3+ dependant children
27. Numbers of male unemployed, 1971
28. Numbers of female unemployed, 1971
29. Numbers of male unemployed
30. Numbers of female unemployed
31. Total household spaces
32. Holiday accommodation – spaces
33. Number of second homes
34. Rooms in hotels and guest houses

Sieve analysis

This is a technique which is used by planning consultants to identify areas where development can take place causing least controversy. It was used by the consultants working for the National Coal Board in selecting sites for new mines. The analysis does not replace the human dimension in the actual decision, but it does help reduce very complex patterns to approachable dimensions. It simply involves identifying factors which would be incompatible with the proposed development and by applying these factors to the area concerned, it sieves out the unsuitable parts of that area. Just as in a nest of sieves used for soil particle analysis, the sieves should start coarse and grade to finer meshes. The following is an example.

Let us suppose that you are seeking the best location for a sports centre and the area is on the edge of town with few buildings. You have maps showing the nature of the geology and its suitability for holding and supporting the sports centre, land use, settlement and infrastructure. You should then decide which factor eliminates most areas. Probably it will be geology. Proceed to eliminate all the areas which are unsuitable for building using the geology map by placing tracing paper over it and shading all the areas unsuitable. You may now have eliminated 50% of the area. You should then only consider the remaining areas. Of these, you should then eliminate the next largest areas, which may be those which have high agricultural productivity, by adding further to your shading on the tracing paper (using the land-use map), perhaps reducing the remaining area by a further 50%. Finally you need to avoid houses, but need to fit the proposed development into the infrastructure pattern, so you should eliminate built-up land and areas remote from roads and services (by using the settlement and infrastructure maps). You may now have reduced the area on your tracing for further investigation to perhaps 10% of its original extent (Fig. T3.1). From this stage, you can now make detailed ground surveys and come up with the best location for your sports centre.

This technique can be used just as well to analyse maps showing different aspects of a country or region and in the text are two examples where you have to analyse maps of the USSR and South-East England.

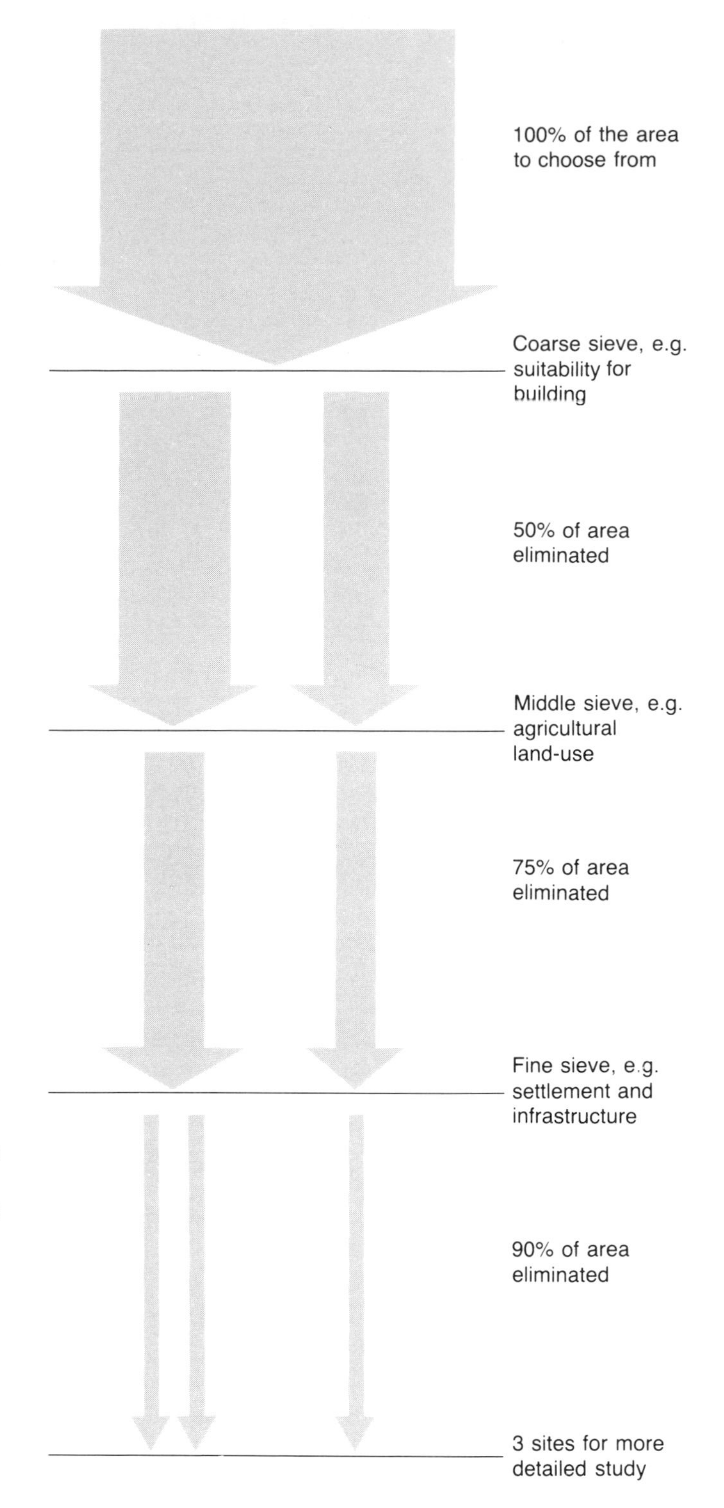

Fig. T3.1 Sieve analysis

Theme 4 *Some Important Issues in Industrial Development*

The World Context

In Great Britain, the Industrial Revolution in the eighteenth century was largely based on the invention and utilization of machinery to increase output. The steam engine was used to provide the motive force for machines that completed almost every stage in the production of garments, machines that made parts for other machines and machines that made finished iron products, for instance. At that stage, many workers were still required by industry because machinery was not as efficient or as self-regulating as it can be now. Another important factor was that the world market for products was growing fast and competition from other nations was small. As a result, many thousands of people migrated from their farms or cottage industries to the new centres of production. This resulted in a major shift of population and a growth of the large industrial towns. In Great Britain's case, therefore, there was a positive link between industrial expansion and industrial employment, with major benefits as far as family income was concerned. There were also, of course, major urban and social problems engendered by rapid industrialization.

Nowadays, industrialization in the less developed countries takes place against the backdrop of enormous competition. Many countries are competing in the same field. This often means that the chosen path for industrial development requires a high degree of mechanization which cuts down labour costs. The same end could be achieved by paying workers very low wages. Table 4.1 shows changes in the distribution of world population from 1750 to 1980 and can be used to give an indication of the shifting location of the world market.

1 Which continent has the largest share of the world's population?

2 Which two continents had a growing percentage of the world's population until 1930, but a declining percentage after 1950?

3 Which continent has had an expanding population for the entire duration of the statistics?

4 Which continent had a decline in its percentage share of the world's population from 1800 to 1930, but an increase since that time?

5 Given that the world's population in 1990 was about 5300 million, what was the population of Oceania for 1990?

6 Draw graphs of the figures for the total world population and the individual continents. From your graphs, predict the constituent figures for the year 2000. (Help on drawing such graphs is found in Techniques T5.2.)

The growth pole theory

One idea that has been widely used in implementing development policies is that of **growth pole** theory (Fig. 4.1). The idea on which it is based is usually called **trickle down** whereby one new development in an area (e.g. a new industry) is supposed gradually to lead to many others in its vicinity. Another term widely used for the idea is **cumulative causation,** since in the example shown the new industry causes an increase in population due to employment prospects, which in turn causes both an increase in the skilled labour force of an area and an additional local demand for goods and services and so on. These results are supposed to accrue from investment in a relatively small way and are thus very attractive to developing countries, which after all, do not have much capital to invest.

Many countries of the developing world were once colonies of European nations. The patterns of growth that had occurred in these countries by the time that they became independent was often one of concentration of industry and commerce in one region, often containing the main sea ports and the capital city. Such a region is usually referred to as the **core** of the country. Away from this in all directions, if that is possible, there is a decline in industrialization, application of new techniques and ideas, and in the standard of living in general. The areas away from the core are known collectively as

Table 4.1 The distribution of world population, 1750–1990

	Total world population (millions)	Europe	N. America	Lat. America	Oceania	Asia	Africa
		% of total world population					
1750	800	19.8	0.1	1.5	0.3	65.2	13.1
1800	960	21.2	0.7	2.1	0.2	65.9	9.9
1850	1240	23.4	2.2	2.8	0.2	63.3	8.1
1900	1650	26.3	5.0	3.9	0.4	56.9	7.5
1930	2000	26.4	6.7	5.4	0.5	53.2	7.9
1950	2517	22.9	6.7	6.5	0.5	55.2	8.2
1960	2990	21.3	6.6	7.1	0.6	55.9	8.5
1970	3626	19.4	6.3	7.8	0.5	56.6	9.4
1980	4432	17.2	5.8	8.4	0.5	57.9	10.2
1990	5300	15.0	5.0	8.7	0.5	59.3	11.5

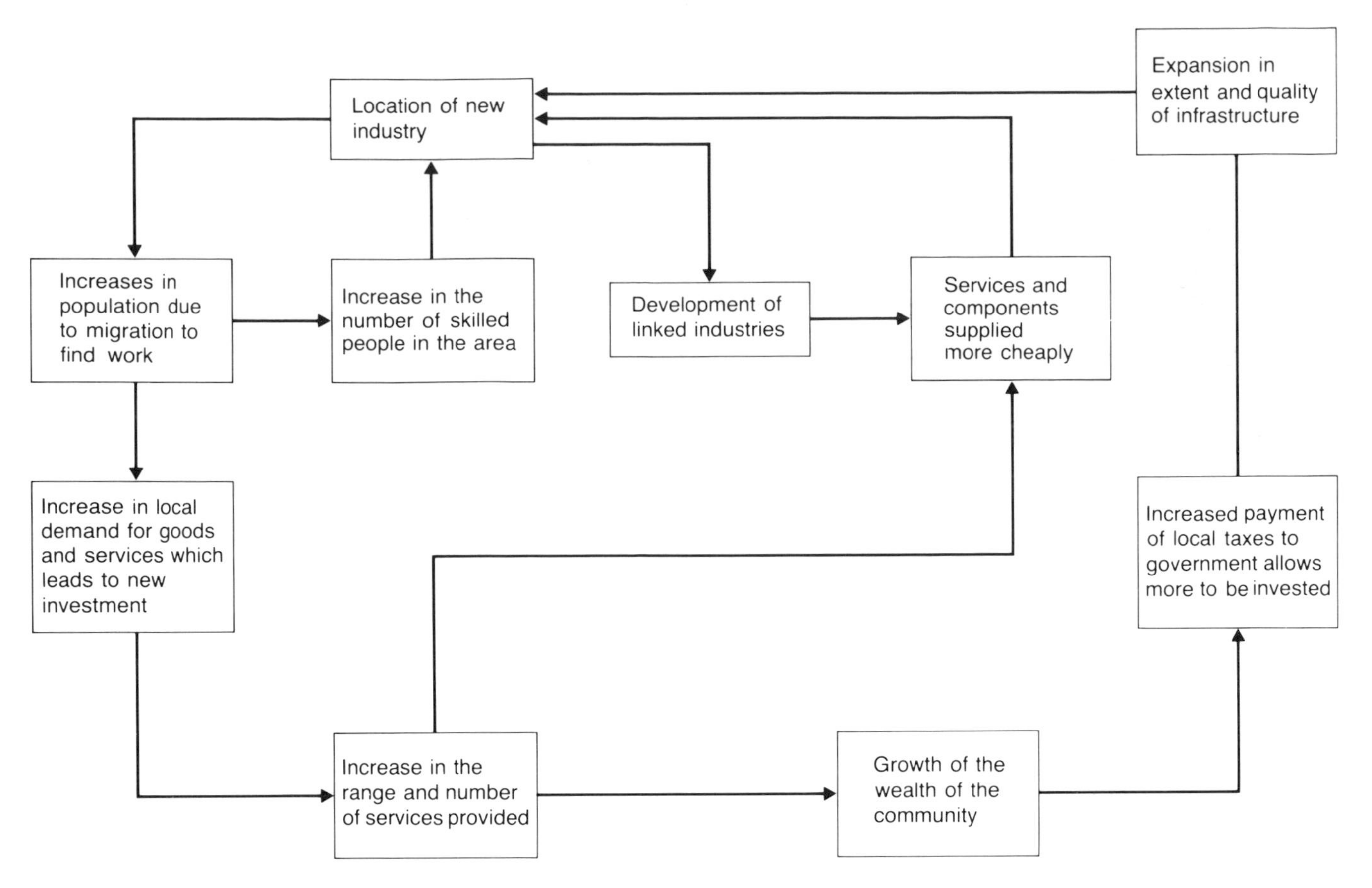

Fig. 4.1 The model of cumulative causation as it explains the growth pole theory

the **periphery** (Fig. 4.2). The growth pole policy is one idea that has been used to even out the distribution of wealth within a country and to spread the kinds of benefits that people have in the core to the periphery.

Fig. 4.2 Core and periphery

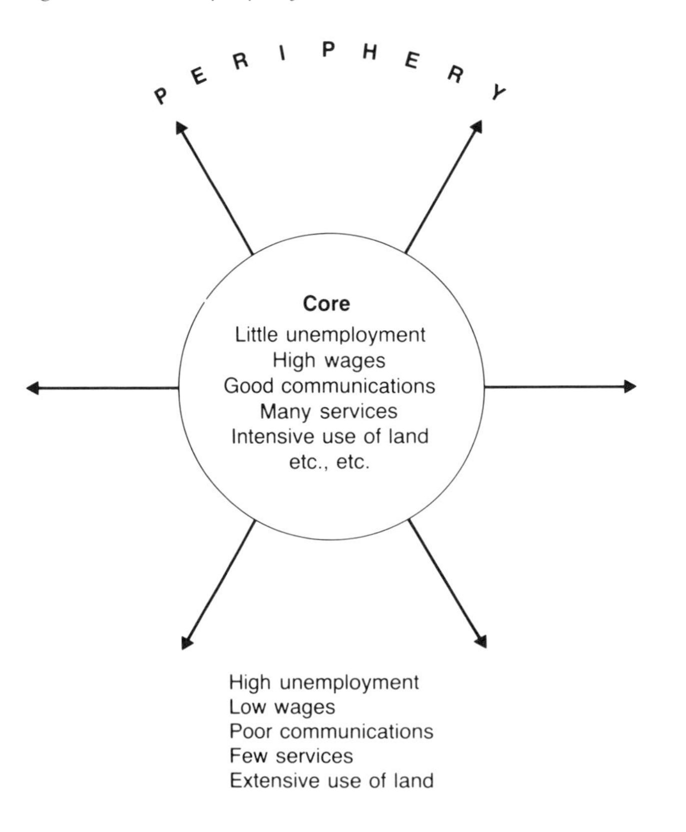

Case-study: development schemes in India

One country in Asia which represents well the dilemmas facing planners of industrial development in the less developed countries is India. In 1982, India's population was some 711 664 000 people with a mean annual growth rate of 2.2%. As such, India has a huge internal market and is equally in a continental location where surrounding markets are expanding strongly. When she achieved independence from Britain in 1947, India had few large industries and most production was concentrated in village workshops and cottage industries. The country was basically an exporter of raw materials and foodstuffs and an importer of machinery and finished goods. Certain very distinct advantages for industrial development were possessed, however. The British left an excellent infrastructure of communications between the main cities, although local roads were mainly dirt tracks. India has some major deposits of raw materials, notably iron ore and coal, and there was a huge labour force willing and able to take on industrial tasks as a way of escaping to some extent from the even more demanding and menial tasks of the farm.

India has attempted to control economic development by the use of Five Year Plans. To begin with, investment was concentrated on the development of heavy industry and infrastructure and attempted to create one major growth pole in the Damodar Valley to the north-west of Calcutta (Fig. 4.3). This cost a very large amount of money and was not entirely successful. After this, the Indian government has invested more in the rural areas, attempting to decentralize and spread the beneficial

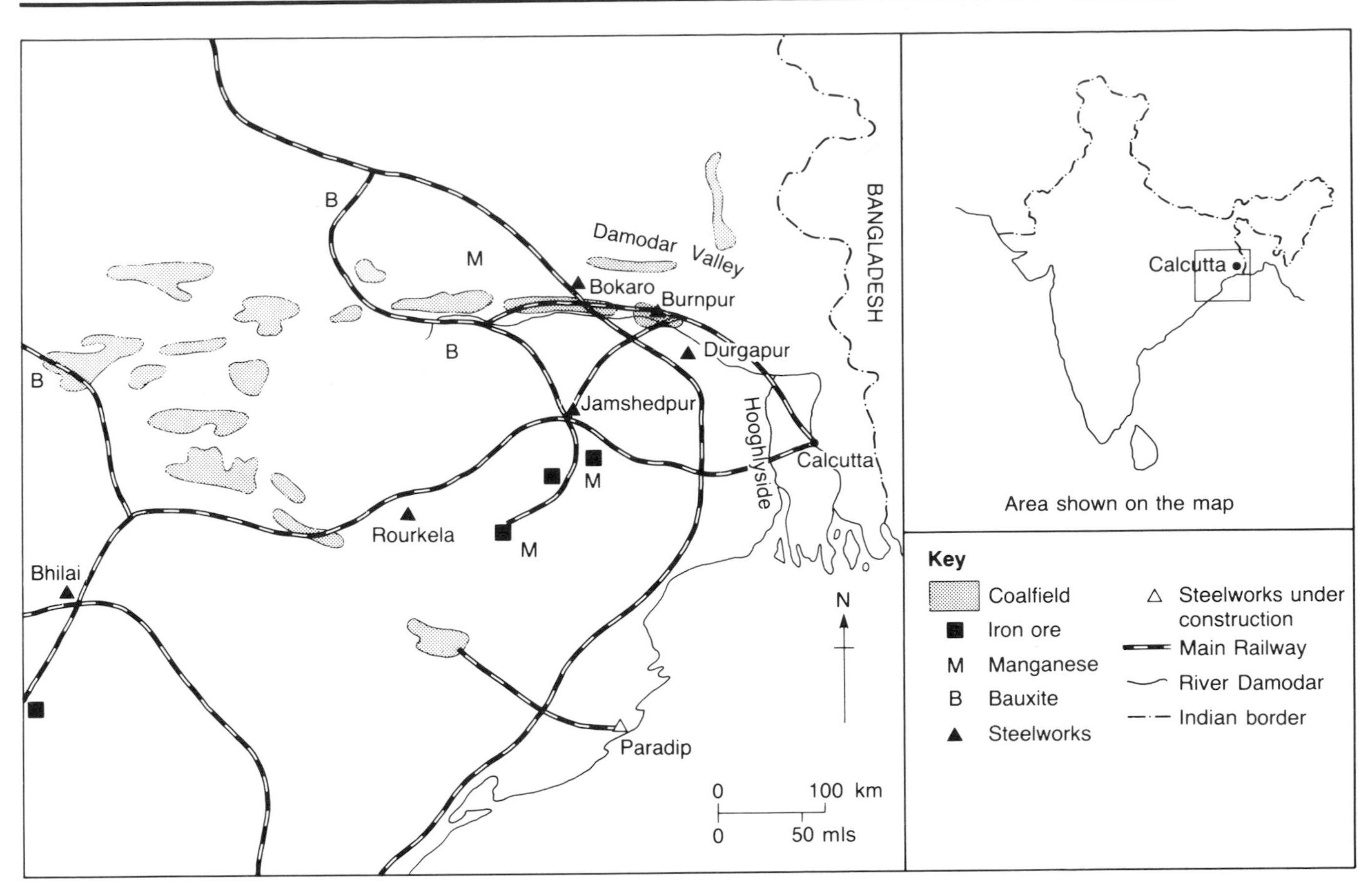

Fig. 4.3 The industrial north-east of India

effects of industrialization further. More appropriate technology was encouraged and the increase of food supplies through the use of better strains of plant, better irrigation techniques and land reform was made a priority. In this way, the Indian government hoped to get the benefits of industrial development to the people. By doing so, they hoped to improve living conditions and encourage a bigger home demand for Indian enterprises.

With such a large and expanding market, India is ripe for industrialization. This has come in two forms. There have been major developments, in part encouraged by the early Five Year Plans, which have led to the development of major industrial growth poles (Fig. 4.4) and to the development of large, highly mechanized industrial plants. Then there has been the counter-movement of the encouragement of more 'human-scale' developments such as village cooperatives.

The Damodar Valley Scheme in north-east India, in the hinterland of Calcutta, was the first area to be selected for large scale industrial development by the Indian government. The area had long been one with serious problems. The monsoon rainfall led to flooding and widespread soil erosion and the farming base declined as a result. Large deposits of raw materials had, however, been discovered in the area and were used along with the control of the river and its development for hydro-electric power as a basis for industrial expansion. Fig. 4.3 shows that heavy industry, based on iron and steel production, was to be the heart of the Damodar scheme. India is now the world's seventh largest producer of iron ore, the sixth largest producer of manganese, twelfth largest producer of bauxite and fifteenth largest producer of crude steel. Despite this, the development of the Damodar scheme was neither as large nor as fast as the Indian government would have liked. It did not have the multiplier effect that they had hoped for and for these reasons, together

Fig. 4.4 Growth poles in India

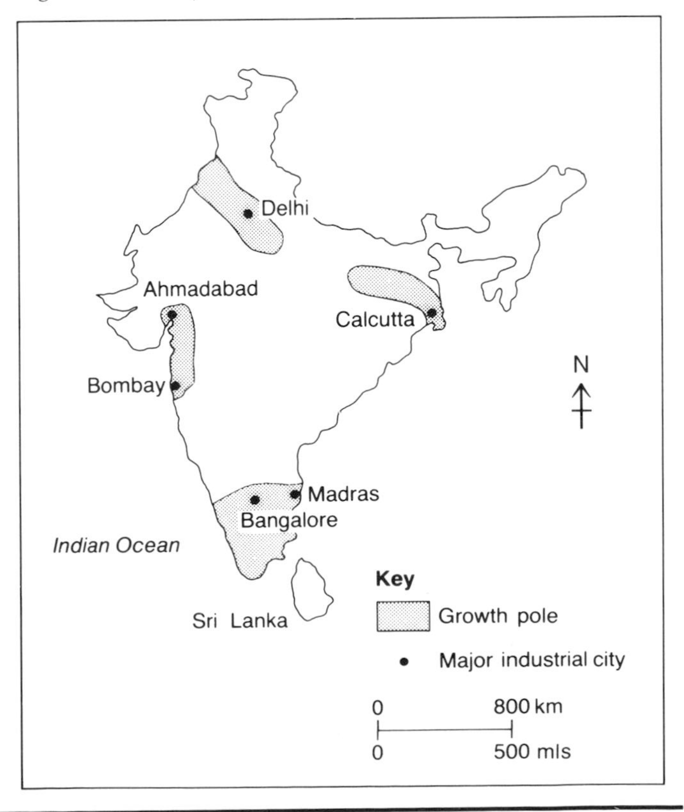

with a series of disastrous harvests which forced them to spend large amounts of money on food imports, they began to invest more heavily elsewhere. The lack of a skilled work force for large technological developments was also a problem, for although India has a huge potential work force, it does not have a very large pool of university or college trained workers. The Damodar Valley Scheme has been successful to a degree, in that it has provided a heavy engineering base which was hitherto largely missing in the Indian economy. However, the main criticism of it was that it was a development that reached only a very few people. Much of the country had seen little or no benefit from the first three Development Plans (Table. 4.2).

In the 1978 Five Year Plan, there was a statement which made it clear that a great change in approach was being applied: 'In the next phase of development it will no longer be appropriate, in the light of our past experience, to formulate the principal objectives of a particular plan period merely in relation to a specified target of growth in the economy. What matters is not the precise rate of increase in the national product that is achieved in five or ten years, but whether we can assure within a specified time-frame a measurable increase in the welfare of the millions of the poor.'

The plan therefore proposed to achieve the following objectives within a period of ten years:

(a) removal of unemployment and the significant under-employment
(b) an appreciable rise in the standard of living of the poorest sections of the population
(c) the provision by the government of clean drinking water, adult literacy education, elementary education, health care, rural roads and housing for the landless and minimum services for the urban slums.

At the same time, it wished to achieve a higher rate of growth of the economy, reduce regional disparities and ensure the country's continued progress towards self-reliance.

Other large scale developments were promoted by individual companies and, although encouraged by the Indian government, were not a direct result of government planning. One example is the Telco truck works in Poona. The factory, started by the Tata family, developed from a site originally chosen over one hundred years ago. Its modern production owes little to the technology of the developed world, as befits a country whose industrial output in fact puts it fourth in world rankings. Most of the machinery has been designed and made in India. The trucks it turns out are very well designed and tested, as they need to be to cope with Indian road conditions. This reliability puts them in great demand in South-East Asia and Tata trucks figure large in India's exports. There are plans to double the production of the factory, which already employs 13 000 people. The Telco company takes the welfare

7 a) Draw histograms to show the approved and actual expenditure for the Plan periods, dividing them to show the concentration per sector.
b) Describe the changes in emphasis you can identify from your histograms.

8 a) Describe the changes in national income and per capita income during the Plan periods. Which periods seem to have had most impact on incomes?
b) 'Inflation helps to make growth in income less than it might seem.' What would high inflation mean for incomes in India from 1977–78 to 1979–80?
c) What problems are there in measuring per capita income in a country such as India?

9 Look at the Objectives for the Seventh Five Year Plan.
a) Write a new column for 1985–90. Read the Objectives for the Seventh Plan and, opposite the correct reference, show whether you think an INCREASE (+) is indicated or a DECREASE (−).
b) How does the emphasis for the Seventh Plan differ from previous attempts?
c) Explain why you support or reject the new priorities.

of its workers seriously. Not only do workers at the company earn considerably more than their counterparts in other industries in Poona, but they are encouraged to take on other responsibilities to help the community in general. Such responsibilities might be running a training workshop or 'hobby school' for the children of Telco workers, or contributing to housing associations. These receive loans from Telco so that a deposit can be put down on a piece of land and building materials bought. In this way, workers at Telco can benefit from building their own improved housing. The feed-back from industrial wages into agriculture in the surrounding area is also great. Workers often send money to close relatives farming near Poona and with this they dig wells, rebuild farm buildings, buy fertilizers and invest in new machinery. A number of small industries has also been encouraged by the Telco management. These often make use of the labour attracted to the truck works, but not skilled enough to be of any use. In one example, 350 men are employed in a workshop making the temporary wooden seating that is used when Telco trucks are sent from the Poona factory to distributors throughout India. The work is not nearly as well paid as at the truck works, but is far better than the employees could hope to get in their villages. In this way, individual factories, if they are large enough, often act as growth poles in themselves.

The situation in India where under-employment and unemployment are endemic is the main reason for the high mobility of the labour force. People may be driven off their land because they get too deeply into debt, or they may decide that a city or industrial

Plan Outlays Expenditure for Public Sector
(By Sector)

(Thousand Rupees)

Sector	First Plan		Second Plan		Third Plan	Annual Plans		Fourth Plan		Fifth Plan		Annual Plan		Sixth Plan 1980–85	
	(1951–52 to 1955–56)		(1956–57 to 1960–61)		(1961–62 to 1965–66)	(1966–67 to 1968–69)		(1969–70 to 1973–74)		(1974–75 to 1977–78)		1978–80		1980–85	
	Plan provision	Actual expenditure	Plan provision	Actual expenditure	Plan provision	Actual expenditure (b)	Approved outlay	Actual expenditure	Approved outlay	Actual expenditure	Approved outlay	Actual expenditure	Approved outlay	Actual expenditure	Approved outlay
1	2	3	4	5	6	7	8	9	10	11	12	13	14	15	16
1. Agriculture and Allied Services and Co-operation	354	72	65	42	125	91	284	300	1238	866	1904	1267	1939	1514	5400
2. Water and Power Development	648	—	117	78	127	92	124	173	459	444	908	618	679	671	11 995
3. Industrial and Minerals	188	73	747	1001	1573	1759	1512	1491	3307	2779	9327	6555	5006	4398	12 771
4. Transport and Communications	570	406	1203	1075	1225	1782	1034	998	2670	2363	5505	3826	2982	2697	11 521
5. Social Community Services	532	112	396	285	350							1541	1468	1215	5302
6. Economic and Gerenal Services and Miscellaneous	86	43	43	52	200	488	487	417	1182	1374	3389 86	101	63	261	
Total	2378	706	2571	2534	3600	4212	3441	3379	8856	7826	21 032	13 893	12 075	10 558	47 250

(Source: Planning Commission, in *Statistical Abstract, India 1984,* No. 27 and *1985*, No. 28 (Central Statistical Organization, Dept of Statistics, Ministry of Planning, Govt. of India)

(1) Includes science and technology.

(b) Between 1966 and 1969 there were three One Year Plans since the effects of the Indo-Pakistan conflict had eaten deep into the Indian reserves. In 1978 the plans changed to 'rolling' plans where re-assessments were made very year, before reverting in 1980 to Five Year Plans again.

Table 4.2 India's Five Year Plan

1 Continued growth, but with emphasis on equality and social justice, self-reliance, improved efficiency and productivity. Acceleration of foodgrain production to increase employment opportunities and raise productivity.
2 Employment to help people stand on their own feet and work with self confidence and self respect, which is the first essential for their participation in development tasks. Increases in cropping intensity and extension of new agricultural technologies to low productivity regions and to small farmers.
3 Emphasis on rural employment. Improve the environment for the workforce such as provide clean drinking water.
4 Increase the number of irrigation schemes.
5 Industrial development. The slow growth of agriculture limits the possibility of non-inflationary industrial growth, which will be based on modernization and the upgrading of technology.
6 There will be a huge investment in infrastructure.
7 Take note of what people say.

Table 4.2a Objectives for the Seventh Plan

Fig. 4.5 Changes during Plan periods

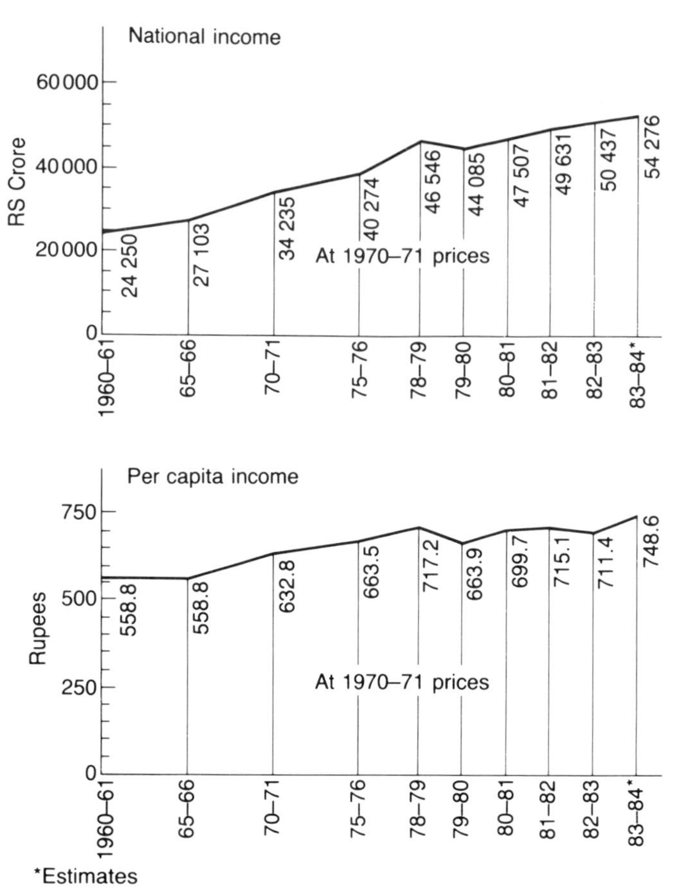

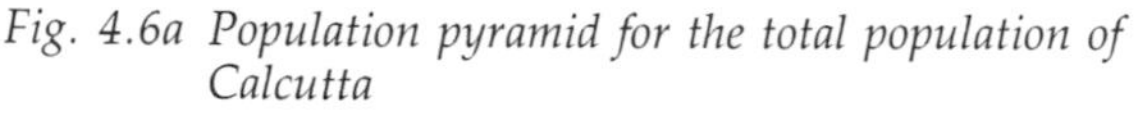
Fig. 4.6a Population pyramid for the total population of Calcutta

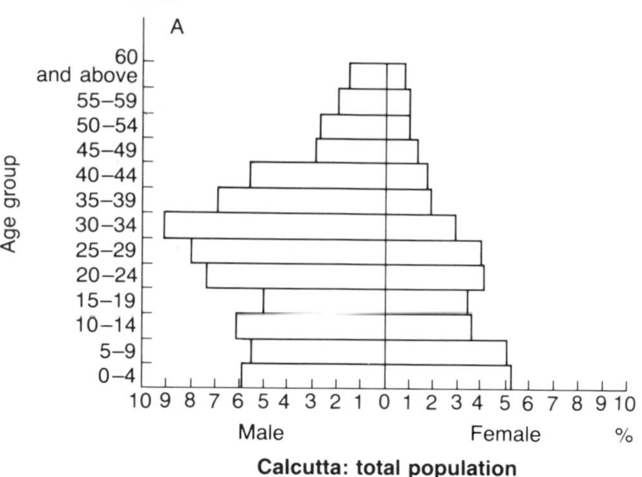

Fig. 4.6b Population pyramid for Calcutta's in-migrant population

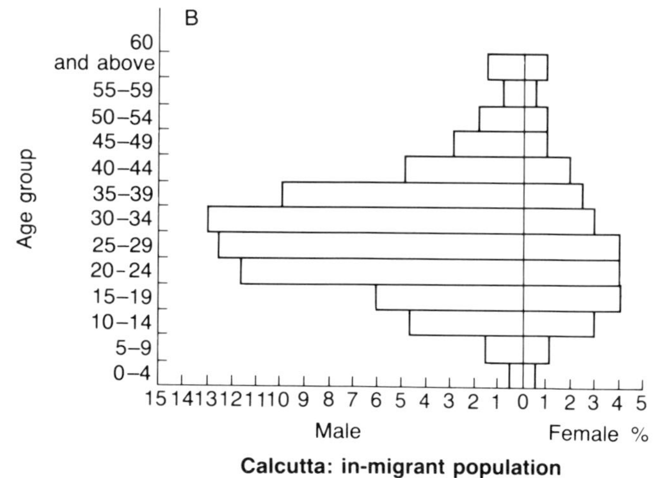

centre offers greater potential for employment. In either case, they may migrate to a place where they perceive there to be greater work opportunities (Fig. 3.28 on p. 97).

This may, in fact, be an illusion, but the fact that they have moved imposes an obligation on them to try to make the move work. In any case, they may find it extremely difficult to return. In many cases, they have 'burnt their boats'. They have nothing to return to. This process, whereby urbanization throughout the developing world is increasing, is illustrated by Fig. 3.28. It may mean that cities expand at a considerable rate in developing countries, so great is the pull exerted by the potential for employment. In many cases this leads to the growth of shanty towns in marginal areas surrounding the cities, such as on steep slopes, in swamps and on land adjoining industrial plants. Such was the case in the example of the Union Carbide pesticides factory built at Bhopal in Madhya Pradesh (Fig. 4.7). The plant had been built away from the suburbs of Bhopal, but at an insufficient distance. Before very long, it was encircled by low class housing and shanties, known in India as bustees. These began virtually outside the wire perimeter fence surrounding the factory. In 1981, a worker had been killed at the factory by an escape of phosgene gas, of a kind used in gas warfare in Europe. It seems that this was not taken as a warning and articles in the local press which also warned of impending disaster were ignored.

10 *Describe* the total population pyramid for Calcutta (Fig. 4.6). Your description should refer to:
(a) percentage of children under 10
(b) percentage of people over 60 (together these form the dependent population for Calcutta)
(c) percentage of males between 10 and 59
(d) percentage of females between 10 and 59
(e) the 'bulge' on the male side at 15–49

11 Attempts to *explain* the pyramid in the light of the structure of the in-migrants. You should mention:
(a) one member of the family often moves to seek work, leaving the rest of the family behind
(b) the most mobile people in India are young male workers. They may be single, or they may leave their family behind
(c) the extended family is important in India (that is perhaps grandparents, uncles, aunts etc. living as one unit).

At around midnight on the 2 December 1984, a major escape of gas occurred. One of the storage tanks used to hold stocks of liquid for later manufacture into Sevin, a branded pesticide, had ruptured. Within minutes, a cloud of dense gas, hugging the ground, had moved through the neighbouring bustee, killing as it went. Survivors said that they felt a terrible stinging sensation in their eyes and coughed violently. Many ran out of their houses and made for a nearby hill. Many did not make it. About 40 km^2 of the city were affected by the gas and nearly 200 000 people were caught in its grip. The final death toll of the disaster was 2850. Of the remainder of those affected, most suffered severe eye problems, which later thankfully improved, but also breathing problems, which did not.

Matters were made much worse by the fact that Union Carbide refused to admit that gases other than methyl isocyanate were involved. That gas was supposed to have no long lasting effects, despite the fact that very little experimental information was available. In fact, the company's medical officer told worried doctors, 'The gas is non-poisonous. There is nothing to do except to ask the patients to put a wet towel to their eyes.' Only later did it emerge that phosgene and hydrogen cyanide had probably escaped with the methyl isocyanate and that the chemistry and effects of the mix were both unknown and deadly.

There is considerable suspicion in many quarters that Union Carbide suppressed information on the gas cloud and made the effective treatment of symptoms of cyanide poisoning much delayed. It was only in late January, about eight weeks after the gas escape, that the Indian Council of Medical Research advised that an antidote for cyanide should be administered under medical supervision. The only place administering it up to this point was a temporary hospital set up in a police bungalow. An earlier attempt to treat people in this way had failed when the government closed down a Health Centre. This had been on Union Carbide's property. Law suits in both India and America only came to court a year later. In the meantime, Union Carbide set up an orphanage and paid some interim compensation, which lawyers advised against accepting for fear of jeopardizing the major claim, which was for over $15 billion.

Many of the reports in the Western press following the disaster concentrated on the amount of compensation to be paid. Fig. 4.8 shows the sort of emphasis that was given. The issues raised were seen in a different light in India, however. In his book *Bhopal, The Lessons of a Tragedy*, (Penguin, India, 1987), Sanjoy Hazarinka stresses that, in future, there has to be a better balance between tighter safety measures and the need for chemicals. He says that, so far, only a beginning has been made and an all-encompassing safety code, implemented by those who are expected to abide by it (the multinationals and other sectors of industry) is sorely needed. It would be hypocritical, he continues, if such a code of conduct did not cover both local entrepreneurs and large, medium and small-scale industries which are perhaps more involved in the development process in their respective countries. He says that there is no doubt that industry would resent being regulated, but it is in its own interest to regulate itself and be more open. He sees the overseers of any code as being industry itself, but also governments, which could offer special incentives for compliance, environmental groups, to maintain and sustain a level of public interest, and universities and colleges, to encourage an awareness of the environment. The multinationals could fund such an education programme. Hazarinka's measured and thought-provoking response reflects one type of Indian reaction. In his book *Bhopal Gas Tragedy – Accident or Experiment?* (Paribus Publishers and Distributors, New Delhi, 1986), B N Banerjee gives an altogether different viewpoint. He says that 'the current trial in New York on Union Carbide's responsibility is nothing but a disgraceful farce. American justice, so quick to deal with human rights campaigners in the USA, shows a strange leniency towards the culprits of the tragedy. . . . The public of India and other countries urgently demand that an end be put to the rapacious activity of transnationals which, in pursuit of profits, endanger the health and often the life, of millions of people. The lessons of Bhopal should not be forgotten: the memory of the Bhopal tragedy appeals to all people.'

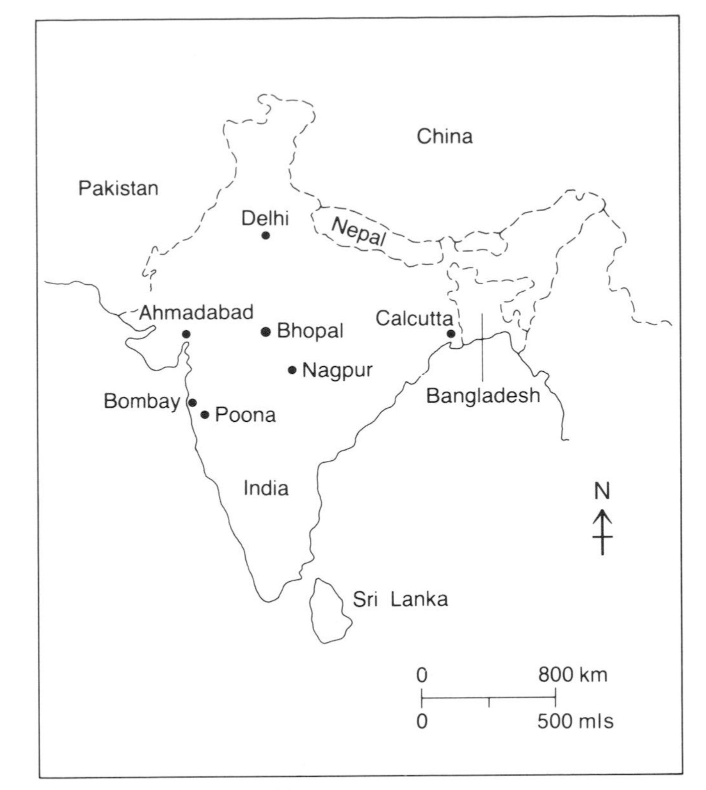

Fig. 4.7 The location of Bhopal

Indian commentators feel, then, that Western governments and the boards of multinational companies are not taking a responsible enough attitude to the control of worldwide industrial activity. An enforceable code of conduct is desperately needed if their faith in large scale industrial development is to be restored.
this scale. Their location on the edge of the factory site itself was due to their original poverty and the need to find work.

$350m for Bhopal victims

From Mark Tran in Washington

Union Carbide and lawyers acting for victims of the Bhopal disaster have reached a tentative settlement worth $350 million, 15 months after the world's worst industrial accident.

The figure is less than thought — the final settlement had been expected to be in the $600 million region. Even more surprising is that the Indian Government, which claimed that it alone represented the 200,000 injured in the gas leak and the relatives of the 2,000 people who died, is not part of the settlement, although it agreed to it.

The settlement, reached last week, covers everyone harmed by the leak, whether or not they have filed a suit. It is subject to final negotiations and requires approval of the New York District Court judge, John Keenan. However, the final figure and conditions are expected to change little, if at all.

Union Carbide previously offered $100 million in compensation. More than 120 personal injury suits have been filed in the US on behalf of 103,000 Indian plaintiffs, and the Indian Government filed its own suit a year ago, with another 4,000 suits filed in India. The claims total more than $100 billion.

The agreement, reached verbally on Thursday after months of bargaining, is now being put in writing.

It appears that some of the $350 million would be put in an interest-bearing fund to help to pay for continuing medical care for thousands of people suffering from lung ailments.

Fig.4.8 Reporting on the financial consequences of Bhopal – a news report from The Guardian

It emerged during the heart-searching following the Bhopal disaster that the escape of chemicals and accidents at chemical plants were rather more common than was supposed. In Germany, an explosion at a chemical plant at Oppan killed 561 in 1921. A similar explosion killed over five hundred people at Texas City in 1947. In 1956, 1100 people died in an explosion at Cali, Colombia and in 1976 an estimated 300 people died of anthrax after an accident in a biological and chemical warfare establishment at Novosibirsk, USSR. An explosion at the Pemex factory in Mexico in 1984 killed at least 452 people when natural gas exploded. Even after Bhopal, five separate leaks of chemical fumes in different parts of India killed four people and injured several hundred. It seems that the location of the Bhopal plant is echoed throughout the world and that stringent measures need to be taken to keep such installations safe.

In the town of Ludhiana in the Punjab, an alternative method of industrialization has taken shape, less by government action than by adaptation to local and national demands. There are now many small factories in the town that have grown up in a spirit of cooperation. In fact, there is a very strong linkage between many of the workshops. One of the largest factories in the town belongs to the Hero bicycle company. Although it was started only in 1956, this bicycle factory is now the largest in the world outside the United States of America. Its present production is over 5000 bicycles per day and still increasing. The company sends its finished product out as kits to be assembled by retailers, but it actually manufactures only the structural parts of the bicycle such as the frame, crank and wheels at its Ludhiana plant. Since its inception, it has encouraged the growth of other small plants which produce such items as seats and pedals. Often these are produced by companies which have been started by former Hero employees.

In the last ten years, Hero's profits have increased by 1500%. This figure has been achieved by very efficient production and low outlay in wages. Despite this, workers are very much better rewarded at the Hero works than in other industries in Ludhiana or in the farming industry of the surrounding countryside. It is not uncommon for a Hero worker to be able to save 200 rupees (£13) per month. This money is often ploughed into either housing or farming improvements. The employment possibilities at the Hero bicycle company are large and increasing. Directly, Hero employs more than 5000 people and will gradually employ more as the production of its home-designed moped begins to increase. There are also plans to build motor-cycles for Honda and this may eventually lead to the manufacture of four wheeled vehicles.

The Hero company is attempting to diversify as the market for its principal product must eventually become saturated. Many less skilled workers, as well as a large number of children, are employed in the smaller works that contribute items to the Hero factory, There is an Indian law forbidding child labour, but this applies particularly to large factories and is not strictly applied in smaller works. This is because the money that young children can earn making small parts for assembly into bicycle seats, for instance, is vital to their families. Often this money means the difference between survival and failure for some farming families with very little land.

The government would find it very difficult to apply the child labour laws in the face of such an arrangement.

The balance between development by encouraging large companies and small ones is well illustrated by another important Indian industry, that of match making. Wimco, the West Indian Match Company (a subsidiary of Swedish Match) made one half of all India's matches in its five factories spread throughout India in 1975. In that year, however, the Indian government made a policy decision to switch the production of matches from large, mechanized industrial plants such as those owned by Wimco, to small units. This fitted in with the aims of the Five Year Plan. This policy has been followed through so thoroughly that now the small units in, for example, Tamil Nadu produce more than three quarters of the matches made in India. A comparison of the lot of two workers, one in each of the types of match production unit, is, however, quite instructive.

At Wimco, a typical worker lives on the Wimco estate near the factory. He is able to walk to the factory where he works from eight in the morning to five at night. He earns about 1300 rupees per month (about £87) of which he is able to save some 300 rupees per month. His children attend the school on the estate. Meanwhile, in Tamil Nadu, one of the 75 000 people employed in match making lives in a small village surrounded by agricultural land. Her husband scrapes a living from an undersized plot of land and she travels to work every morning by the bus which tours the villages picking up employees. She leaves home at 7 o'clock and starts work at 8 o'clock. Working through until 6 o'clock in the evening alongside her six-year-old daughter, she earns six rupees per day (the equivalent of 40p). In fact over 45 000 match workers in Tamil Nadu are under 15 years of age and 10 000 are under five! Again, the child labour laws are not enforced. The Indian government has been successful in encouraging production in smaller, rural units but the cost has been great in some respects.

Decision-making Exercise I: The Choice of Development System for Industrial Development in India

Part A

You are to decide on the type of industrial development that will be implemented in the next Indian Five Year Plan. An outline of two possible development models is given below. Use the information below along with the instructions at the top of p. 112 to try to devise a Five Year Plan for India.

	Village development model	*Ahmadabad factory model*
Organization	Run on a cooperative system. Cost of raw materials kept down through bulk-buying. All processes carried on in village (spinning, dyeing, weaving). Cotton cloth sold by cooperative. About 400 workers employed, earning perhaps 40p per day. Children can work alongside parents	Modern factory using chemical feedstock-produced fibre (polyester). Factory is highly modernized, using mainly Indian produced machinery. Four hundred people employed in Asia's largest design studio, but far fewer on factory floor. No child labour. High wages
Other Benefits	Village school with free school books for children of cooperative workers. Doctor available if they fall ill. Paid holidays	Health benefits available at factory
Finances	Supported by Gandhigram Trust. Profits go to cooperative, not to middle-man	Company profits ploughed back into increased mechanization and higher production
Market	Mainly among lower class. Cotton goods of high quality but low price are produced	Volume production of 'modern' fibre means huge middle class market can be supplied with fashion goods as well as providing large quantities of material for further manufacture and export

Part B

How will you oversee the type of industrial development you suggest? Devise a code of conduct for the companies which set up industries under your plan. Write six 'rules' that they will have to abide by and suggest how you would ensure that they are adhered to.

You are in charge of the finances for the next five years of India's 'rolling plan'. You are particularly keen to encourage the production of textiles and you have a distinct choice to make. Should you encourage the building of a large new factory in a previously undeveloped part of the northern Deccan near Nagpur, or should you encourage the growth of more small cooperative units around the country?

As background for your decision, you are presented with two models. One is of a small village, famous for its production of high-class textiles. The other is of a large modern factory at Ahmadabad. You should ensure that you have read the section preceding this exercise on match making and you can use evidence from that section in making your decision.

Write two paragraphs summarizing the information in support of each of the two models of development. Finally, write a third paragraph explaining which method of development, if either, you would wish to support.

The Growth Pole Method of Development Using the Tourist Industry

Many governments have attempted to encourage development within their borders by instituting one large scheme. It is hoped that by so doing they will encourage other developments, without being totally responsible for those developments themselves. In other words, they hope that after they have started the process, individuals or groups will take advantage of the new conditions offered and will add to that development.

The theory of this is explained in Fig. 4.1 and Fig. 4.9. In the example shown, a dam has been built in the country. Such was the case in Ghana, where the Volta Dam was built, in Nigeria (the Kainji Dam), in Egypt (the Aswan Dam), and in many other parts of the world. The example of the James Bay hydro-electric scheme shows that the policy is not confined to developing countries but is also used by countries of the developed world to help redistribute industrial capacity and lessen regional disparities. The diagram shows that the dam that has been built leads to the formation of a lake. This lake provides a source of drinking water, encourages the development of a fishery, gives water for irrigation purposes, encourages the development of water-borne transport and creates power for the growth of industry. In the case of the Volta scheme, the principal industry developed was the Tema aluminium refinery. Other industries were, however, drawn to the same site because of the infrastructure created for the Tema works. This infrastructure consisted of a new deep-water harbour, electricity supply, an industrial site provided with water, drainage and sewerage, improved road links and the development of a skilled industrial workforce. Other industries to take advantage of the conditions offered were a fish processing works, a dry dock, a cement works, a steel works employing electric arc furnaces, a textile works, a motor assembly works and a cocoa storage and processing plant. All of these to some extent depend on the electricity developed in the first place solely for the aluminium refinery. None of the other developments mentioned was instituted by the government. The conditions were created to encourage industrial growth and private concerns took advantage of them, thus creating a larger scale industrial growth. In the case of Ghana, this created a point of growth within the country – a growth pole. Other countries have used this idea to help development in the poorest or least developed areas. One such example is found in Tunisia.

Fig. 4.9 The 'spread effect' of a dam development

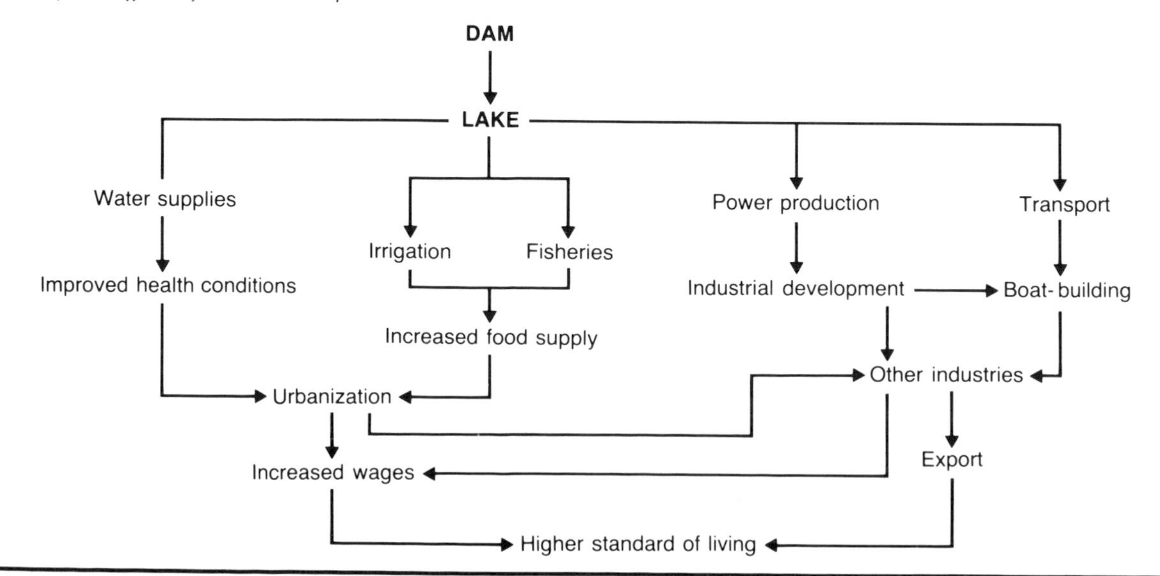

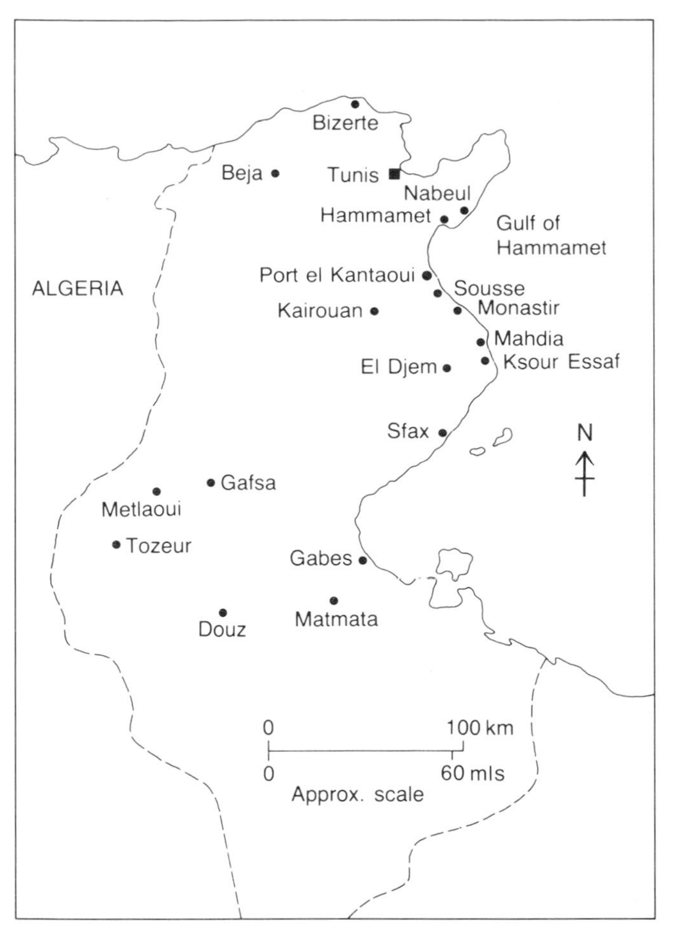

Fig. 4.10 A map of Tunisia showing the location of Port el Kantaoui

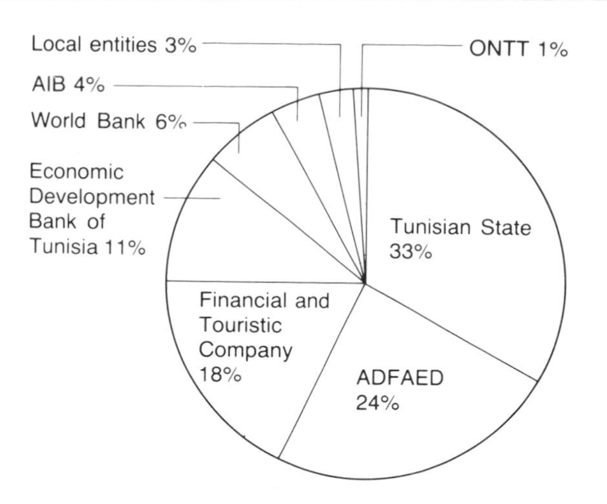

ADFAED Abu Dhabi Fund for Arab Economic Development

AIB Arab International Bank

ONTT Office National du Tourisme et du Themalisme

Fig. 4.11 Capital structure of the Société d'Études et de Développement de Sousse-Nord

Fig. 4.12 The hotel belt of Sousse

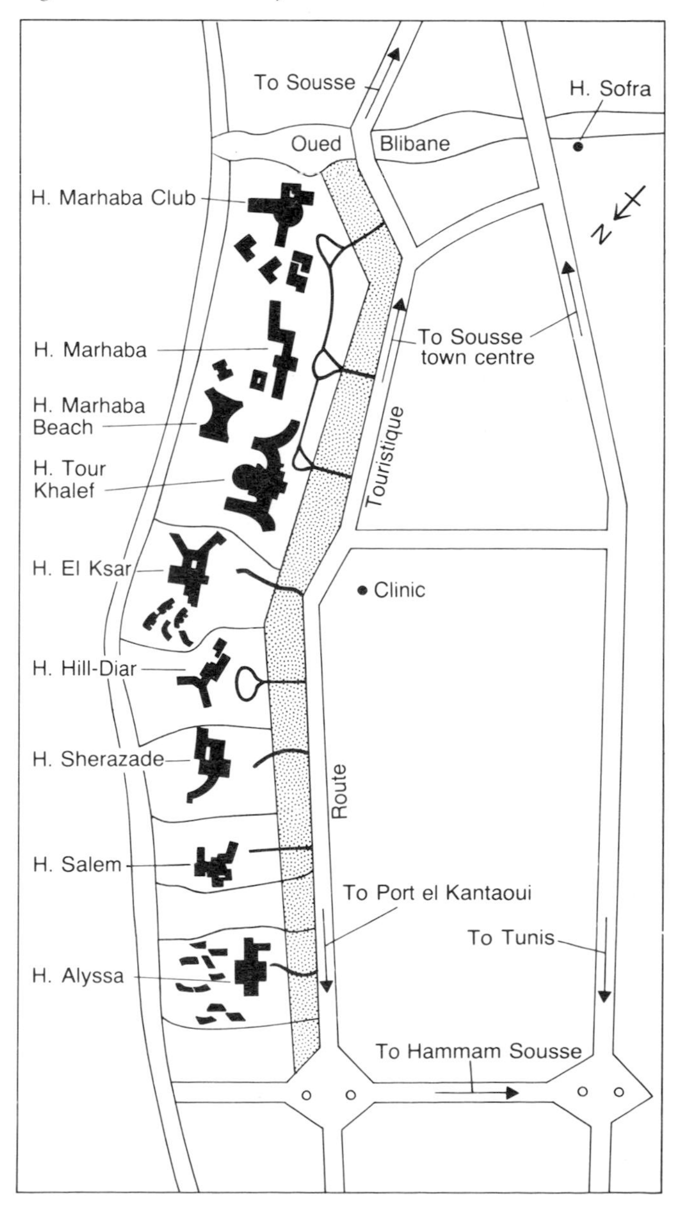

Case-study: Port el Kantaoui

The Port el Kantaoui tourist complex was conceived by the Tunisian government to encourage development in precisely the same way as the schemes discussed above. It is located on the coast, 6 km north of the town of Sousse, the third largest town in Tunisia (Fig. 4.10). Like the Volta Scheme, finance for the development was not available in the country, for huge amounts were needed. In Ghana, private funds were raised from the Kaiser Aluminium Company, but in Tunisia the funding was arranged by the government in association with the Société d'Études et de Dévéloppement de Sousse-Nord (SEDSN), a research and development company set up to control the project. The funding was derived from the government, the Abu Dhabi Fund for Arab Economic Development, various banks including the World Bank, and tourist development concerns. Funding was originally to cover the estimated cost of US $100 million (Fig. 4.11), but by 1985 the cost had risen to at least US $750 million. A master plan was drawn up to accommodate several important design ideas and building was due to be completed in 1987, although through modifications and additions some revision of this seems likely.

The idea of an integrated tourist complex came about as a natural consequence of the development of the tourist industry in Tunisia from 1960–1975, years which can be described as the boom years of Tunisian tourist development. Before 1960, the only hotels that there were in the Sousse area were small

Fig. 4.13 Questionnaire for holiday-makers in Tunisia

1 Is this your first visit to Tunisia?	Yes 79%
2 Have you been abroad before?	Yes 97%
3 Would you return to Tunisia?	Yes 58%
4 Would you recommend Tunisia to your friends?	Yes 71%
5 What do you like about your holiday?	The desert safari, local culture, good hotels
6 What do you dislike about it?	Food, the local people, hotels

A sample of 32 people were asked. Respondents came from England, Australia and Germany.

units largely in the French quarter of the town next to the old Arab walled town, the Medina. Between 1960 and 1975 there occurred a massive growth of hotels along the coastal zone to the north of the city (Fig. 4.12). In conjunction with this, a new international airport at Monastir was constructed to deal largely with the package holiday flights. The problem with package holiday developments for a developing country is that much of the money generated by the developments is siphoned away from the host country. Often the holiday companies pay very low rates for rooms and transport because they guarantee such a large trade. In addition, they organize their own flights and often encourage the import of products and goods that they know their clients will want. In this way, the spread effect referred to above more often affects the country where the tour operator comes from rather than the developing country concerned. Other problems of this somewhat haphazard coastal veneer of hotels were revealed by two pieces of fieldwork carried out by a Sixth Form geography group in and around Sousse in 1985.

A questionnaire was arranged so that the questions were short and to the point and not too intrusive (since the respondents were, after all, on holiday). These were aimed at ascertaining the acceptability of package holidays in general, and Tunisia as a base for such holidays (Fig. 4.13).

Secondly, a bi-polar analysis of the hotels immediately to the north of Sousse was completed (see Techniques p. 157). This was done before the new development at Port el Kantaoui had been visited. The students were asked to place a tick in one of five boxes between two extreme statements about the hotels. In order to analyse the responses, each of the boxes was given a value between one and five. Totals were made for each row and the range of responses noted, so that the statements about architectural style achieved, for instance, an average score with a range of between one and four. This incidentally shows the widely differing views that people have of the same thing, varying perhaps because of sex, age, previous experience and many other factors. Later another bi-polar analysis was made of the complex at Port el Kantaoui. The results are shown in Table 4.3.

Table 4.3 Bi-polar analysis of the hotel belt of Sousse

	1	2	3	4	5	
The overall plan is pleasing to the eye		√				The plan is cluttered
The architecture has a 'Tunisian' feel				√		The architecture is European
There is a comprehensive range of shops			√			The range of shops is very narrow
Buildings are blended well with the environment		√				The buildings are intrusive
The development is very tidy	√					The development is cluttered
The provisions for young children are excellent			√			The provisions for young children are poor
The provisions for teenagers are excellent		√				The provisions for teenagers are poor
The provisions for adults are excellent		√				The provisions for adults are poor
There is excellent provision for a wide range of nationalities		√				Few nationalities are catered for
Directions are fully provided					√	There are few directions
Gardens are well kept		√				Gardens need much attention
There is excellent provision of yacht berths					√	There are few yacht berths
Transport is fully available		√				There is poor transport availability
There is a comprehensive range of services	√					The range of services is poor

(The analysis shows one person's response. The overall score from the entire party was 602. The ranges of scores for each point were large, generally being 1 to 5)

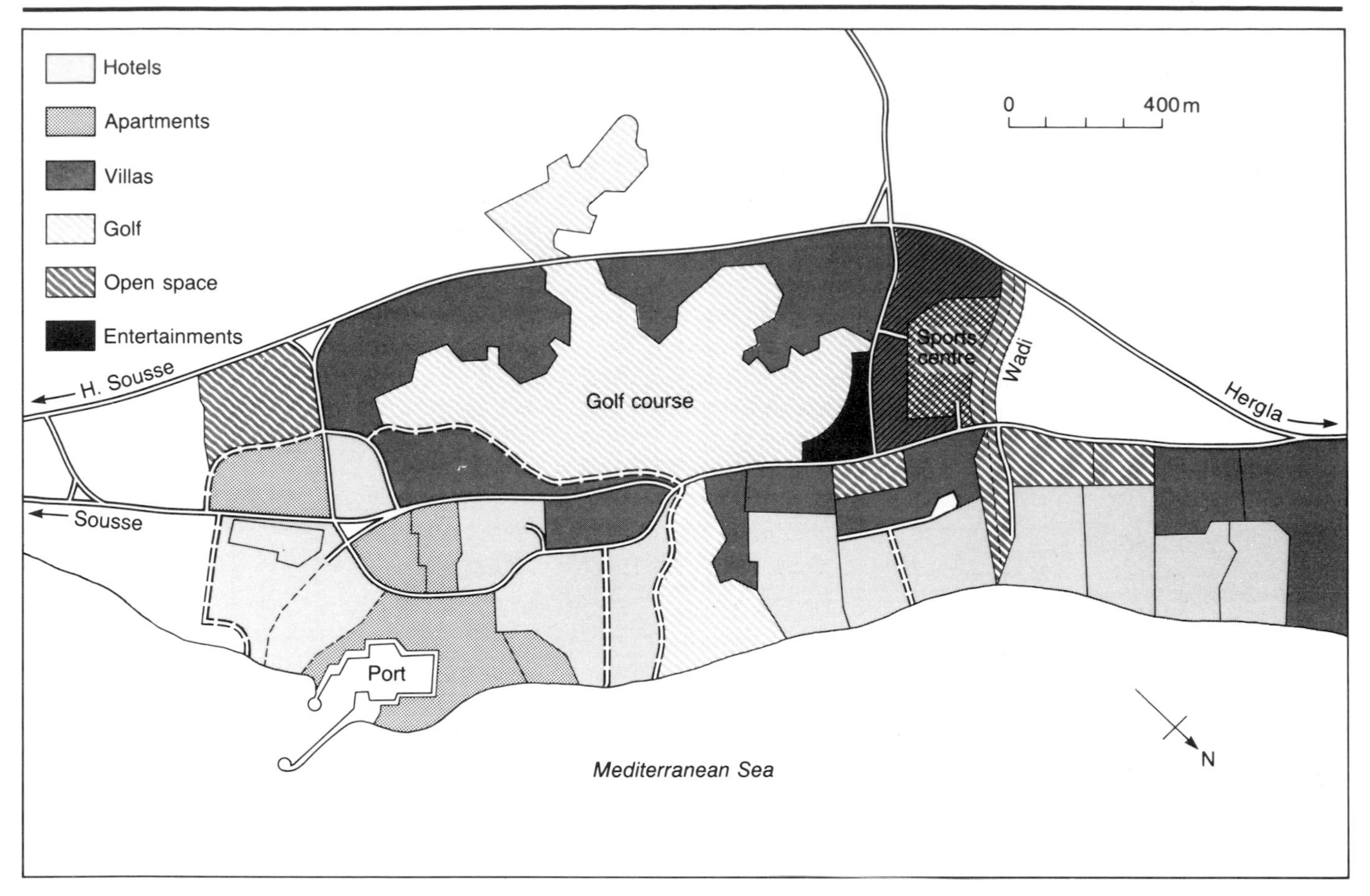

Fig. 4.14 Original master plan for Port el Kantaoui tourist complex

The fieldwork helps to confirm some of the points that were made by SEDSN prior to the development of Port el Kantaoui. It was felt that the linear development of hotels was rather inconvenient. This was simply an adjunct to the town of Sousse. Some people apparently disliked the location of the hotels which meant that they had to travel in to Sousse for purchases such as medical items, and many people disliked the experience of the old Arab town, the Medina, with its narrow souks, or shopping streets, in which people were expected to barter for items. They were also shocked by the relative poverty it brought to their notice. The questionnaire revealed that some of these items still worried people in 1985, and the worrying percentage of people who would not return to Tunisia for another holiday points to the necessity of improving the image.

Port el Kantaoui was developed as an integrated tourist complex. It unashamedly caters for the 'top end' of the market and there is no significant presence of tour operators in the hotels there. The original master plan (Fig. 4.14) shows that somebody visiting the complex for a holiday would have no need to leave its confines unless they really wanted to. The shops and apartments have been designed on the most successful local models. Certainly the bi-polar analysis reveals that the development has been much more successful in most respects than the coastal hotels north of Sousse (Table 4.4).

The other main aim of the development was to make the **spread effect** work for the national and local community much more than it had done in the past. To this end, only local building firms were employed (Fig. 4.15). Traditional decoration was

Fig. 4.15 Board at Port el Kantaoui showing predominance of local builders

Table 4.4 Bi-polar analysis of the Port el Kantaoui tourist complex

	1	2	3	4	5	
The overall plan is pleasing to the eye	√					The plan is cluttered
The architecture has a 'Tunisian' feel			√			The architecture is European
There is a comprehensive range of shops	√					The range of shops is very narrow
Buildings are blended well with the environment			√			The buildings are intrusive
The development is very tidy	√					The development is cluttered
The provisions for young children are excellent			√			The provisions for young children are poor
The provisions for teenagers are excellent			√			The provisions for teenagers are poor
The provisions for adults are excellent			√			The provisions for adults are poor

	1	2	3	4	5	
There is excellent provision for a wide range of nationalities	√					Few nationalities are catered for
Directions are fully provided		√				There are few directions
Gardens are well kept	√					Gardens need much attention
There is excellent provision of yacht berths	√					There are few yacht berths
Transport is fully available	√					There is poor transport availability
There is a comprehensive range of services	√					The range of services is poor

(The response of the same person as in Table 4.3. The total score for the same group was 490. The ranges were generally between 1 and 3, showing less disagreement than the analysis for the area near Sousse)

Fig. 4.16 Port el Kantaoui

Fig. 4.17 New hotel building, Port el Kantaoui

used, employing local stonemasons and carpenters. Furniture, fixtures and fittings were all bought locally. A large hotel school is included as part of the complex and the eventual employment figures of 7000 (a ratio of one person per two beds) emphasizes the importance of this to the economy of the local area. One of the most noticeable things about the fringes of Sousse is the proliferation of new buildings (Fig. 4.16). The employment of vast numbers of people on the many construction sites means that money is being fed into the local economy and is often being used to upgrade the standard of living of the people by additions to houses or the construction of new ones. Added to this is the increase of transport requirements both to the construction sites and from the hotel complex to Monastir, Sousse and Tunis. This is satisfied by the very large taxi fleet and by the regular bus service along the coast. An interesting development has also been the construction of a film studio, originally for the making of one film, 'The Pirates', but now geared up for both internal shows and to use the picturesque locations provided by the port itself and the interesting, traditional-looking narrow streets of Port el Kantaoui. This is one example of how the original master plan has been modified, but in fact most of the plan has been put into effect. A view from the escarpment overlooking the complex shows that, to date, about one third of the complex is complete and construction work is still progressing apace (Fig. 4.17).

The tourist complex, like other examples of big developments in the Third World, has provided work in other areas of the economy. The ratio of employment in the complex to employment locally is estimated in this case to be 1:4. People in the local area are employed not only in taxis and construction, as shown above, but also in the manufacture of tourist items to be sold in the hotel shops and those of the souks. The fact that the development is meant to be as self-contained as possible, however, may eventually mean that the spread effect is not as widely felt in the surrounding area as it would be for a more traditional tourist development. Things like entertainments, car hire, restaurants and cafés are all provided at Port el Kantaoui. There is no need to venture outside the confines of the development to find them. On the other hand, the type of tourist likely to be attracted to Port el Kantaoui will be rather richer than the package tour people going to the hotels near Sousse. Perhaps this will ensure that some of the wealth generated will still find its way directly into the local community. Alternatively, a more negative effect might be fed into the surrounding area as was the case with the Volta Scheme (Fig. 4.20).

The results of the attempted use of the growth pole idea in Tunisia are awaited with interest, not only by Tunisia but also by other countries of the Mediterranean and the Third World which have an interest in promoting development by means of tourism.

12 Attempt to draw a spread effect diagram to illustrate the influence of the Port el Kantaoui complex on the surrounding area.

13 Look at the negative aspects of the spread effect diagram for the Volta scheme (Fig. 4.20). What negative aspects might there be for a tourist development such as Port el Kantaoui? Attempt to draw a similar negative aspects diagram for the tourist complex.

14 Do you think the use of tourism as a development focus is justified or not? Explain your answer.

Fig. 4.18 The 'local' market at Sousse

Fig. 4.19 The tourist influence on Sousse market

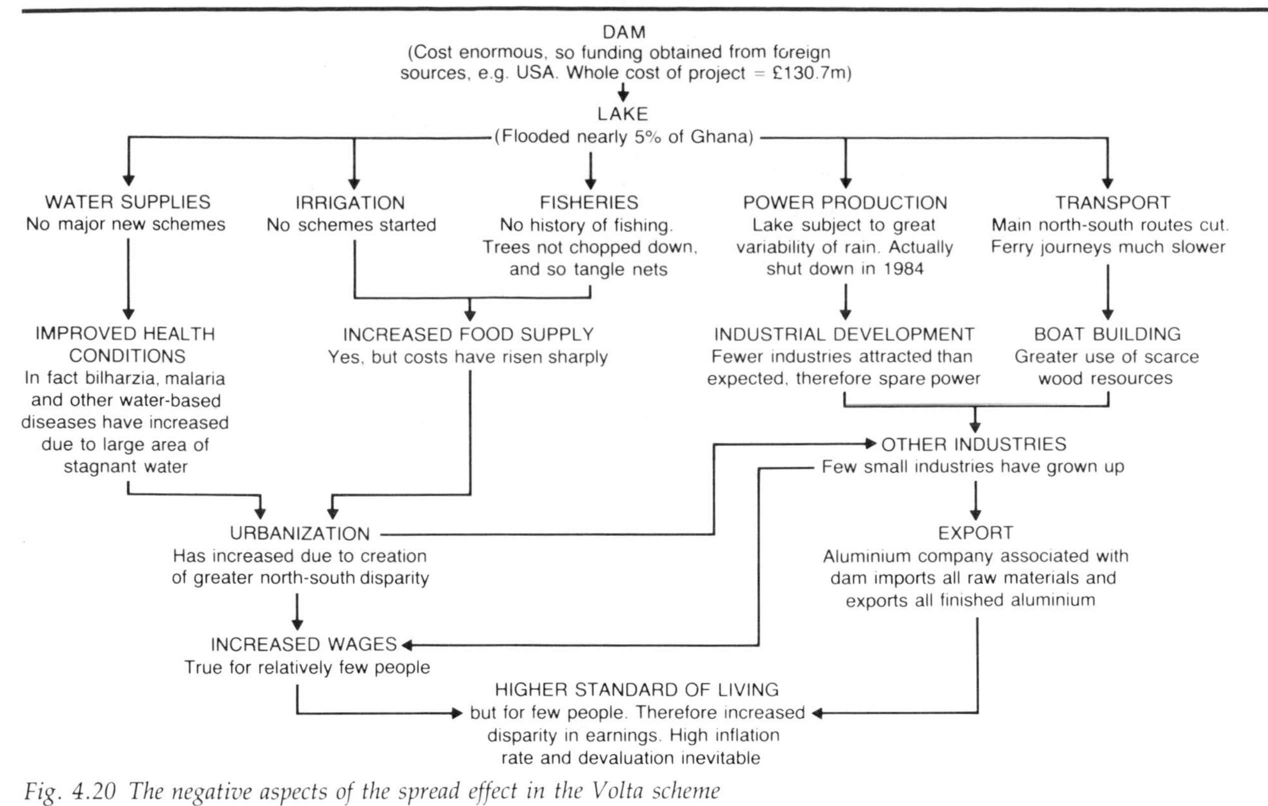

Fig. 4.20 The negative aspects of the spread effect in the Volta scheme

Decision-making exercise J: Tourism and Development

J

Tourism can be a major foreign currency earner and can provide the stimulus for development. It can also have detrimental effects on the society in which it occurs.

You are a consultant called in by the Algerian government to advise on the potential of increasing tourism in three centres (shown on the accompanying map, Fig. 4.21). Your written report should begin by looking at the *present contribution* of tourism to the economy. It should look at the *potential* as compared with Tunisia or Morocco, two countries which have developed tourism to a far greater extent. It should also look at the possible detrimental effects of tourism on a traditional Islamic state.

The three centres named are:

Algiers
(1 503 000) – the capital of Algeria. It has a long history of trading and as such has a port, handling imports and exports with European countries, especially with France. It is perhaps more accustomed to Western customs and influences than anywhere else in Algeria (with the possible exception of the oil towns). It has superb sandy beaches and a ready-made infrastructure to cope with a tourist influx. There is an international airport in Algiers.

Constantine
(350 000) – a farming and industrial settlement about 75 km inland in the foothills of the Atlas Mountains. It has a large Berber population. The Berber are the earliest inhabitants of the area who traditionally tend sheep and goats and grow olives and cereals. The nearest international airport is south-east of Annaba, some 200 km away. Constantine is a considerable rail focus with routes to Algiers, Skikda on the coast, Tunis and the oasis settlements to the south.

Tamanrasset
– a small oasis settlement in the Hoggar massif in the Sahara Desert. It is reached by one road from the north, but it may take three days solid driving from the coast to reach it. There is no airfield nearby and no rail link. Marketing local produce remains the most important focus of life in the area. Dates, vegetables and cereals are all produced in the irrigated oasis fields while camels are by far the most important livestock. Nomadic herding is an important way of life.

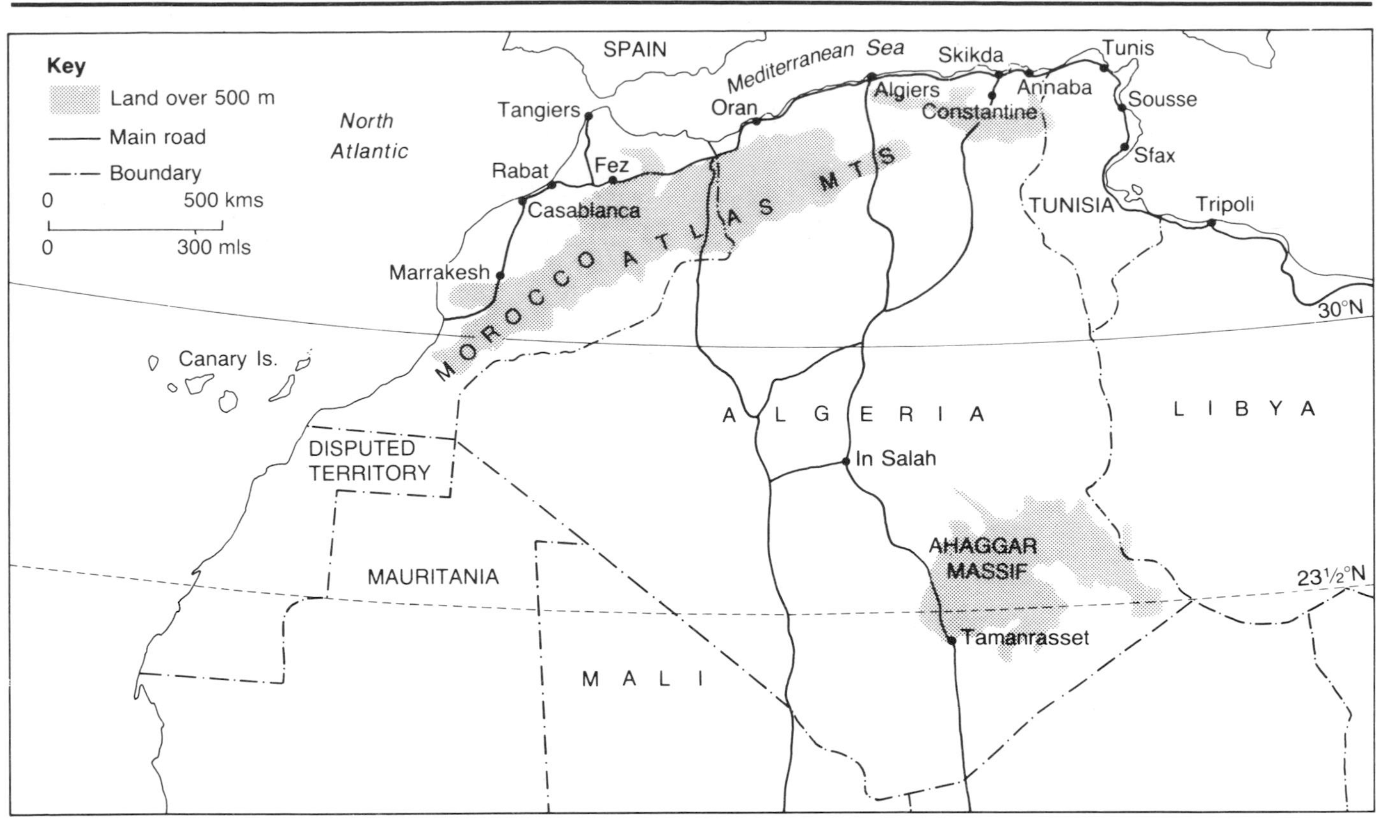

Fig. 4.21 North-west Africa

Fig. 4.22 Cartoons warning of problems caused by tourism

Tourism

. . . may increase begging

. . . may place stresses on local customs

. . . may lead to stress in agriculture whereby goods are produced for the tourist market, leaving production of staple foods short

. . . may lead to an unrealistic change in expectations

. . . may lead to an increase in dependence on one sector of the economy

Table 4.5 Data for Algeria, Morocco and Tunisia

	Algeria	Morocco	Tunisia
Area (000 km²)	2383	447	164
Total population (000)	20 293	21 667	6672
Density of population (per km²)	9	48	41
Balance of trade (1981) (US$ 1982)	+3052	−2253	−1275
GDP (million $ 1978–81)	31 359	12 426	8728
GDP per capita (million $ 1978–81)	1724	657	1370
Origin of GDP (%)			
Agriculture	7	14	14
Mining and manufacturing	40	23	25
Trade, tourism, transportation	18	19	18
Others	35	44	43
Consumer price index numbers 1970 = 100	258	193	155
Tourist receipts in $US 1985	137	800	488
Tourists visiting, 1986	596 000	2 186 000	1 542 000

Fig. 4.23 World tourism

J

The latest estimates for total world travel in 1983

	Arrivals		Receipts $m	
AFRICA	**6 700 000**	**2.3%**	**2000**	**2.1%**
East	1 200 000		435	
Middle	160 000		65	
North	4 150 000		1175	
South	390 000		47	
West	800 000		278	
AMERICAS	**50 928 000**	**17.6%**	**23 261**	**24.2%**
North	33 978 000		13 661	
Central and South	10 200 000		5700	
Caribbean	6 750 000		400	
E. ASIA AND PACIFIC	**23 250 000**	**8.1%**	**9300**	**9.7%**
EUROPE	**196 150 000**	**68.5%**	**57 000**	**59.2%**
East	30 500 000		1250	
North	22 450 000		9750	
South	64 200 000		20 000	
West	78 000 000		25 000	
MIDDLE EAST	**7 000 000**	**2.4%**	**3300**	**3.4%**
SOUTH ASIA	**2 450 000**	**0.9%**	**1250**	**1.3%**
WORLD TOTALS	**286 478 000**	**100.0%**	**96 211**	**100.0%**

Source: *WTO Regional Economic Statistics*

World tourism has grown dramatically over the last 30 years. It slowed recently during the recession but started to pick up again in 1983/84.

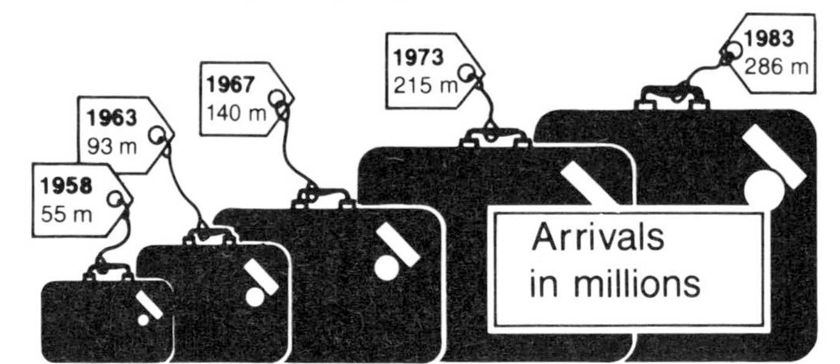

SHARE OF TOURISM IN 1982

Developing countries are now the destinations of 17 per cent of international tourists.

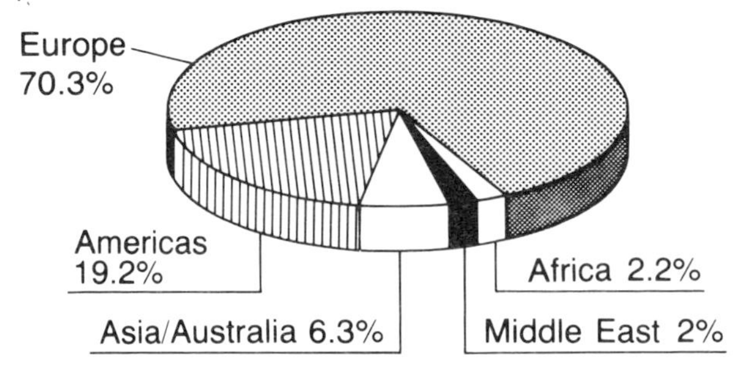

30 million tourists visited Third World countries in 1982

Costs and benefits of tourism

Tourism can be attractive for developing countries. But there are social, economic and environmental disadvantages as well as advantages.

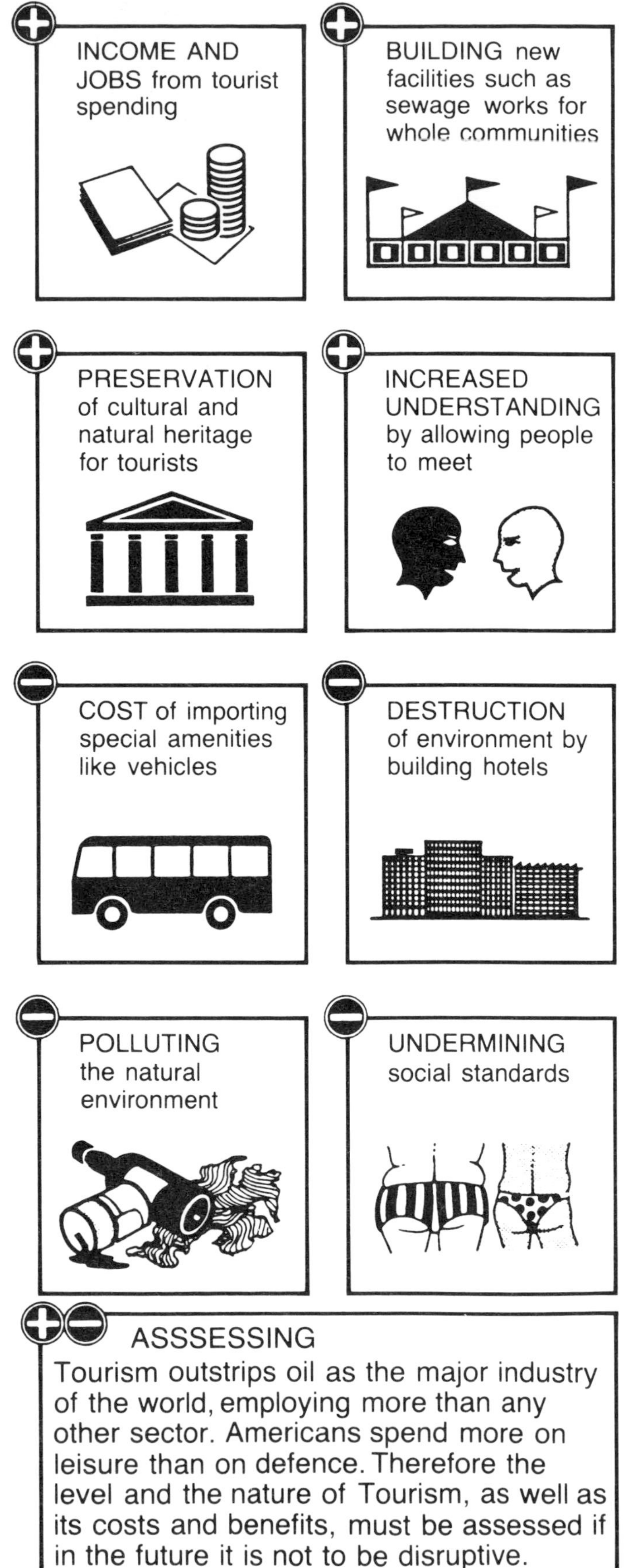

J

A journey through Algeria

In 1979, Tony Gribben made an overland journey from Tunisia, westwards through northern Algeria to the Trans-Saharan highway and then south to Tamanrasset and the border with Niger. This is the transcript of a conversation on the journey in order to provide some background for your decision.

The border between Tunisia and Algeria is rather a problem. I met one German missionary who had been stopped there for four weeks as he tried to drive his Mercedes lorry south to the Central African Republic. I actually saw a hundred or more wrecks between the two border posts and I can only think that they had been let across from Tunisia and refused entry to Algeria, but also refused re-entry to Tunisia! On the whole though, there is less problem on this border than there is to the west between Algeria and Morocco, a border that was closed in 1979. On my journey I was not pestered at all to buy things, as I had been in Tunisia. On the contrary, the only problem I had like this was people trying to buy my shirts, jeans etc. There is a decided lack of consumer items in Algeria, and people generally seemed to have a fair amount of money to spend, but little to spend it on. For instance, on my journey I did not see one sign for Coca Cola, and I only managed to find one place that sold it, a sure sign of the lack of commercialism in the country. The journey from the Tunisian border at Nefta to El Oued and Taggourt in Algeria took me across a very sandy part of the desert where the road was covered with sand to a depth of two to three feet following the sand storm they had just had. We had to dig several people out as we drove along this section. The road was surfaced, but treacherous on account of the sand.

Between Ouargla and Ghardia, the desert was stony rather than sandy and the scenery was somewhat boring. Ghardia is a real oasis town and the main one before reaching El Golea, the last reliable source of food before Tamanrasset in the south. It is here that you pick up the Trans-Saharan highway, a tarmac road to the south which has in parts been broken up by the extremes of temperature, which falls to several degrees below freezing at night but may reach 40°C by day. I remember arriving at El Golea at about 2 o'clock in the afternoon and waiting until about 5 o'clock for the town to wake up after its siesta! When it did, I was able to buy bread and provisions for my journey south. The Tademait plateau south of El Golea is stony and rocky, with mushroom rocks sticking up on either side of the road. A friend who was also on the journey told me that it was like Arizona without the MacDonalds! We came to In Salah, the windy city, which was again very sandy. The houses had regularly to be dug out of the sand! Perhaps this explains why there was only one café to be found there. Nevertheless, In Salah appears on the atlas map and, despite its size, it is an important settlement in that it is the only one for some 200 km before Arak. Again, Arak is on the maps, but in fact it is a collection of some dozen buildings and a tea shack. The scenery is spectacular here. Arak is situated in a gorge cut by a stream which intermittently flows from the Moug Dir mountains. Not far away is Moulay Liscine, where used to live a holy man who blessed trucks on their journey south to Tamanrasset. When he died, a shrine was erected to his memory, and trucks now drive round it seven times to ensure his blessing! The road has suffered as a result.

It is then 400 km south to Tamanrasset, which is on the major caravan crossroads in southern Algeria. The town is very small, and although you have to spend any remaining dinars before you leave the country to the south there is very little to spend them on. I went to the one somewhat basic hotel and managed to buy a drink. It was a good job that I did not want water because it was not working when I was there. I spoke to several others at the hotel who told me about the things to see and do in the area. The list was not a long one! One piece of advice that I am glad I took, however, was to go up into the Hoggar mountains to Assakrem in the early morning to see the sunrise over the desert. You need a four-wheel drive vehicle to negotiate the mountain road, and the journey is an extremely cold one, but you cannot fail to be impressed with the view of the sun rising over the deeply dissected rocks. Before it comes up, the sky changes colour several times and then the giant ball of fire rises in the sky. It is a most spectacular sight. You can climb the last 200 m or so on foot. At the top is a shrine to the memory of Charles de Foucauld, a French hermit, killed by the Tuareg in 1916. A number of French monks live there still. We were also lucky enough to encounter a group of majestic, blue-robed Tuareg on the

J

summit of the mountain. They are an amazingly tall race, something accentuated when you see them on camels. To the north of the Hoggar is an area known as the Tassili Plateau, where there are some spectacular cave paintings and frescoes, but this area is only reached with difficulty from Tamanrasset. When I eventually left the town from the hotel, I went to the border post. You have to clear customs both at Tamanrasset and at the Niger border, and at each your money declaration form has to tally.

Algeria, especially in the south, is relatively unspoilt. The potential for tourism here seems to be more for the 'rough it' brigade, although there is perhaps the possibility of two centre holidays, part spent by the coast and part inland. Compared with Tunisia, it is undeveloped but has at least the same potential for tourism, especially from the French, but eventually from other European nations as well.

Fig. 4.24 View from Tademait Plateau

Fig. 4.25 Arak Gorge tea hut

Fig. 4.26 View from Assakrem

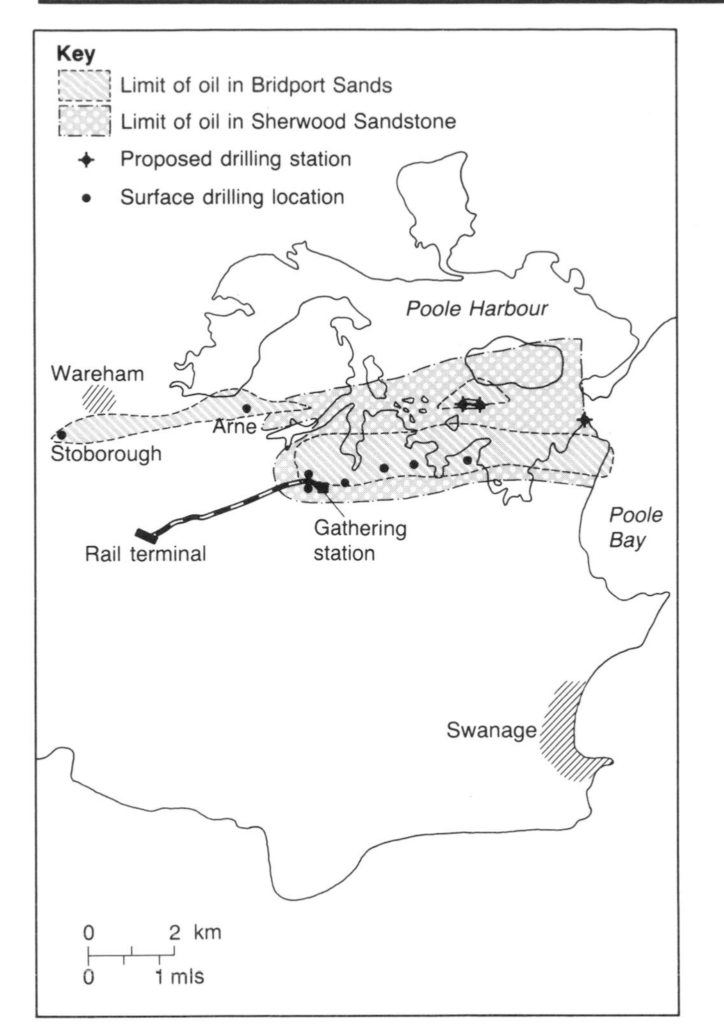

Fig. 4.27 Wytch Farm oilfield limits

Industrial Development and Conservation

Case-study: the Wytch Farm oilfield development

Much has been written about the United Kingdom's vast oil finds in the North Sea. These finds have led to major onshore developments at sites such as Aberdeen, Peterhead and Teesside. Other industries have received a major boost through the growth of the oil industry. Firms supplying drilling equipment, diving and shipping supplies and steel for pipelines have had to increase their output and, in many cases, their location. Added to this, the increase in services required by the large labour force on the drilling rigs and shore depots has led to the growth of shops, banks, restaurants and other associated services. The spread effect of the growth of the offshore oil industry has been great.

Less has been heard of Britain's onshore oil finds, though these have been significant. One of the areas where oil has been produced for some time but which has now been proved to have larger reserves than hitherto thought is in the Purbeck area of Dorset, known collectively as the Wytch Farm oilfield (Fig. 4.27). A production licence was granted to British Gas and British Petroleum in May 1968. Exploratory drilling led to the discovery of reserves in 1973 and by 1980 production started from the field. The great length of time between 1973 and 1980 is explained by

Fig. 4.28 Planning constraints in the Poole-Purbeck area

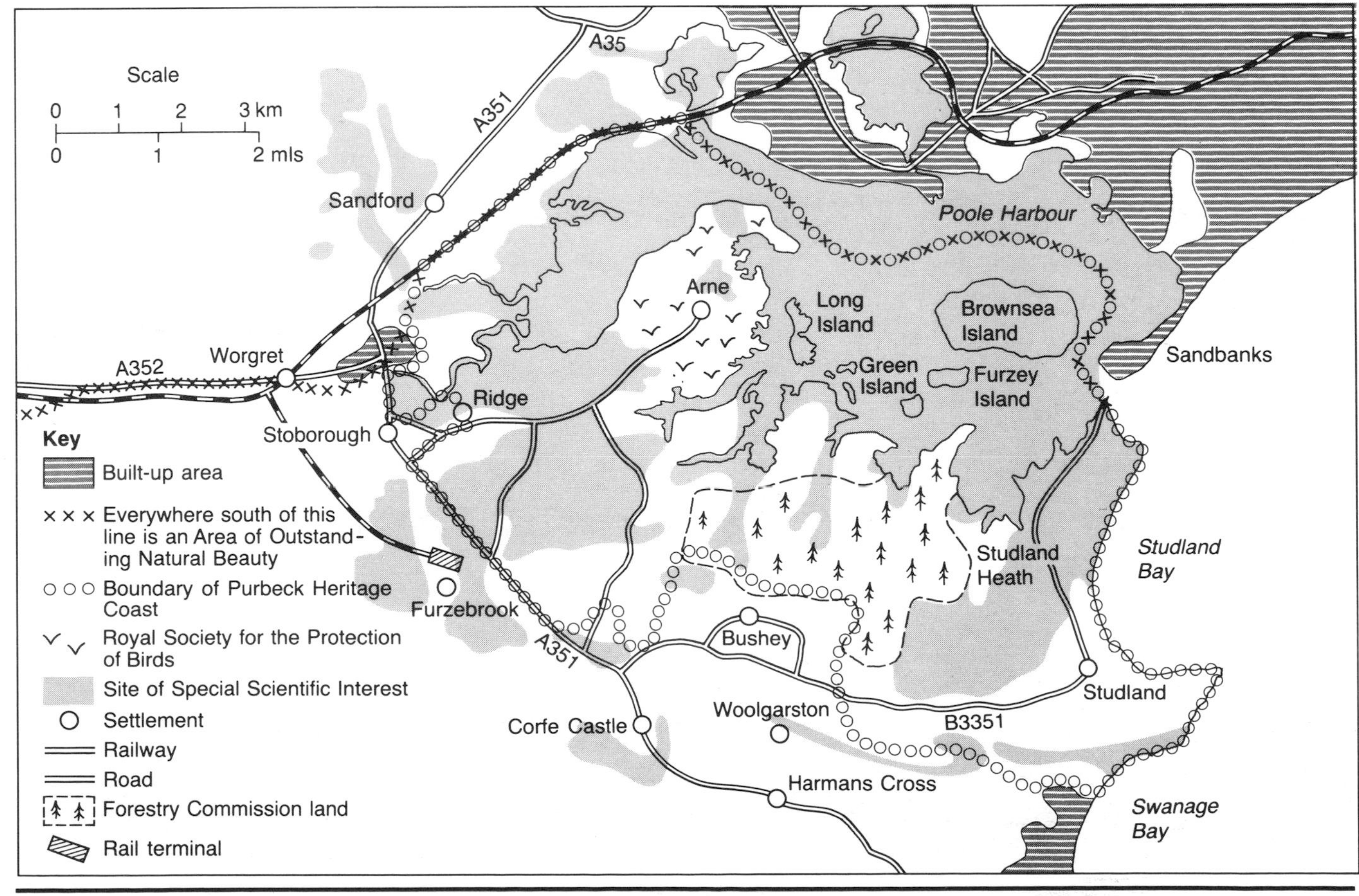

the fact that the oil was found in an environmentally sensitive area. Land in the vicinity of the drilling sites is owned by the Royal Society for the Protection of Birds and several endangered species occur on its Arne reserve. As well as this, the entire area is an Area of Outstanding Natural Beauty (AONB), with several Sites of Special Scientific Interest (SSSIs) studded within it. Surrounding the development is the Purbeck Heritage Coast and there are several local nature reserves where endangered species such as rare orchids, amphibians and reptiles are found. The Forestry Commission also owns a large area of land here (Fig. 4.28). The planning of the first development stage was therefore extremely carefully done with the environmentally conscious companies producing a large amount of literature for the consultation process. The eventual production start-up in 1980 was to accommodate the output of some 4000 barrels per day, which was collected from the various wells at the gathering station at Wytch Heath and then piped to the railhead at Furzebrook. From Furzebrook, the oil was taken by rail to the refinery at Fawley on Southampton Water.

During the development of the first stage, deeper exploratory wells were sunk and a much larger oil reservoir was located. Fig. 4.27 shows the extent both of the Bridport Sands Reservoir, which produces the 4000 barrels per day (520 tonnes) and the newly discovered, deeper Sherwood Sandstone Reservoir which could increase the production by ten times. In so doing, new developments will have to be made. These include two new drilling sites on Furzey Island and one at Studland Point, new pipelines to link these with the Wytch Heath Gathering Station and a new method of getting the oil from here to a refinery.

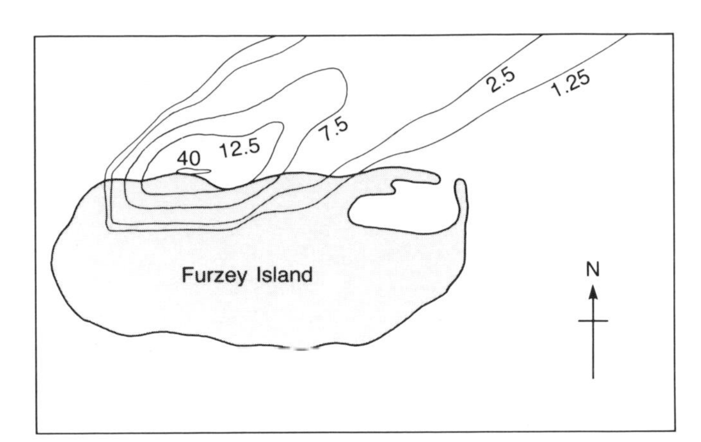

Fig. 4.29 Predicted peak ground level concentrations of NO_2 ($\mu g/m^3$) due to flaring when well-testing

The development of Furzey Island presents great problems. The island is low-lying and in the environmentally sensitive Poole Harbour. Within the harbour there are several species of plants, birds and crustaceans which would be decimated if there were to be an oil spillage. For this reason, BP, which

Table 4.6 Wytch Farm expansion feasibility study: conceptual programme

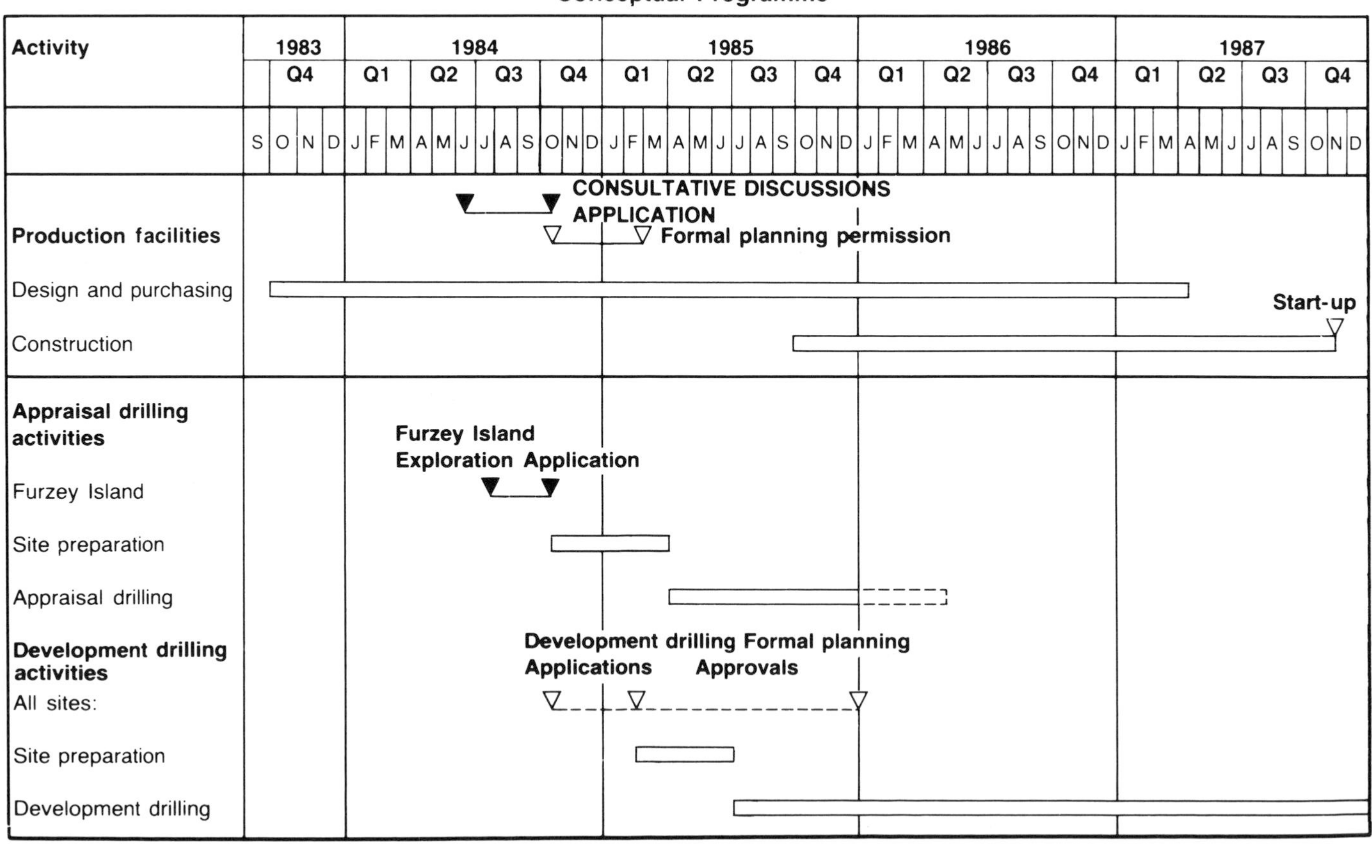

took over responsibility for operating Wytch Farm following the sale into private hands of the oil interests of British Gas, issued several consultative documents. One looked at the existing environment of Poole Harbour and the Isle of Purbeck and another looked at Furzey Island by itself. A third document dealt with the justification of the choice of two well sites on Furzey Island, another with visibility studies on the well sites and yet another was a visual impact analysis of development at Furzey Island. The company left no stone unturned in its efforts to get the development accepted by the various interested parties, which included Dorset County Council, Purbeck District Council, the bodies which administer the various reserves and sites on the Isle of Purbeck and the Department of the Environment which has the ultimate say. The case it had to put up had to be one supported by detailed scientific information. It was not enough that the development would be in the national interest by supplying a significant amount of oil. It also had to be shown that the proposed developments would not injure the environment to any significant degree, both during development and after its completion. Maps were produced by BP to show the location of archaeological sites on Furzey Island, possible routes for boats bringing drilling equipment to the island and roosting sites for the various wildfowl found in the area. Tables were produced to show the noise to be expected from drilling operations and subsequent workings. Further, maps were drawn to show the predicted concentrations of sulphur dioxide and oxides of nitrogen during drilling and after it (Fig 4.29). The extremely sensitive nature of the application meant that BP was only hoping to start up the mainland part of the operation by the end of 1987, having begun in 1983, and on Furzey Island some time after that. The procedures planned by the company are shown in Table 4.6.

An interesting aspect of the application for development on Furzey Island is the visual impact analysis. Since the island is so low lying and is overlooked from the hills on the mainland of the Isle of Purbeck, it was essential that any development could not be accused of spoiling the skyline or the view. For this reason, BP set up a study which aimed:

(a) to identify the characteristics of the Study Area and of Furzey Island
(b) to examine the physical components of the development
(c) to study the visual impact of the development
(d) to identify ameliorative measures to minimize the visual impact
(e) to formulate landscape and management proposals for development.

BP estimated that the lifetime of the oil field would be approximately twenty years and that at the end of this time the landscape should bear no scars. It prepared a map of the island to show the 'visual

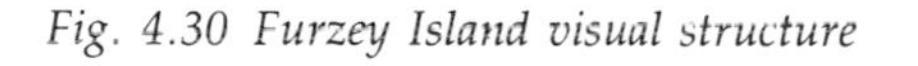

Fig. 4.30 Furzey Island visual structure

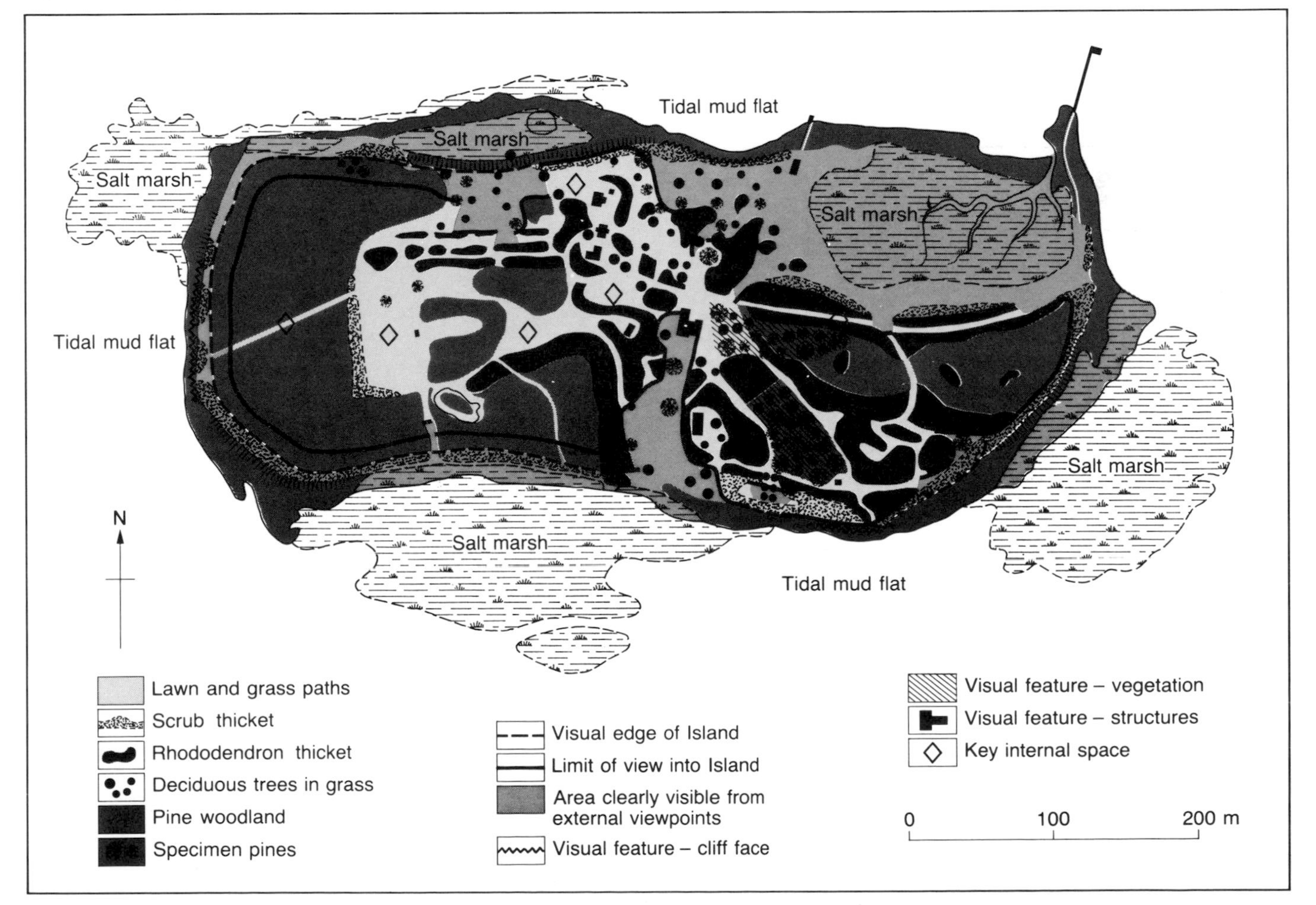

structure', marking on the key elements that could be seen from the highest surrounding viewpoints (Fig. 4.30). This enabled the company to isolate 'key internal spaces' where it could consider development without any great visual impact, and led to the proposal for two well sites as shown in Fig. 4.31. To assess the visual impact of developing these sites, tethered helium balloons were floated up to tree-top level so they could be seen from a nearby viewpoint. Photographs were then taken and a modified profile of the island produced from the positions marked by the balloons. The results for one viewpoint are shown in Fig. 4.32. On the new profile both the areas to be cleared and those in which the density of vegetation would alter are shown. All this information was made available by BP to support its application.

In 1989 a large deposit of oil was confirmed in an extension of the field under Poole Bay (the seaward extension of Poole Harbour). Discussions then ensued about how this oil might be recovered in this highly sensitive area. Proposals were made to find methods of exploiting the oil. They were to build an artificial island from which drilling could take place, with the oil being transferred ashore by pipeline, to site an onshore well at the end of the Studland Peninsula, or to use one of a variety of offshore drilling rigs located in the bay.

In 1990, BP was also given permission to exploit a new field which had been found to the west of Wareham and contained some six million barrels of recoverable oil.

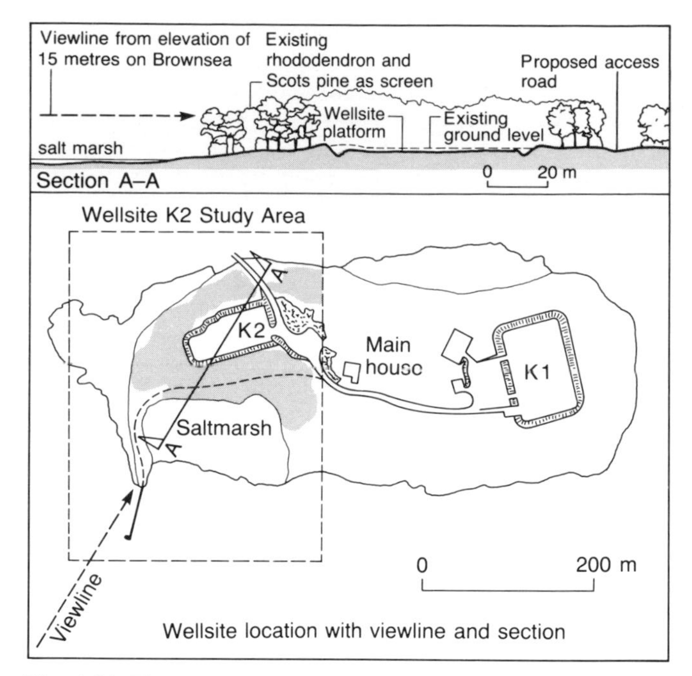

Fig. 4.31 The proposed development of two wellsites on Furzey Island

There remained one problem and this was how to get the oil from the gathering station at Wytch Heath to the refinery. To this end, a further consultative document 'Oil From Dorset, A Consultative Document' was prepared. The document presented four options which it asked the various interested parties to consider.

Fig. 4.32 A sketch of the view from the Baden Powell Memorial, Brownsea Island

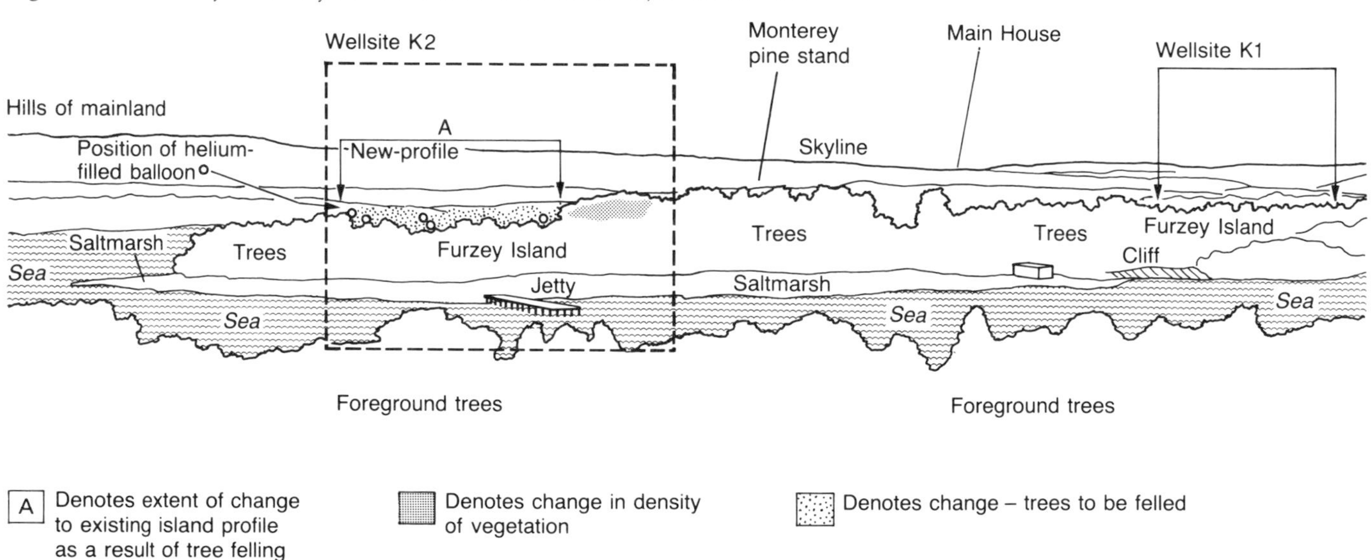

Decision-making exercise K: Oil from Dorset

You are presented with the four possible alternatives given by BP for getting oil from the Wytch Farm development to an oil refinery together with background information about the transport routes involved. You are to take the part of five people in turn, giving the alternatives each a score out of 20. Thus the Harbour Master of Poole Docks might give Option D a score of 20, Option A 5, Options C and B 1 each. At the end of this exercise, a total should be achieved for each option. It is not necessarily the option which scores the highest that will be

chosen, for the views of some people may carry more weight than others. Which of the five people represented will carry most weight in this case? You could use this decision to multiply the score given by the most important person by some number to make the scores more realistic. If you are working in a group, you could each play the part of one of the five. Each should make a statement supporting the case to be made by that person. A discussion following the statements could then arrive at a weighting score by which the totals could be multiplied. If you are working on this exercise by yourself, try to decide the order of importance of each of the five by placing them in rank order.

The rôles to be played are:

(a) The Harbour Master of Poole Docks, who is anxious to bring new investment to the docks and to encourage commercial shipping. There is already a very flourishing yachting business in the area, but the cargo facilities are not as fully used as they could be.

(b) The Financial Director of BP, who is primarily concerned with obtaining the best financial return for his company, although it is his company's policy to fit any developments into the environment as carefully as possible. He will take into account the cost of the development but will be most impressed by the cost of operating the chosen option over the next twenty years. He will also be aware that any further oil finds in the area will need to be accommodated by the option chosen.

(c) The Chairperson of Dorset County Council will have eyes on the biggest money earners in the council's area, which are farming and tourism. There is a great unemployment problem in some areas of the county and the oil developments will be seen as a chance to do something about that, though the possibility of any development affecting employment in farming and tourism will also be borne in mind.

(d) The President, Royal Society for the Protection of Birds. Arne Reserve is noted for its rare birds, including Dartford Warbler and Nightjars. Nearly all Britain's few nesting Dartford Warblers occur here. Added to this, Arne Heath has a population of sand lizards and smooth snakes. Britain's natural heathland has been decreasing alarmingly in the last few years as it is ploughed up for farming, built on or used for forestry.

(e) The Controller, Southern Region of British Rail. There has been a considerable debate about the possible electrification of the rail line from Bournemouth to Weymouth. Any extra traffic on that line would strengthen the case for electrification. At peak production, about twenty trains per day would be needed from Furzebrook.

Table 4.7 Scoring options

	OPTION A		OPTION B		OPTION C		OPTION D	
	Raw score	x w (weighting)	Raw	x w	Raw	x w	Raw	x w
Harbour Master, Poole Docks								
Chairperson, Dorset CC								
Financial Director, BP								
Controller, S. Region, BR								
President, RSPB								
TOTAL								
WEIGHTED TOTAL								

Information about options

Option A: by rail from Furzebrook

Oil will be processed at Wytch Heath, the gathering station, so that the lighter fractions of butane and propane as well as the gas will be separated from the oil. This is known as **stabilizing** the oil.

The rail loading facilities will need to be enlarged to cope with 60 000 BPD of oil as well as the Liquid Petroleum Gases (LPG). This will mean creating more sidings, inevitably eating into the Stoborough Heath Site of Special Scientific Interest. Crude oil storage tanks would need to be located between the gathering station and the Furzebrook depot, in addition to a new pipeline between Wytch Heath and Furzebrook.

The rail route from Furzebrook is already constructed. It could take products to BP's terminal at Hamble, where new rail sidings, crude oil storage tanks, a connecting pipeline and other facilities would need to be built. A jetty is already in existence which can cope with cargoes up to 28 000 tonnes, a relatively small amount compared with other oil ports. Oil tankers capable of moving this amount economically are in short supply. The rail route also runs to Esso's refinery at Fawley, where additional facilities would also have to be built.

The environmental impact of this option would be limited to increased movements of rail traffic on existing lines. When peak production is reached, a total of twenty trains per day would run between Furzebrook and Hamble or Fawley. This would include night-time movements and would compare with the present movement of one or two trains per day. British Rail have tried to calm public disquiet about noise levels by stating that the increased number of movements should not lead to unacceptable noise levels.

Financially, this would be a cheap option to install, but operating costs would be higher than a pipeline option (Fig. 4.33).

Fig. 4.33 Option A: by rail from Furzebrook

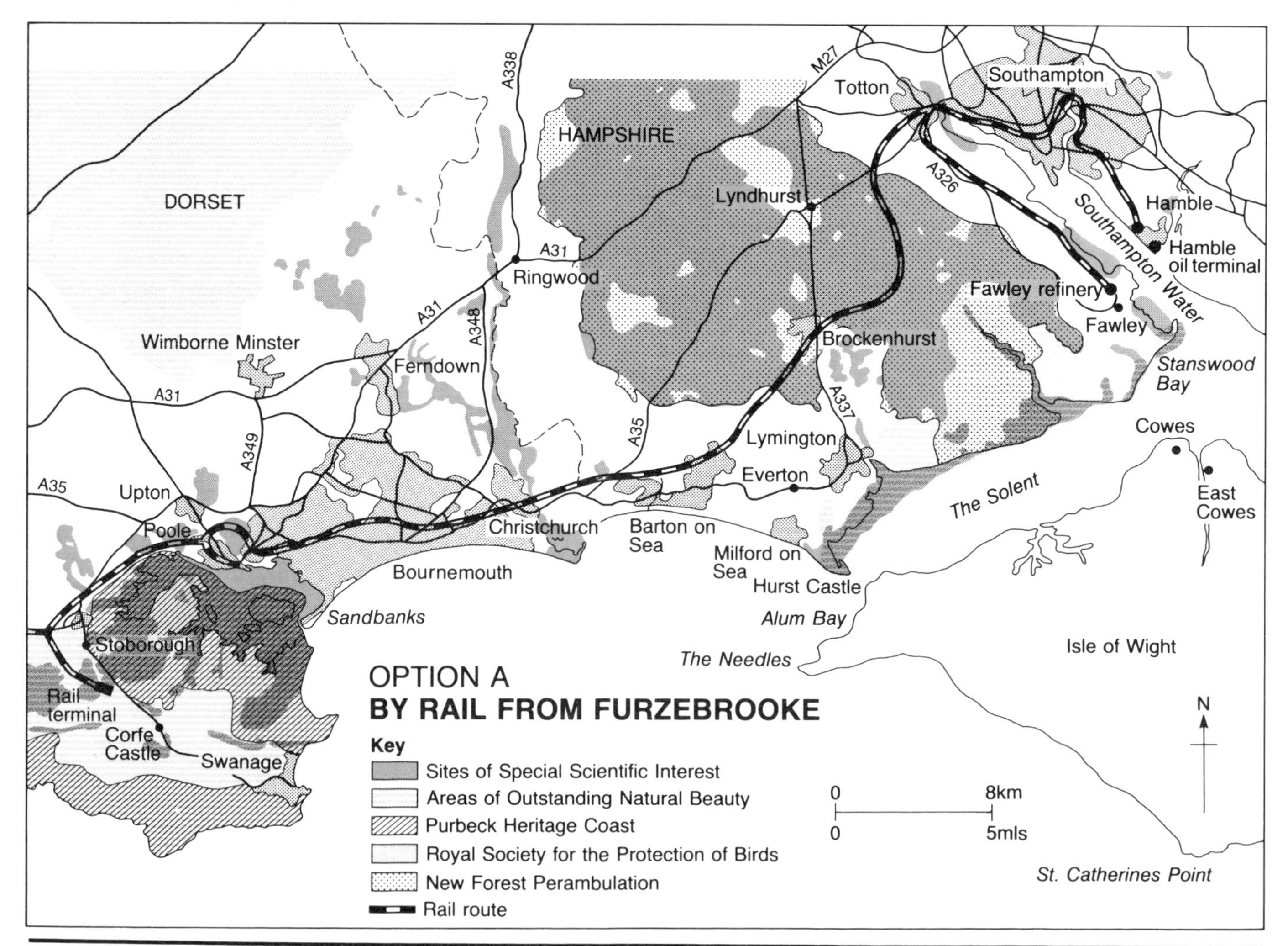

Option B: by pipeline to Portland Harbour

Oil would be moved from Stoborough Heath 50 km to a new marine terminal in Portland Harbour via a twelve-inch-diameter buried pipeline. The Ministry of Defence own the harbour concerned and a proposal has been put to them regarding the development.

The pipeline route (see Fig. 4.34) would largely cross farming land, running around the edge of Weymouth before entering the harbour and ending in a permanent loading facility. The pipeline would need two valve stations, each of which would require the fencing of an 8 m × 8 m square.

The loading facility would consist of a berthing structure with a floating storage vessel moored on one side, itself consisting of a modified tanker capable of storing about 100 000 tonnes of crude oil and large amounts of ballast water. It would also have ballast treatment facilities.

Tankers loading from this berthing structure would be capable of carrying on average 55 000 tonnes. One vessel would be needed per five to seven days and would load in about twenty-four hours.

The short pipeline route would have minimal environmental impact and it could be installed at such times as to avoid holiday traffic and holiday sites, especially near Weymouth. Reinstatement of the land over the pipeline would be included in this plan and the land should recover quickly, even where it is agricultural land. The pipeline would then be almost invisible and definitely silent.

There is, of course, a worry about oil pollution of Portland Harbour, but experience at other terminals shows that any spillages could be very quickly and effectively dealt with. Strict operational procedures would cover the harbour water quality, local conservation areas and bathing beaches. The low volume of tanker traffic together with the strict control of movements by the Queen's Harbour Master would further minimize the possibility of spillages.

Fig. 4.34 Option B: by pipeline from Portland Harbour

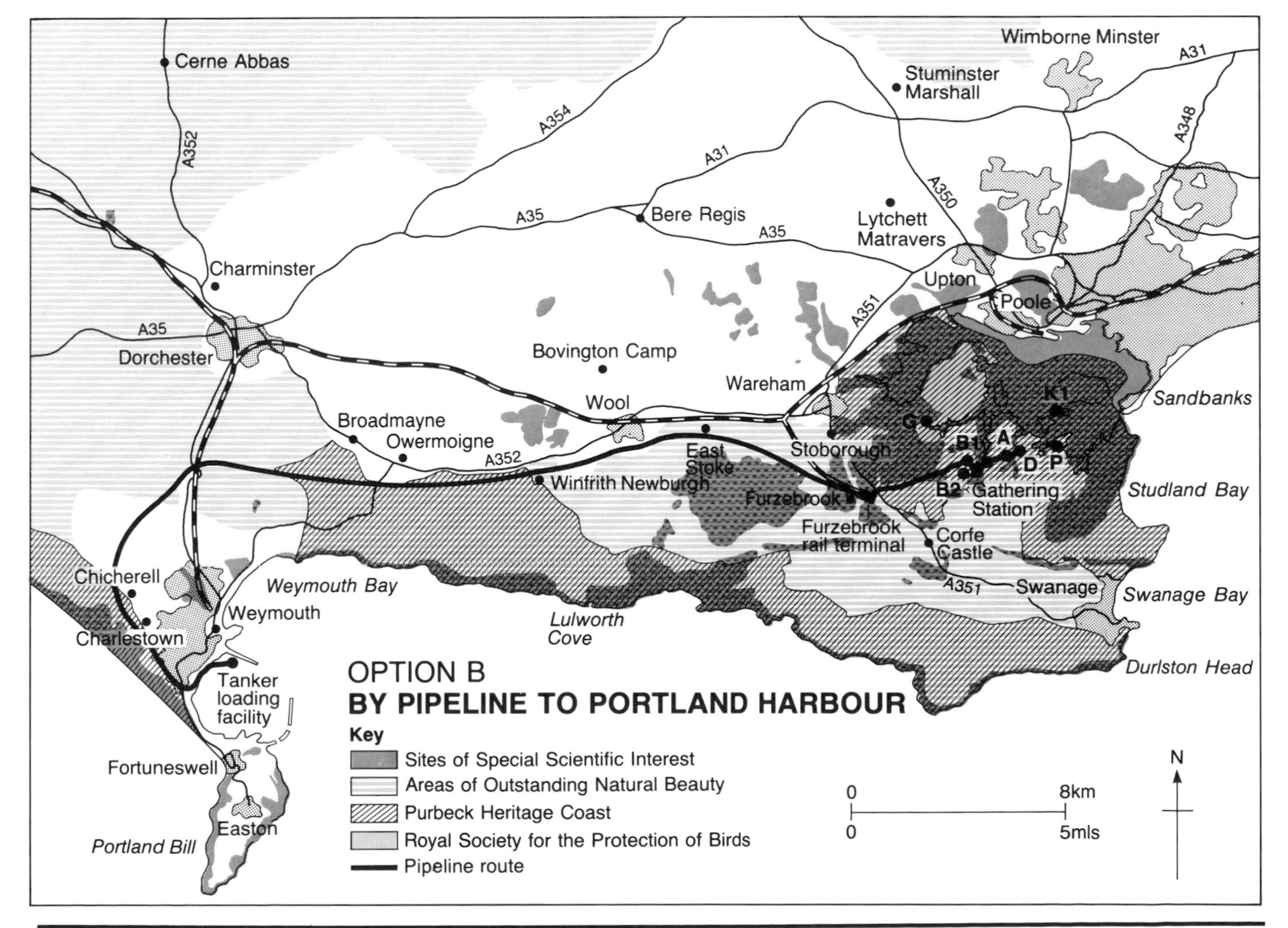

Tankers to be loaded would normally carry segregated ballast, which is clean water kept separate from the oil cargo. During stormy conditions, it is possible that further ballast, possibly oil-contaminated, would be carried. The storage vessel would have facilities to clean this water prior to discharge. This storage vessel would be larger than the vessels at present using Portland Harbour, but its visual impact would be kept to a minimum.

Option C: by pipeline to Southampton Water

A buried sixteen-inch diameter pipeline, 75 km long, would be installed from Wytch Heath to Fawley and Hamble, where facilities to deal with it already exist (Fig. 4.35). The suggested route, which is subject to further debate and consultation, would avoid built-up areas where possible, but would cross the New Forest Heritage Area for about 20 km. It would avoid the New Forest Site of Special Scientific Interest and cross within the New Forest boundary proper for only 6 km. Four valve stations, each with an 8 m × 8 m fenced area above them would be needed.

There is a 'strong presumption' against such a development in the planning policy for this area and the location of the exact route would inevitably be subject to considerable debate.

The main benefit of this development would be the possibility of making use of existing facilities on Southampton Water where oil refining and shipping movements are already well developed. If Hamble were chosen as the destination, the existing jetty could be used, but this is currently restricted to tankers of 28 000 tonnes, making it potentially less attractive to customers.

The potential for oil spillage obviously exists in Southampton Water but the increase in risk is not significant with an increase in tanker movements of about two per week.

Fig. 4.35 Option C: by pipeline to Southampton Water

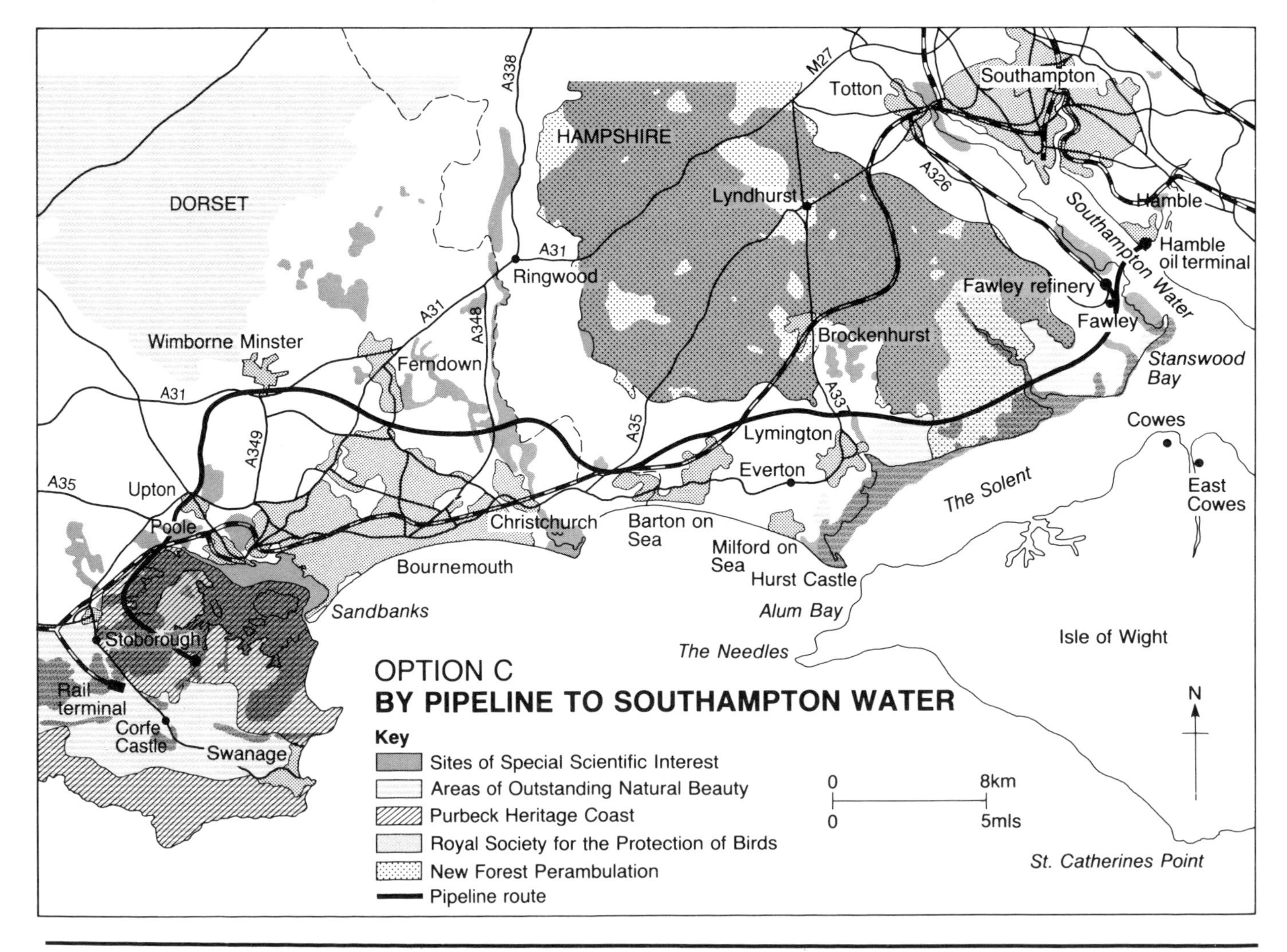

K

The route of the pipeline would be dealt with sensitively. It would be fully restored, and where required near habitation, where working times would be limited, construction would fit in with the current land use.

Option D: by pipeline to Poole Harbour

Originally, this option was rejected by BP but was included in the discussion document because of the strong case made out by the Poole Harbour Commissioners. Fig. 4.36 shows that the pipeline would not need to be long, but would cross several very sensitive areas. In particular, it would run through several Sites of Special Scientific Interest, through Arne RSPB Reserve and underneath Poole Harbour itself at Wareham Channel. The tanker loading facility would need to be located on reclaimed land at Lower Hamworthy and would include a new jetty extending into Wareham Channel. Oil storage tanks and a ballast water treatment plant would also be required. Poole Harbour is shallow, and to overcome this drawback, Poole Harbour Commissioners propose the building of wide beam tankers capable of carrying about 10 000 tonnes of oil and segregated ballast water. Each tanker would cost more than the entire pipeline to Portland and at least three would be needed. If these were not built, the largest tanker that could be used would be of around 2000 tonnes, requiring some 30 loadings per week to move the quantity of oil envisaged. In an area of considerable environmental value and in which very large numbers of small craft move daily, strict operational procedures and contingency measures would need to be adopted.

This option would keep the new jobs created in the Poole area and would generate port revenue.

Fig. 4.36 Option D: by pipeline to Poole Harbour

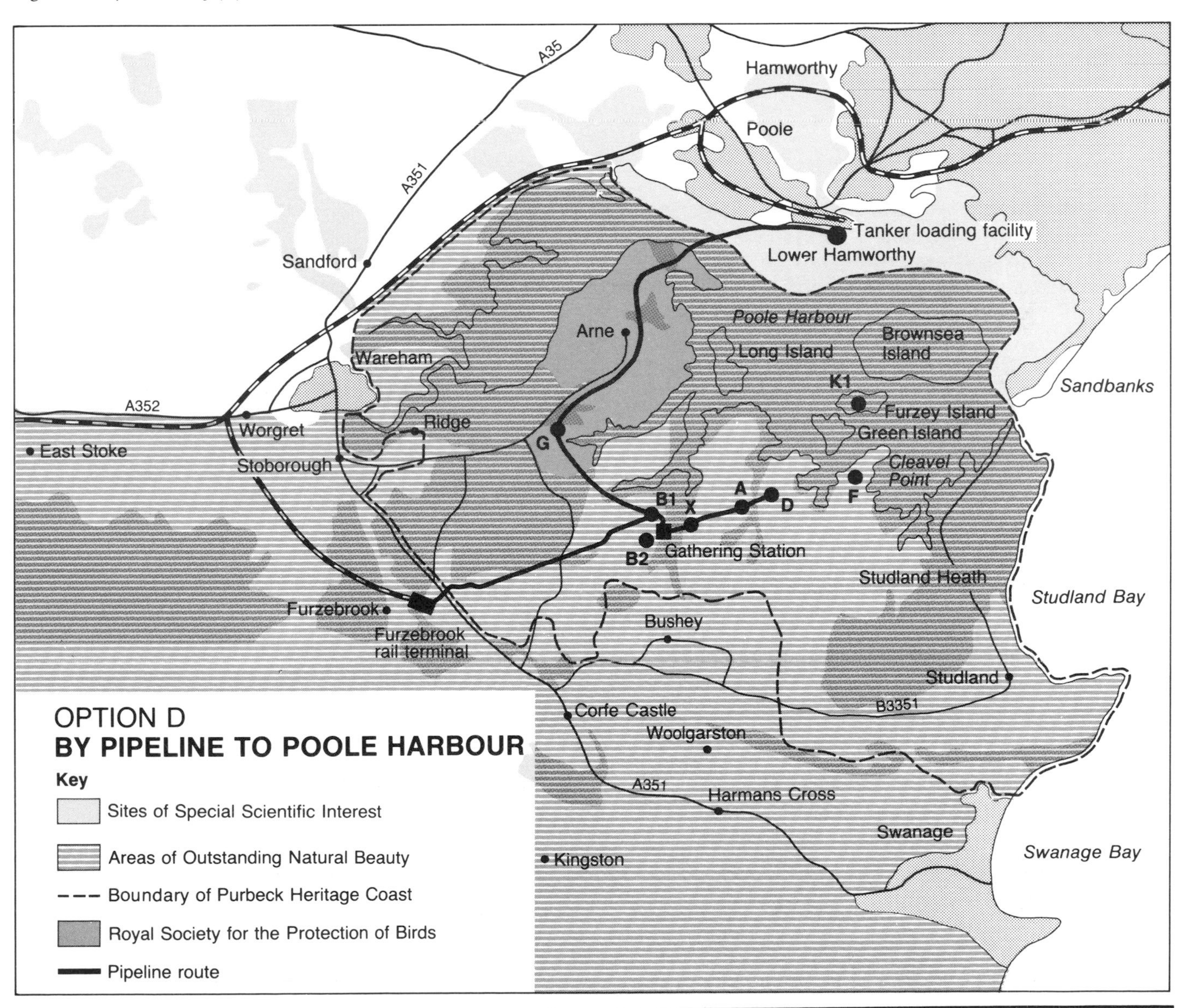

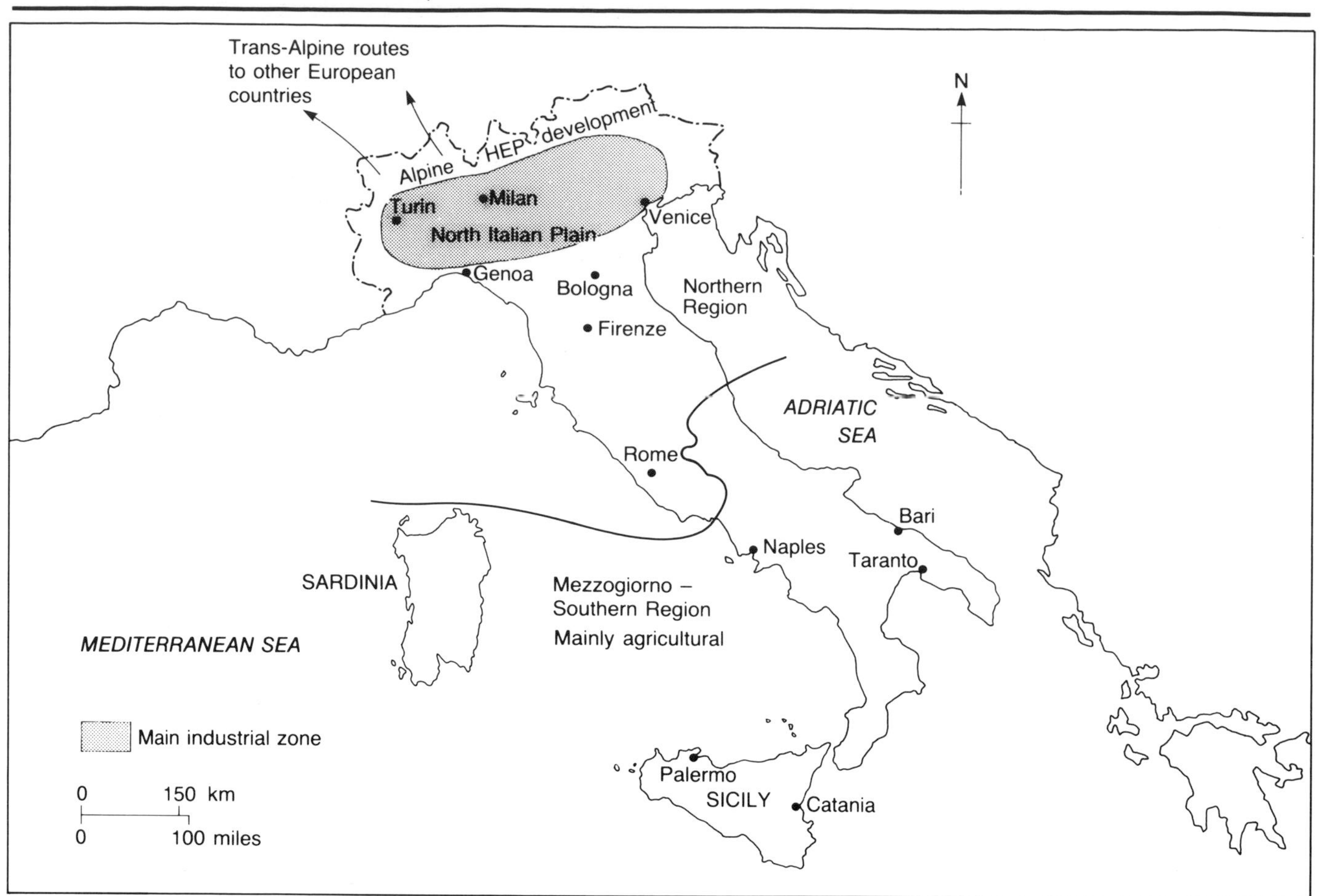

Fig. 4.37 The Mezzogiorno in Italy

Case-study: unemployment and regional aid in Italy

Like many countries of reasonable size, Italy has areas of high industrial activity and areas of economic backwardness. The principal division is between the relatively industrial and prosperous north and the agricultural and poor south of the country, known as the Mezzogiorno (Fig. 4.37). This division is a long-standing one. Over the centuries, the south has become impoverished by various agricultural malpractices. The harsh Mediterranean climate with its long period of intense heat and little rainfall has long given problems to non-irrigated farming. Land holdings in the south have also been the source of poverty. Absentee landlords have traditionally run huge estates. These 'latifundia' practised monoculture, the production of one crop such as olives or wheat, and employed a low-paid peasant workforce which had no incentive to improve the quality of the land or its production. Soils throughout the south are poor and thin and soil erosion is common. Deforestation over centuries of use has left soils on steep slopes exposed to the devastation of violent rainstorms which are common in the south. For all these reasons, there has been a history of emigration from the south towards the more prosperous regions of the north (Figs. 4.38 and 4.39). When industry set up in Italy, it did so mainly in the large cities of the north where power could be obtained from the Alpine streams, where routes through the Alps meant that this was the most accessible of Italian regions and where various natural resources such as oil and natural gas were found. This left Italy with a major regional disparity

Fig. 4.38 Migration of population from the Mezzogiorno, 1951–1961

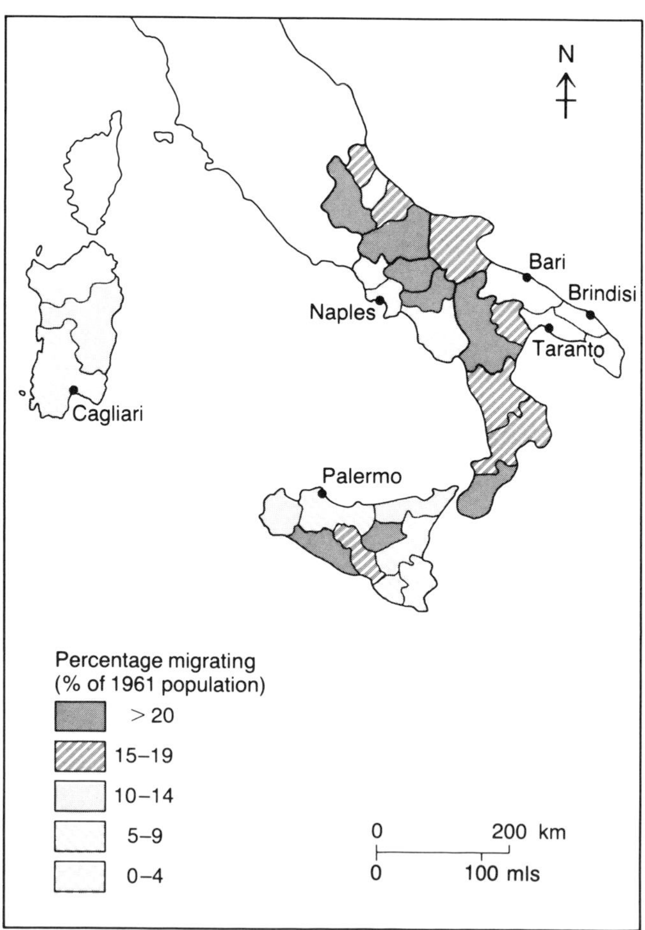

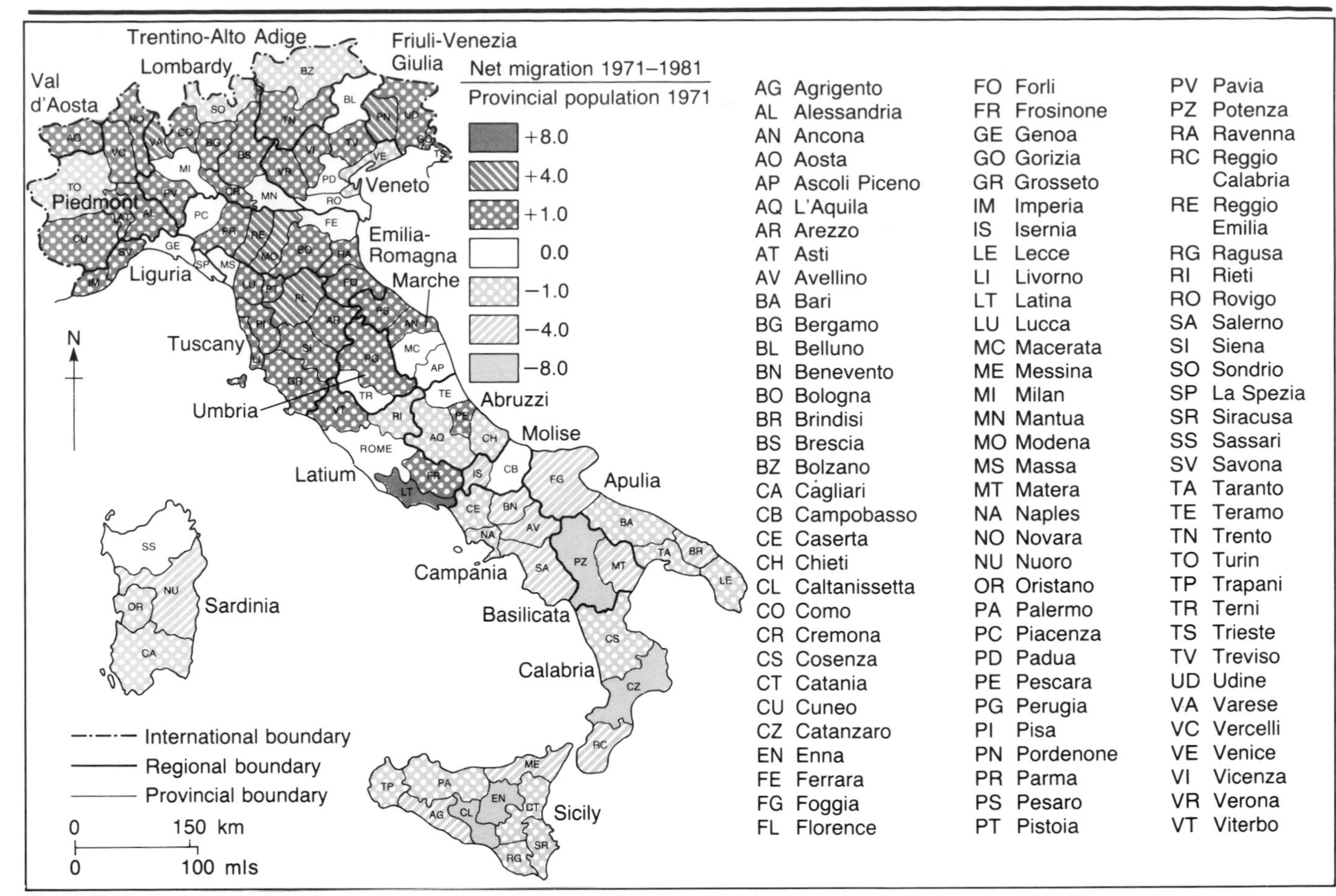

Fig. 4.39 Italy: net migration 1971–1981, by province

and it has been one of the main aims of the government to overcome these differences.

The regional disparity was so great after the Second World War that whereas 21% of the working population in the north was employed in manufacturing industry, the figure for the south was a mere 8%. As well as that, efficiency was much greater in the north so that value added per employee was 40% higher than in the south. Industries of the north were mainly large concerns such as the Fiat car works in Turin and the shipyards and oil refineries of Genoa. In the south, there were many small industrial units, often catering for local needs rather than earning money outside the region. As a result, development plans were drawn up to decrease the inequality in Italy, largely by developing the Mezzogiorno. One means by which the problem was tackled was the introduction in 1950 of the 'Cassa per il Mezzogiorno'. This was set up as an executive body answering to the Committee of Ministers for the south, which included representatives from the ministries of agriculture, forestry, industry and commerce, public works, labour and social welfare, transport and the treasury. It was granted funds (at first a thousand thousand million lire, or £600 million) to instigate a ten-year improvement plan, but it was also encouraged to attract funds from elsewhere. The World Bank, for instance, provided nearly a third of the Cassa's total investment during its first ten years of operation.

The first aim of the Cassa was to improve the agricultural situation and the infrastructure, especially communications. One major way in which it pursued the agricultural aim was to expropriate the land from absentee landlords. The huge 'latifundia' were therefore available for redistribution. A minimum size of plot to allow self-sufficiency in the south was worked out to be about five hectares, so the 'latifundia' were broken up into units of about this size for redistribution. Land capable of irrigation or of good quality could support a family on about five hectares, but poorer quality land had to be reallocated in larger plots. The redistributed holdings are known as 'poderi'. About 48 000 of these had been redistributed by 1982. Smaller plots known as 'quote', were given where the family also had another income, from industry for example, or where some land was already owned. About 65 000 of these were distributed by the same date. The concentration on wheat and olives was broken by this move and crops such as vegetables and citrus fruits increased greatly in output, together with the number of livestock kept.

Expenditure between 1950 and 1965 shows that 56% of the Cassa's budget was spent on farming improvements, but the next largest slice, about 22%, was allocated to improvements in the general infrastructure. The improvements in farming by 1965 had been so great that from 1966 to 1969 the percentage allocated to farming improvements dropped to about 25%. In fact only about one tenth of the total land reform area was actually expropriated (Fig. 4.40) and costs for what some critics call a small improvement were astronomical. It has been estimated that the cost of resettling one family averaged £5300, but at least, after years of neglect, the ball was rolling. Considerable improvements were also made during this period to watershed protection by instigating channel modifications, afforestation and soil erosion schemes. At the same time, the expenditure on infrastructure led to improved roads, drainage, sewerage, water and electricity supplies, from which the farming system also benefitted.

Migration from the south has been a problem for some time in that it is usually the most ambitious, active and involved members of society who make such a move. Certain villages in the south had indeed lost the best part of their working population by the mid-1960s. Taken from an overall viewpoint, however, the problem is perhaps not as great as it may seem, for the density of population was such that there was no hope of land reform in some areas providing enough land for people to make a reasonable living. It is only by encouraging individuals to make profits that other sectors of the economy can be promoted. Since farming was the way of life of the great majority of southerners, it was necessary for them to derive a reasonable income in order to pay for services, which then provided work for more people. This effect was seen best in the manufacturing industry sector.

At first the Cassa concentrated its funding on agriculture and the infrastructure. In 1957, it reassessed the situation, looking seriously at the encouragement of industry. The Industrial Areas Law was passed in that year, authorizing the Cassa to support the setting up of industrial zones in which both infrastructure and industry would be supported by groups of companies under the guidance of the Cassa. From 1959 some very important measures were taken to encourage industrial growth in the south. These are summarized in Fig. 4.41.

In addition to the background improvements listed in Fig. 4.41, it was made obligatory for the large government-sponsored firms in the country to invest at least 60% of their new investment in industry in the Mezzogiorno. This ruling was steadily strengthened, so that 40% of all investment, not just new investment, was to be in the south by 1964. By 1971–75, 80% of new investment and 60% of total investment had to be in the south. Despite these measures, the designated growth poles of the south only boasted offshoots of northern plants by 1980 and very few smaller plants had been spawned as a result. The **multiplier (or spread) effect** had not begun to work to any significant degree, but the major growth poles were in a far stronger position for 'take-off' to be achieved than before the Cassa took hold of the situation.

The growth poles, which were allowed to evolve rather than being imposed, are shown in Fig. 4.42. Some of these contained major industries by 1970. Near Taranto, for instance, there is now a huge iron and steel plant, a cement works, an engineering works, a brewery and a Shell refinery, together with several light industries. Taranto itself has received an extensive face-lift with improvements to its port facilities and large additions to its built-up area. Employment in the manufacturing sector more than doubled between 1951 and 1982 in the Taranto area. In Naples, state-run industries have been introduced

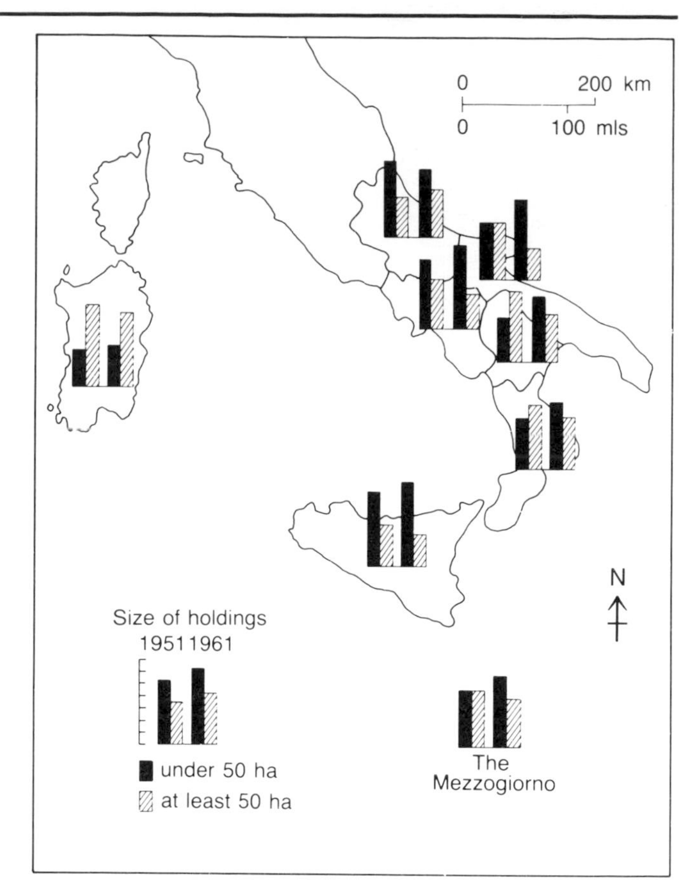

Fig. 4.40 Size of agricultural holdings in the Mezzogiorno, 1951 and 1961

Fig. 4.41 Measures introduced in the Mezzogiorno under the industrial law

- Authorities in designated development zones encouraged to work together to improve infrastructure (especially rail and road communications, water and power supplies)
- Authorities enabled to expropriate land for industrial and infrastructure improvement purposes
- Buildings were to be offered for sale or rent to attract industries
- The modernization of old industries in all but the largest cities was encouraged by the allocation of Cassa funds. The Cassa could provide up to 85% of the total capital cost of such improvements for small and medium sized industries
- In the case of large industries, grants could be made by the Cassa as long as they were in the designated growth areas
- There was to be a ten-year exemption from tax on industry in the Mezzogiorno
- Any equipment imported to modernize industry or start new industry was to be free of import tax. Local authorities were free to decide whether local tax should be applied in such cases or not
- Rail freight concessions were granted to Mezzogiorno industries
- The Cassa encouraged credit institutions to make loans to new industries
- The overall aim was to improve this industrial environment of the south

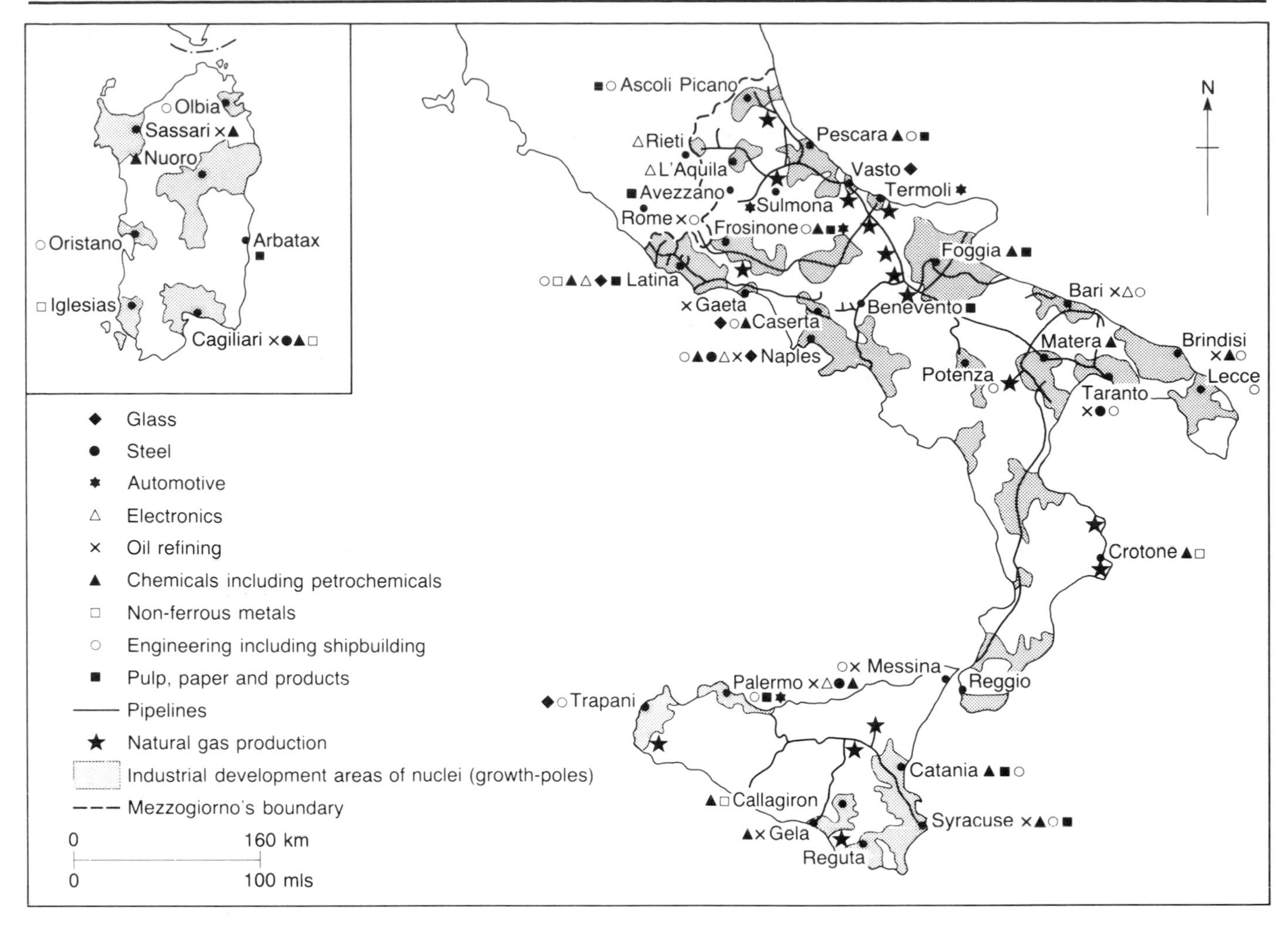

Fig. 4.42 Industry in the Mezzogiorno

so that it now has an iron and steel works, an Alfa Romeo car plant and a cotton textile factory. Other industries near Naples include an electronics factory, an oil refinery and an aircraft fabrication plant. The determined location of the Alfa-Sud factory (a subsidiary of Alfa Romeo) near Naples was a ploy to attract component manufacturers to the area. This has not occurred to any great degree so that the increase in industrial employment in the area has fallen far short of the planned total.

In fact, since the economic recession following the collapse of the oil market in 1974, unemployment in southern Italy has increased steadily. The great strides that had been taken, for instance between 1951 and 1961 (Figs 4.43 and 4.44), have not led to as great an improvement as would have been predicted before 1974. Expectations have been raised in the south but not satisfied. For example, there were over 50 000 applications for jobs in the Taranto iron and steel works for just 5000 vacancies. A new law in 1976 in fact took away from the Cassa a lot of its influence as a promoter of industrialization, a recognition perhaps of the failure of planning policies to provide the stimulus required for the south to begin a self-generating cycle of industrialization. The emphasis on capital-intensive industries has certainly created several major potential growth poles, but these have been remarkably unsuccessful in generating the labour-intensive offshoots so necessary if the south is to break out of the mould

Fig. 4.43 Decline of the working population engaged in agriculture in the Mezzogiorno, 1951–1961

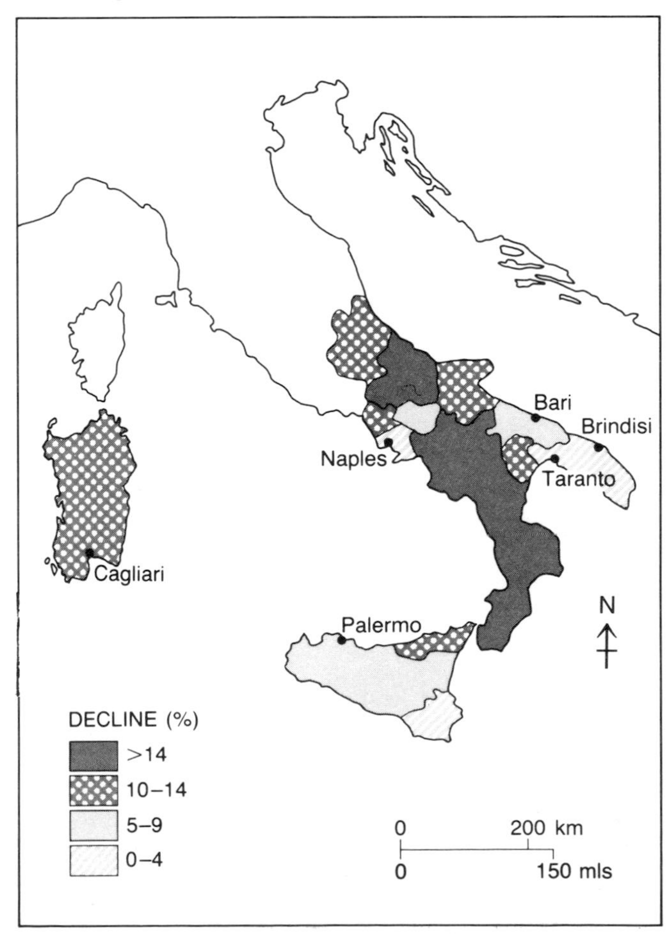

of being a labour supplier for the major industries of the north. One labour-intensive industry of potential in the Mezzogiorno is tourism and this could be an important contributor to the labour market in the future. (See the decision-making exercise following this section.)

15 If you were an Italian industrialist in 1960, why would you have been keener to locate your new plant in the north of Italy rather than in the south?

16 In what ways did the agricultural reforms of the Cassa 'set the scene' for industrial improvements?

17 Why is it essential to improve the infrastructure before industry can be coaxed into an under-developed area?

18 Do you agree with the idea of the government telling industry where it must expand? Explain your answer.

19 Refer to Figs. 4.38 and 4.39. Has there been any change in the emigration pattern from the Mezzogiorno between the two periods shown?

20 Why do you think the growth pole idea has not so far worked to any great degree in the Mezzogiorno?

Fig. 4.44 Changes in manufacturing employment in the Mezzogiorno, 1951–1961

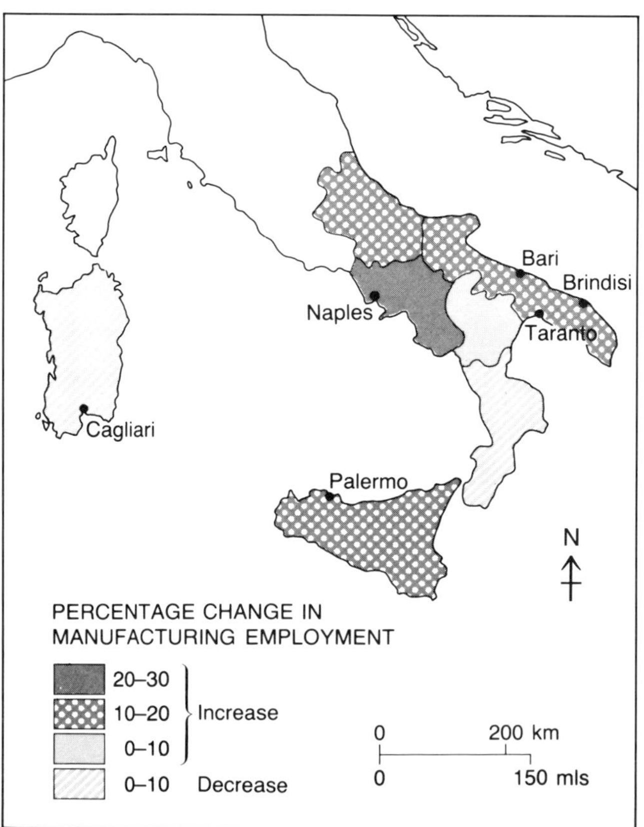

Like farming output, average income and the distribution of industry, there is a great difference between northern and southern Italy as far as unemployment is concerned. Table 4.8 shows the relationship between population and unemployment in the regions of Italy in 1984.

21 (a) Use the technique of drawing Lorenz curves (see Techniques p. 159) to illustrate the data for the distribution of unemployment in Italy in 1984 and 1980 (Table 4.8).
(b) To what extent do the measures implemented in the Mezzogiorno seem to have worked?

22 Produce a choropleth map of Italy on the base map provided (Fig. 4.45) to show the distribution of unemployment in Italy in 1984 using the data in Table 4.8. Which regions of Italy have percentage shares of unemployment well above average? Attempt an explanation of the distribution you have found. (Refer to Techniques p. 160 for information on choropleth maps.)

In an attempt to deal with concentrations of unemployment such as in the Mezzogiorno, the European Economic Community, of which Italy is a member, has two regional funds. The first is the European Social Fund. This was set up under Article 123 of the Treaty of Rome which states that the Social Fund is supposed to improve the employment opportunities for workers in the Common Market

Table 4.8 Population and unemployment in Italy, 1984 and 1989

Region	% share of Population 1984	% share of Unemployment 1984	% share of Unemployment 1989
Italy	100.00	100.00	100.00
Piemonte	7.65	6.00	6.90
Valle d'Aosta	0.20	0.07	0.10
Liguria	3.26	2.50	2.30
Lombardy	15.32	7.57	11.10
Trentino	1.70	0.46	1.60
Veneto	7.62	4.00	5.28
Friuli-Venetia Giulia	2.29	1.02	1.60
Emilia-Romagna	6.92	3.67	5.42
Toscana	6.29	3.12	5.50
Umbria	1.58	1.03	1.53
Marche	2.62	1.46	1.70
Lazio	8.77	10.85	8.47
Campania	9.62	23.73	14.80
Abruzzi	2.30	1.58	2.39
Molise	0.77	0.74	0.41
Puglia	6.88	8.35	8.60
Basilicata	1.15	1.63	2.00
Calabria	3.65	4.83	4.70
Sicilia	8.57	13.03	10.61
Sardegna	2.84	4.30	4.90

(Source: *Eurostat Basic Statistics of the Community*, 1985, 1990)

and contribute to raising their standards of living (by) rendering the employment of workers easier and increasing their geographical and occupational mobility'. From what we have seen about the Mezzogiorno, this region is an ideal recipient of funding. Most of the fund now goes to help finance training, retraining and job creation schemes and many such schemes have been started in southern Italy. Forty per cent of the total fund is spent in the priority areas of Northern Ireland, Eire, Greece, the Mezzogiorno and the French Overseas Departments. The other source of finance to combat the unemployment problem is the European Regional Development Fund which came into operation in 1975. This works on a system of national quotas, but the importance of Italy in this scheme is revealed by the figures for 1982 (Table 4.9).

About one third of the quota is used to finance industrial, tourist and service sector projects and about two thirds is used to help finance infrastructure projects run by public authorities. In this way, the European Regional Development Fund is used to support improvements made by the member states themselves.

How successful have all the various methods of support for industrialization and development been for the Mezzogiorno? One way to find out would be to look at the distribution of certain basic industries throughout Italy to see whether the regional disparities are as great as those described above, where for instance it has been pointed out that most of the large manufacturing industries before 1950 were in the north and that, whereas 21% of the working population in the north was employed in manufacturing industry, the figure for the south was only 8%.

23 Work out the industrial location quotients for textiles, chemical and metal articles for the regions of Italy from the data in Table 4.10 (for 1980). Plot these on copies of Fig. 4.45 using a choropleth technique. (For information on location quotients, see Techniques p. 161.)

24 (a) What do your maps show? Was there still a disparity in 1981 between industries in northern and southern Italy despite all the measures that have been taken?
(b) Looking at Tables 4.8 and 4.11, would you expect the distribution of industrial employment by type to have been similar in 1990? Explain your answer.
(c) How does Table 4.12b support your answer?

Your maps will no doubt have shown that great differences still exist despite all the efforts made to overcome the problems. The consequences for Italy are very great, but not as immediately obvious as normal unemployment figures might suggest.

Fig. 4.45 The regions of Italy

Table 4.9 National quotas (%) under the European Regional Development Fund for 1987

	Range for Grants (%)
Italy	21.62–28.79
UK	14.50–19.31
Greece	8.36–10.64
France	7.48–9.96
West Germany	2.55–3.40
Belgium	0.61–0.82
Netherlands	0.68–0.91
Luxembourg	0.04–0.06
Ireland	3.82–4.61
Portugal	10.66–14.20
Spain	17.97–23.93
Denmark	0.34–0.46

Table 4.10 Employment statistics for Italy by region, 1981

Region	(Figures in 1000s)			
	Total industrial employees	Chemicals + man-made fibres	Metal articles	Textiles
Italy	6595.3	289.8	408.7	479.6
Piemonte	779.8	26.8	66	48.1
Valle d'Aosta	18.5	0.8	0.5	0.6
Liguria	177.2	9.7	14.0	4.2
Lombardy	1590.8	101.1	135.3	127.1
Trentino Alto Adige	92.2	2.6	4.3	3.3
Veneto	554.6	22.7	43.9	51.4
Friuli-Venetia Giulia	163	3.1	12.4	7.4
Emilia Romagna	502.3	19.6	37.5	27.8
Toscana	463.1	15.7	23.1	55.3
Umbria	104.2	7.2	6.2	8.5
Marche	172.3	2.4	8.6	18.5
Lazio	424	22.5	12.7	20.0
Campania	451.1	10.5	15.6	31.5
Abruzzi	111.7	3.1	2.7	8.6
Molise	25.4	0.1	0.5	2.6
Puglia	307.9	7.3	8.9	23.2
Basilicata	58.5	3.4	0.7	3.1
Calabria	136.6	2.0	2.1	5.6
Sicilia	349.2	18.8	9.1	28.3
Sardegna	112.7	10.2	4.3	4.5

Unemployment has not struck equally in Italy, as your analysis of Table 4.8 has already shown. However, it has also affected some elements of the population more than others as Table 4.11 reveals.

25 Construct further maps of Italy to show the percentage of female unemployment and youth unemployment in 1983. Use a scale which shows four classes between 35% and 75% for both types of unemployment.

26 What pattern do your maps show? Is it the same pattern as that revealed by the distribution of industry?

27 Why are female unemployment figures likely to be less reliable than those for males?

Table 4.11 Regional variations in unemployment in Italy, 1987

Region	Number of people unemployed (in 1000s)	% of total unemployed (female)	% of total unemployed (under 25)
Piemonte	169.7	62.70	64.40
Valle d'Aosta	2.3	56.70	50.00
Liguria	56.6	50.97	68.01
Lombardy	272.7	61.41	65.85
Trentino	24.3	56.13	64.15
Veneto	131.2	55.89	62.77
Friuli-Venetia Giulia	40.9	59.86	59.86
Emilia Romagna	133.7	67.17	52.86
Toscana	135.3	65.67	55.02
Umbria	37.5	53.04	53.04
Marche	43.1	65.25	50.90
Lazio	210.3	50.55	64.19
Campania	363.8	54.12	61.07
Abruzzi	59.5	55.32	60.64
Molise	10.3	58.97	55.77
Puglia	210.8	50.15	59.53
Basilicata	50.3	57.14	54.33
Calabria	115.7	50.69	50.25
Sicilia	260.8	47.76	56.74
Sardegna	121.8	51.96	64.50

The consequences of unemployment are great for any country. Long-term unemployment leads to lack of motivation of the work force and is a drain on the nation's resources as far as social security payments are concerned. In effect, the dependent part of the population is increased and unless new and more profitable industries are introduced to provide greater national income the result is that fewer financial resources are spread ever more thinly. There have not been many industrial successes in Italy, for its index of manufacturing output has only risen to 119.6 in 1984 from a base of 100 in 1975. At the same time, unemployment has risen from less than 1.3 million to 2.7 million in 1984. The slight increase in revenue from taxes and levies on industry has been more than matched by the number of people needing unemployment benefit. The disaffected part of the population is a breeding ground for unrest and many people have put down the rising crime rate and terrorist activity in Italy at least in part to unemployment.

28 From Table 4.12, construct an unemployment pyramid as you would for a population pyramid. Use a vertical scale to show age groups. Show males to the left of this central axis and females to the right. Use a percentage scale on the base of the pyramid.

29 From your answers to questions **25** and **28**, attempt to say what you think are the main problem elements of the employment spectrum in Italy.

30 In view of your answer, should there be a different emphasis in development strategies for southern Italy?

Table 4.12a Unemployment rates in Italy by age groups, 1989

Age group	(Percentages) Males	Females
14–19	21.8	20.4
20–24	35.8	33.8
25–29	16.5	19.0
30–34	7.2	9.5
35–39	4.5	6.7
40–44	3.5	4.0
45–49	3.4	3.5
50–54	3.4	2.2
55–59	2.9	0.7
60–64	0.7	–

Table 4.12b Movement of population, 1985

Region	Net migration (per 1000)
Piemonte	−0.4
Valle d'Aosta	3.7
Liguria	3.2
Lombardia	0.5
Trentino-Alto Adige	0.6
Veneto	1.5
Friuli-Venetia Giulia	1.8
Emilia-Romagna	2.2
Toscana	2.8
Umbria	4.2
Marche	2.0
Lazio	3.0
Campania	0.5
Abruzzi	3.3
Molise	1.4
Puglia	1.0
Basilicata	−1.9
Calabria	1.0
Sicilia	1.5
Sardegna	2.1

Decision-making Exercise L: A New Car Plant for the Mezzogiorno

Part A
Refer to any relevant data in the previous section and the data in Fig. 4.46. You are to look for the general potential for new development in the Mezzogiorno. You should produce a well argued statement of not more than 1000 words to explain why it would be advantageous to site a new light engineering plant in Potenza. This statement should be illustrated with any relevant data, which could include material on the success of similar new industrial developments in 'peripheral' areas elsewhere.

Part B
You are also provided with data on a particular site on the outskirts of Potenza which has been recommended for development. The Ford Motor Company is interested in it as the site of a new engine plant.

Write a report for the Ford management in which you put forward the main arguments for building on the site. You should refer to:

(a) Locational factors (Are there problems in having a number of different plants? See Fig. 4.47. Do some areas of Europe produce more than others? Why might this be the case? See Figs. 4.48 and 4.49)
(b) Company structure and links (Fig. 4.47)
(c) Local economic advantages (ERDF Grants, Fig. 4.9)
(d) Potential markets (Figs. 4.14, 4.52, 4.53)
(e) Social factors (Figs. 4.51, 4.12, 4.11, 4.8)

You should also highlight any disadvantages you see with developing the new plant.

Make your overall recommendation to the Board as to whether to locate at Potenza or not.

Part C
With reference to Table 4.12b and Figs. 4.38 and 4.39, suggest how successful recent developments have been in reversing the trend of regional decline in the Mezzogiorno.

L

Table 4.13 Total employment in Italy, 1984

	Total	Agric., forestry & fishery	Fuel & power	Ind. prods.	Bldg & constr.	Mkt servs.	Non-mkt servs.
Italy	20 826	2410	194	5112	1645	7832	3633
North west	2667	206	33	773	153	1080	423
Lombardy	3659	132	33	1375	229	1409	481
North east	2431	228	21	648	175	948	410
Emilia-Romagne	1676	213	13	441	106	643	260
Centro	2241	233	23	614	155	814	402
Lazio	1874	114	18	280	138	862	462
Campania	1689	276	13	321	170	594	315
Abruzzi-Molise	550	119	4	101	59	165	101
Sud	2103	520	15	302	243	640	384
Sicilia	1451	278	14	197	162	520	280
Sardegna	487	90	8	62	54	157	116

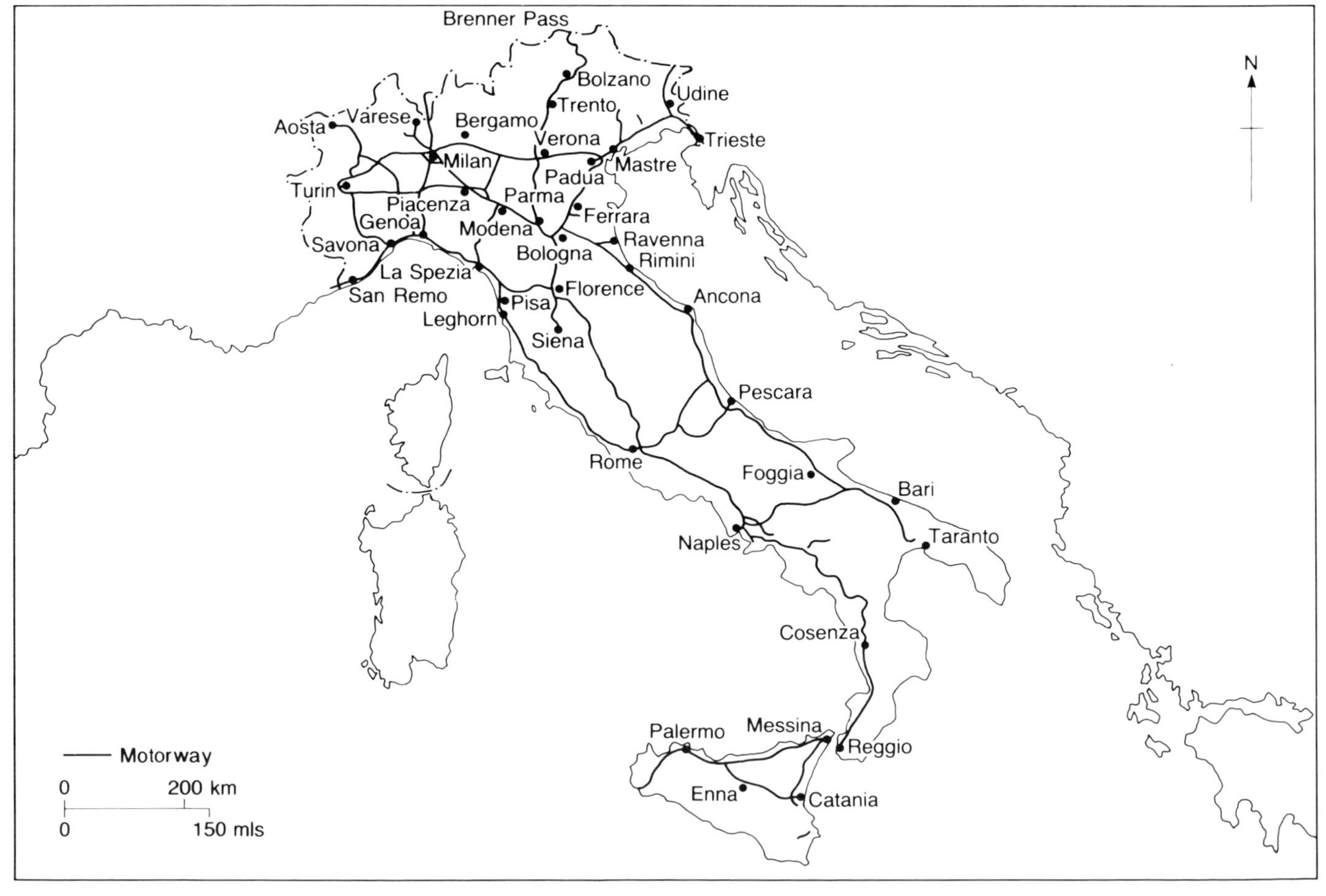

Fig. 4.46 Autostradi of Italy

Table 4.14 Motor vehicles, production and assembly, in Italy (with comparative figures) 1986

	Passenger cars Production	Assembly	Commercial motor vehicles Production	Assembly
Italy	1663	–	179	–
Belgium	230.8	1064	54.5	5.4
Germany	4269	–	297	–
Spain	1282		251	
France	3029		507	
Netherlands	119		24	6.6
Portugal	–	59	–	28
UK	1018	–	227	–
Japan	7810		4450	

Fig. 4.47 Ford Motor Co. Europe – The Fiesta.

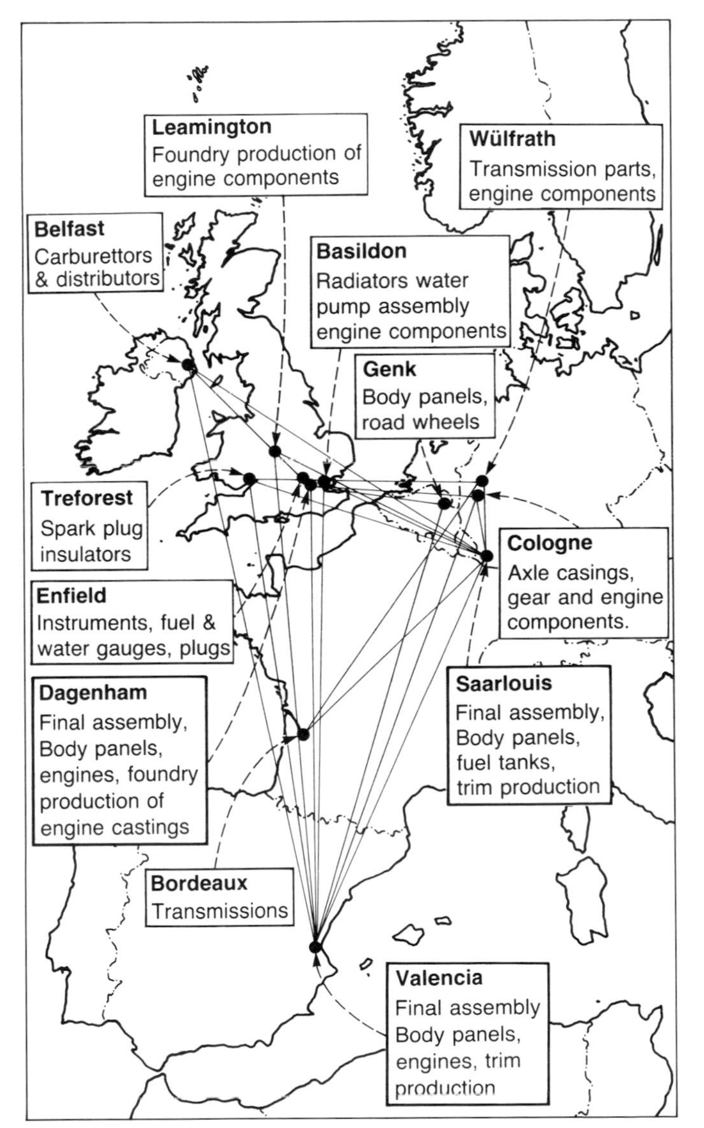

Fig. 4.48 Hours per car

Plant	1988
Valencia – Fiesta	33
Genk – Sierra	40
Dagenham – Fiesta	57
Dagenham – Sierra	67
Halewood – Escort	59
Southampton – Transit	79

(Source: Ford internal management report 1989)

Fig. 4.49 Vehicles per employee

Plant	1988
Dagenham	22
Halewood	29
Southampton	23
Genk	34
Cologne	34
Saarlouis	42
Valencia	38

(Source: Ford internal management report 1989)

Fig. 4.50 A potential site for a car factory

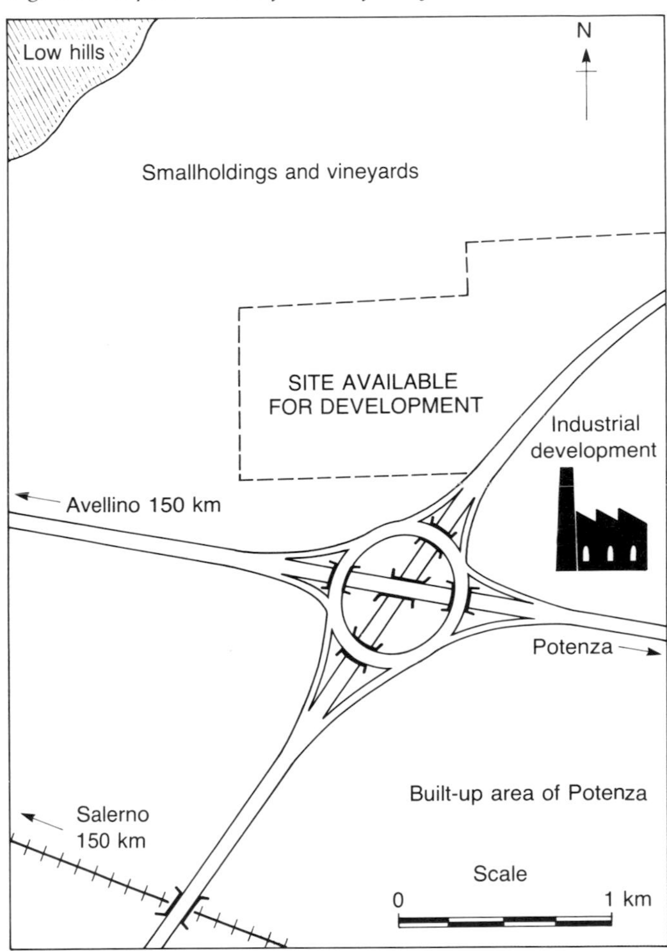

Fig. 4.51 Comparative wage rates in the car industry (1986)

Country	Hourly rate (US$)*
US	$19.88
West Germany	$16.91
Canada	$13.45
Sweden	$12.75
Japan	$11.97
France	$11.22
Italy	$10.55
Australia	$ 9.56
UK	$ 8.66
Mexico	$ 2.66
Taiwan	$ 2.44
Korea	$ 1.90
Brazil	$ 1.73

* includes average rate plus benefits and bonuses.

Fig. 4.52 Top ten car companies, 1985

Company	Total production (millions of cars)
General Motors (US)	7.1
Ford Motor Co. (US)	3.8
Toyota (Japan)	2.6
Volkswagen (W. Germany)	2.1
Nissan (Japan)	2.0
Renault (France)	1.6
Peugeot/Citroen (France)	1.6
Chrysler (US)	1.3
Fiat (Italy)	1.2
Honda (Japan)	1.1

Fig. 4.53 Japan, Inc

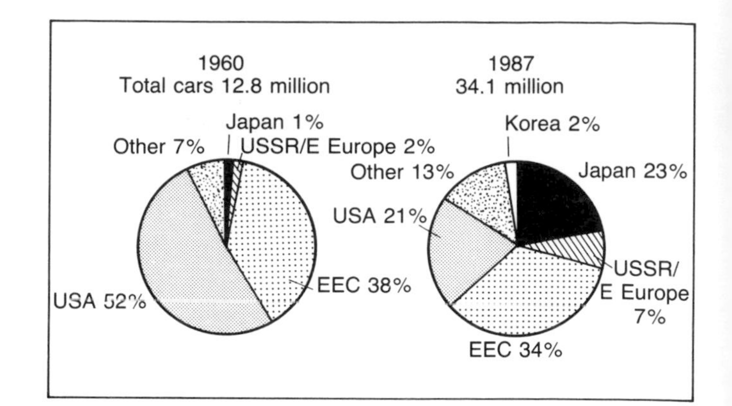

Fig 4.54 A major car plant in Northern Italy – the Fiat factory at Turin

Reclaiming Derelict Land

Case-study: the Lower Swansea Valley

Industrial decline has been widespread in many of Western Europe's old-established heavy industrial areas. This decline has brought with it the problems of unemployment and subsequent migration, but also it has left large areas scarred by the activities of the past. Derelict buildings, disused communications and waste tips of various sorts blot the landscape in such areas as the Sambre-Meuse coalfield in Belgium, the Ruhr in West Germany, the north-east coalfield of France and in several places in the United Kingdom. One example of industrial dereliction and what has been done about it is found in South Wales, centred on Swansea.

The Lower Swansea Valley (Fig. 4.55) is one of Britain's oldest established industrial areas. It was developed particularly for smelting non-ferrous metals from 1717 onwards and by the early nineteenth century it had a large number of important works linked by canal, road and railway to the port of Swansea, 2 km to the south. These works were developed without taking care of the environment which at one time had been described as one of the most beautiful in Wales. That was hardly the case by 1900. Factories were belching out noxious fumes, spewing liquid wastes into the filthy River Tawe and generally despoiling the landscape.

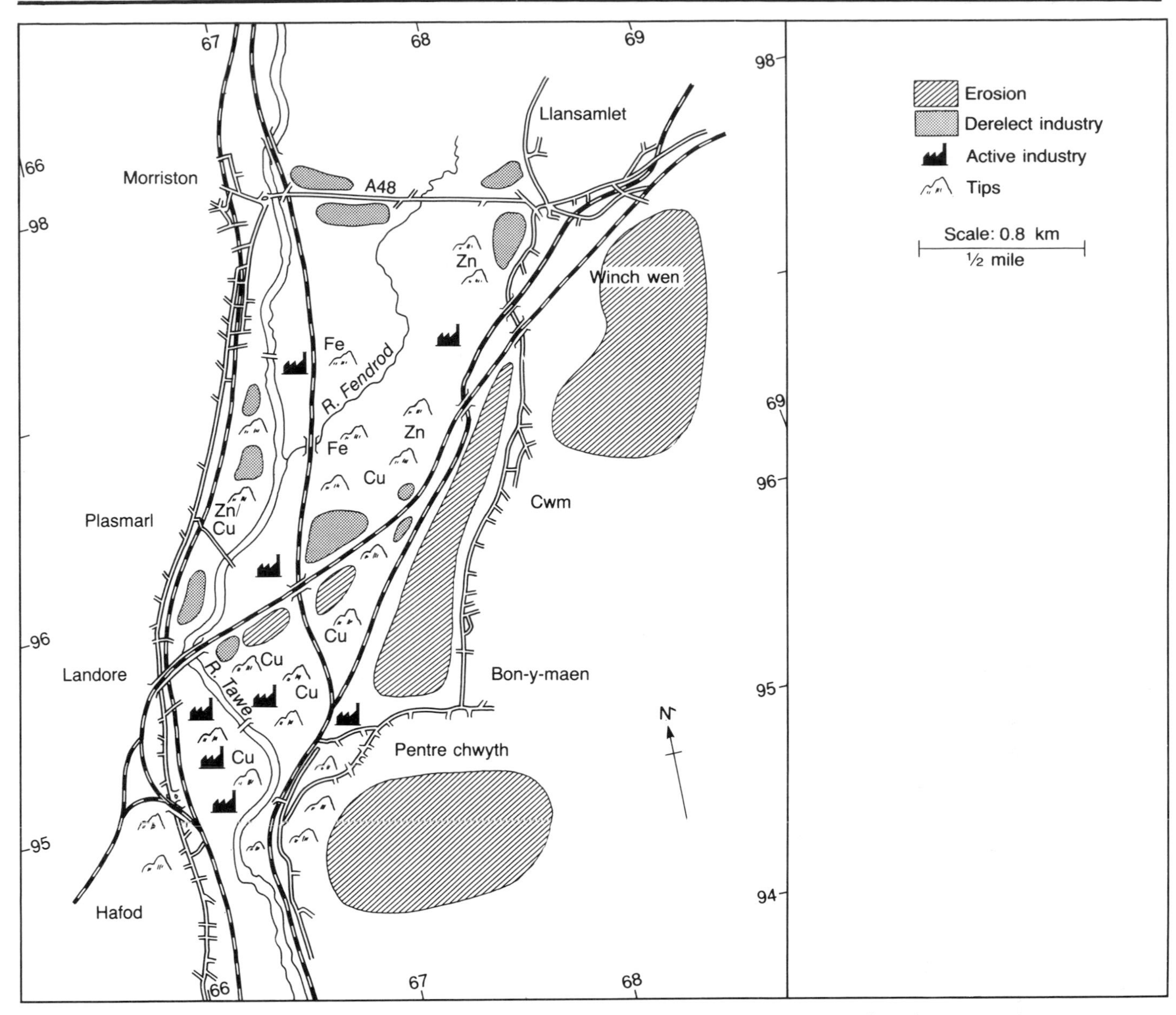

Fig. 4.55 Map of the Lower Swansea Valley in 1960, showing the distribution of tips, dereliction and erosion

Worse, if that is possible, was to follow. The metal works in the Lower Swansea Valley began to close down and by 1918 the majority of them were finished. The closure of the coal mines in South Wales was one influential factor here, but so was the peripheral nature of the industry which was unable to compete with those in the West Midlands which were nearer their markets. Transport costs became prohibitive for both the import of raw materials and the export of finished products. When the factories closed, they quickly fell into disrepair. Vandalism and a certain amount of looting of building materials hastened the process, so that by 1969, John Barr in his book *Derelict Britain* was able to say: 'Nowhere in derelict Britain is there a more dismaying example of man creating wealth while impoverishing his environment than in the Lower Swansea Valley . . . (which) . . . has been often called in the past the most concentrated and uninterrupted area of industrial dereliction in Britain.' The position in 1960 is summarized in the map (Fig. 4.55) which shows extensive spreads of slag heaps, waste tips, derelict industry (including old mine shafts), and transport media as well as areas where the surface vegetation had been killed by poisonous fumes, leading to rampant soil erosion. It seemed a hopeless situation. The land was in the ownership of many individuals and companies and was divided by old railways and canals.

The overwhelming view that nothing could be done was ignored by one person, Robin Huws Jones, the then director of courses in social administration at University College at Swansea. He became convinced that the situation was not hopeless and his conviction was such that his ideas led to the setting up of the Lower Swansea Valley Project which appointed its first director in 1961. The project was based on the University which took the lead in research into both physical and socio-economic aspects of the area. The study area covered 475 ha. A detailed geological map was produced after the physical survey and various experimental efforts were made to vegetate different types of derelict land. The socio-economic survey revealed a distinct lack of open space, schools and good housing in the area. Reports were published in 1967 and thereafter various schemes were implemented to begin the reclamation of the area. One of the most important first steps was to secure the land in the ownership

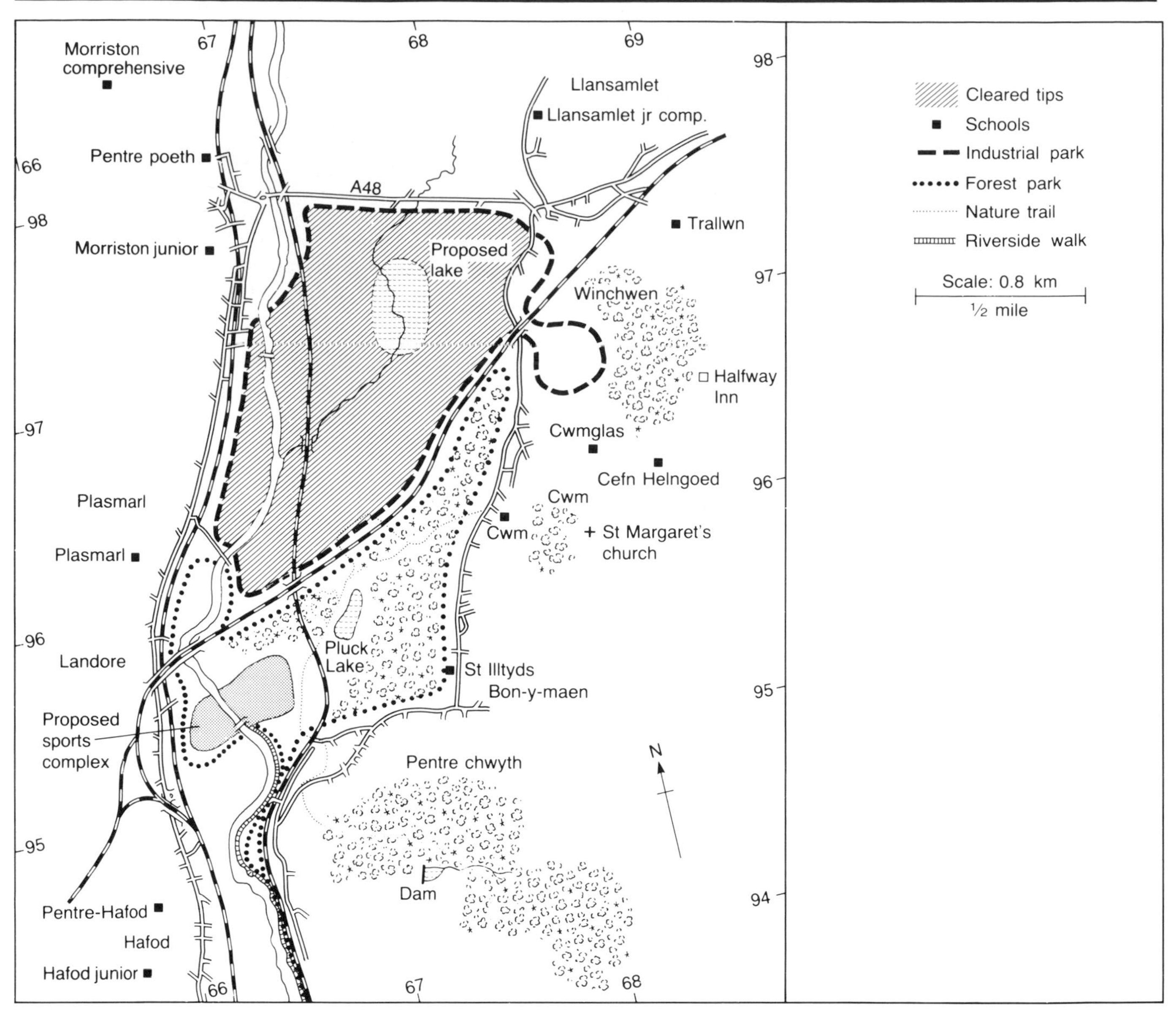

Fig. 4.56 Map of the Lower Swansea Valley in 1980 showing land-use proposals (schools involved with the project are shown)

of one body. The county borough council only owned just over one hectare of the area concerned in 1964, but by purchases, some of them by compulsory orders, it owned nearly all of it by 1981. Grants from the Welsh Office and the Regional Development Fund of the EEC allowed both the purchase and the redevelopment of the land.

The experimental planting of trees on land polluted by different materials or badly eroded had resulted in many of them being broken or uprooted. This led to those running the project reaching an important conclusion. It was decided to involve the community in the reclamation of their area. This especially meant the involvement of school children. Early clearances of derelict industrial buildings were done by locally based army units, which meant that costs were kept to a minimum, but later the removal of slag heaps was found to be more of a problem than had been expected. Explosive charges failed to remove the Pentre-Hafod tip, and it was finally achieved by the continued dropping of a heavy weight from a crane, a long and laborious job. The cost of this one operation was some £400 000. However, within three years of the start of the scheme, a new comprehensive school called Pentre-Hafod was built on the site of the former tip. (Fig. 4.56).

Afforestation was possible on the badly eroded soils, but there were considerable problems with establishing trees on sites polluted with metallic waste. Such sites had been treated with sewage sludge and every effort had been made to improve the soil prior to planting. These efforts failed and only a few species of grass and shrub could be made to establish themselves on the toxic waste. Experiments with different species of tree led to the selection of six types for afforestation schemes. These were Lodgepole and Corsican pines, Japanese larch, Norway spruce, birch and alder. The areas now afforested are shown in Fig. 4.56.

The Lower Swansea Valley Project's recommendations included this important sentence: 'The visual improvement of derelict land is an essential first step towards its eventual redevelopment for industrial, residential and amenity use.' Certainly the project has greatly improved the visual impact of the area and this has certainly started to bring benefits in the three categories mentioned.

An industrial park has been established in the north of the project area and has already attracted a number of industries such as warehouses and wholesalers. The buildings have been very carefully designed to fit in with the new environmental aspirations of the area. The small River Fendrod which runs through the park area has provided problems in the past, since it has tended to flood. To overcome this, a large area has been set aside in the centre of the industrial park to act as a flood reservoir. Parts of the River Fendrod have been culverted. There are even plans to build a barrage on the River Tawe to even out the flow and to keep the level high for most of the time, thus creating a recreational amenity and a visually more pleasing scene than the mudflats of the tidal river. Even before this improvement and before any major change in communications, the area has managed to attract a number of industries. The improvement of the environment seems to have been enough.

Or has it? The Swansea **enterprise zone** had attracted 170 firms by 1986. Whereas 22 of the companies attracted in the first 18 months were manufacturing companies, now 70% of the industries are in the service sector. They include several Do-It-Yourself hypermarkets and furniture warehouses. As well as this, many of the 170 firms are in fact ones which have closed down their operations in other parts of South Wales and simply relocated in the Swansea enterprise zone in order to make use of the grants and incentives offered. As a whole, development areas in Great Britain follow this trend, and it has been said that, in fact, development areas would have to attract ten times the number of businesses attracted so far in order for them to qualify for the description 'successful'.

Residential development in the area covered by the Lower Swansea Valley Project has included the upgrading of the terraced housing fronting on to the enterprise zone. Without looking out on to the waste tips and belching chimneys, the properties have become much more desirable. The urban area has also been provided with a number of new schools, such as the Pentre-Hafod school referred to earlier. The schools associated with the project are shown in Fig. 4.56.

The development of a large sports complex at Landore has provided a focus not just for the local area but for the larger urban area of Swansea. Nature trails have been established through the newly forested areas and around Pluck Lake, created from an ugly scar between two spoil heaps. Other developments, shown in Fig. 4.56, have certainly made the project a great success, and an inspiration to other areas of Britain afflicted with such problems. If the area with the greatest concentration of waste land and derelict buildings can be reclaimed, then the same should be possible for many other areas.

Fig. 4.57 Enterprise Zones in Britain, 1990

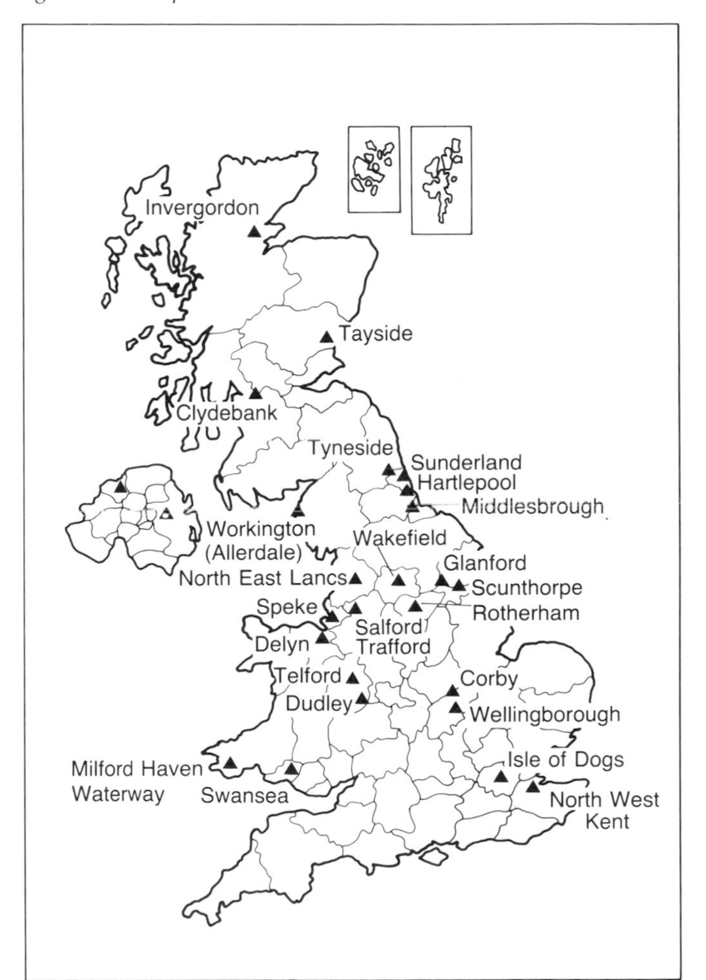

Fig. 4.58 An advertisement from the Observer

31 Using the evidence in Fig. 4.55, describe as vividly as you can a walk from Pentre chwyth to Morriston in 1960, mentioning all the types of dereliction and pollution that you come across on your route.

32 Do you agree with the project that it is important to improve the visual impact of an area before it can start to regain its industrial past? Explain your answer.

33 Design a publicity 'blurb' for attracting new companies to the Lower Swansea Valley Industrial Park after the improvements shown in Fig. 4.56. The area is also now an enterprise zone (Fig. 4.58), which means that government assistance is given to any industries moving to the area. They also get relief from local rates. Other enterprise zones are shown in Fig. 4.57.

Pollution and Weathering

Weathering is the break-up of a material in contact with the atmosphere and in the same place as it was originally (*in situ*). The material may be solid or broken rock, or man-made materials and objects. In the natural environment, weathering results in the production of a **regolith** (a layer of broken particles overlying the original material), and the top part of this regolith may further develop by weathering into what we call a **soil**. The main factors affecting weathering processes are climate, material (e.g. rock) and local factors (e.g. vegetation, depth of soil and speed of erosion).

Objects and materials made by people are also broken up and damaged by weathering, which can therefore be a very expensive process. The weathering of materials can also be hastened through the addition of damaging pollutants to the air, or by the exposure of fresh surfaces by quarrying or building.

Fig. 4.59 The fate of the Leeds Town Hall lion

The rate of weathering in any area is very difficult to measure with any meaning, since it is a function of so many variables which may themselves change over time. For example, an area of granite may begin to decompose under the influence of chemical weathering, but as a soil develops this may hold water which will be in contact with the underlying rock for longer periods. This may hasten chemical decomposition. However, when the soil and regolith are very deep, they may contain a permanent supply of water which may lead to a slowing down of weathering, since some processes require the presence of air to proceed (see Table 4.16).

Table 4.16 shows one possible way of classifying weathering processes. These processes seldom operate independently. For instance, granular disintegration, hydrolysis, the growth of plant roots and solution may well proceed together in the humid tropics and on certain rock types. The net result of weathering processes is either the **disintegration** or the **decomposition** of the rock.

Weathering processes can be hastened by human activity. For instance, quarrying exposes fresh faces of rock for weathering processes to attack. Urban areas contain many 'fresh faces' of concrete, brick, cement and stone; at the same time the urban air is often infused with potentially damaging pollutants that may hasten weathering. Nor is this a minor point. An OECD report in 1976 suggested that in costing the effects of weathering, the following points have to be considered:

(a) the reduced life of materials (including installation costs to replace them)
(b) the decreased productivity or use of materials
(c) the research to find suitable substitutes for the affected materials
(d) losses due to inferior substitutes
(e) the need to protect the affected materials from pollution
(f) the cost of extra maintenance (including cleaning)

Recently, the weathering of the limestone of St Paul's Cathedral has been monitored and the rates of break-down and removal have been seen to vary greatly due to extent of exposure of surface, amount of direct rain impact, amount of standing water and many other factors. Since it is known how long parts of the building have been standing, it is possible to work out average rates of weathering and these vary between 0.0095 mm/yr and 0.391 mm/yr. The result of this is the disfiguring and rounding of surfaces. A similar fate has befallen the limestone lion outside Leeds Town Hall (Fig. 4.59).

Estimates of the cost of increased weathering due to pollution are difficult to find, and must in any case be subject to much guesswork, but they would probably run into many millions of pounds. The decrease in property value as a result of air pollution in the USA was said to be $5000 million per year in 1973. Experts believe that increased weathering through pollution in the last twenty years has done more damage to European art treasures than weathering did in the previous 2000. Since many of

Table 4.16 A classification of weathering

Group	Process	Effects of process	Examples
Physical	Exfoliation	Sheets of rock *peel* off the surface due to repeated heating and cooling (the latter perhaps by dew)	Arid areas in the tropics, e.g. around Ayers Rock, N. Territory, Australia
	Granular disintegration	Rock *disintegrates* to particles the size of individual mineral constituents due to pressures set up by different rates of expansion and contraction of minerals. (There is some doubt about the importance of this process)	Acid and semi-arid tropical area, especially of granite rocks, e.g. northern Nigeria, near Zaria
	Freeze-thaw	Rocks *shatter* as water freezes (and expands) in joints. Pressures may be very great	High latitudes or altitudes, where temperatures fluctuate about 0°C and where there is sufficient moisture, e.g. N. Scotland
	Mechanical collapse	Rocks *break apart* due to undercutting and subsequent collapse	Overhanging cliffs or river banks, e.g. chalk cliffs in S. England
	Crystal growth	Rocks *shatter* because of internal pressures	Growth of ice crystals within porous rocks, and of salt crystals in weathered rock (especially in tropical semi-arid areas)
Chemical	Hydrolysis	Layers *peel* away from spheroidally-weathered rock since some products are bulkier than the parent material. Process due to ion exchange between water and minerals	The most important chemical process, most active in humid tropical areas and on basic volcanic rocks, e.g. Grenada, W. Indies
	Oxidation/reduction	*Change* in chemical composition with the addition/removal of oxygen (often in the presence of water)	Mottled colouring in clay river banks or cliffs
	Solution	Some minerals or weathering products are *dissolved*	Caves in rock salt areas, e.g. Bavaria
	Hydration/dehydration	*Chemical change* and often *volume change* with the adding/subtraction of water	Swelling of shales e.g. Ravenscar nr. Whitby; shrinking of clays, e.g. British drought, 1976
Biotic	Carbonation	*Chemical change* of carbonates to soluble products	Karstic scenery, e.g. Eastern Transvaal
	Growth of roots	Joints in rock are *widened*	Trees growing in shallow soil
	Animal burrowing	Aeration of soil *aids* other processes New surfaces *exposed*	Prairie Dog colonies in N. America
	Enhanced solution	*Solution* aided by the presence of respired Carbon Dioxide in the soil	Areas and seasons of active plant growth
	Solution by humic acids	*Solution* by acids from decaying plants	Deep regolith beneath peat bogs, e.g. New Forest

these treasures are made of marble, the belief is not difficult to support. An American economist reported in 1973 that visitors to the Lincoln Memorial in Washington claimed that the whole building 'fizzes like a giant Alka Seltzer' in the sulphuric rains! Cleopatra's Needle, in Central Park, New York, has lost the hieroglyphics on two of its four granite sides. It was given to the city in 1882, with hieroglyphics on all four. A similar obelisk, erected on the Embankment in London in 1878, was restored in 1979 by applying a layer of detergent paste to remove the thick layer of grime without damaging the granite or inscriptions beneath. This restoration alone cost nearly £5000.

Several of the pollutants shown in Table 4.17 are the cause of what is now called **acid rain**. As well as the effects of this process on weathering, it is also having a devastating effect or water life and vegetation, especially in those areas downwind from concentrations of industry.

Table 4.17 Some important pollutants and their effects on weathering processes

Pollutant	**Non-natural source**	**Materials affected by subsequent weathering**	**Extent to which materials resist weathering**	**Weathering effect of pollutant**
Particulates (any airborne material, solid or liquid, larger than a single molecule)	Combustion (especially coal and urban incineration plants), stone crushing, agriculture, iron and steel production etc.	Metals	Alloy steel P Aluminium G	Abrasion, chemical reaction
		Building materials	Stone F Brick F	Soiling
			Cement and concrete G	Abrasion
		Paint	P	
Sulphur Oxides	Combustion of fossil fuels – more from coal than oil	Metals	Alloy steel P Aluminium P	Corrosion
		Building materials	Stone P Brick G Cement and concrete P	Chemical reaction
Carbon Dioxide	Combustion of fossil fuels	Building stone	Sandstones G Limestones P	Formation of soluble carbonates
Nitrogen Oxides	Combustion of fossil fuels, especially by cars, power stations and central heating units	Building materials	Stone E Brick E Cement and concrete E	Chemical reaction
Hydrogen Sulphide	Decay of organic matter (rapidly converted to sulphur oxides)	Lead-based paints Metals	Alloy steel P Aluminium P	Darkening Tarnishing
Hydrofluoric Acid	The manufacturing process in brickworks and aluminium	Glass Metals	G Alloy steel P Aluminium G	Etching
Ozone	Action of sunlight on hydrocarbons and nitrogen dioxide	Rubber	G	Embrittlement
		Dyes	G	Fading, discolouration

Key: E = Excellent
G = Good
F = Fair
P = Poor

(Source: Various articles in A. C. Stern (ed.), *Air Pollution*, Vol. III Academic Press Inc., 1976)

34 Table 4.16 is only one way of classifying weathering processes. Using the headings **Processes leading to Disintegration** and **Processes leading to Decomposition**, make two lists of the processes shown in Table 4.16. Which classification do you consider the more useful, and why?

35 With reference to Table 4.17, comment on the effect of the following on the weathering of buildings and structures (such as bridges) in an urban area:

(a) the burning of coal in open fires in a densely populated area
(b) the incineration of rubbish outside the urban area
(c) heavy pollution from car exhausts
(d) pollution from a nearby brickworks.

36 What effects would you expect the following to have on weathering in the urban environment?

(a) the establishment of smokeless zones
(b) an increase in road traffic
(c) greater recent use of concrete, cement and steel in construction.

37 **Future trends**. In view of the following figures on atmospheric pollution in the UK would you expect the future weathering rates in cities to be higher or lower than they are now? Explain your answer.

Atmospheric pollution in the UK

	1900	1930	1960	1970	1980	1982
SO_2	2.8	3.2	5.6	6.0	4.67	4.0
NO_2	0.68	0.72	1.35	1.64	1.79	1.67

(in millions of tonnes per annum)

Sources of pollution
Power stations: SO_2 65%; NO_2 46%
Vehicles: NO_2 29%

Case-study: one consequence of industrial activity

Acidification of rainfall is one possible consequence of industrialization. The processes by which rainfall becomes acidified are not as yet very well understood, leading to claims that, on the one hand, acid rain is not the prime result of industry and, on the other, that industry must clean up its act before nature is killed off completely. Given that acid rain exists, and nobody argues against this, what effects does it have? The following are two extracts, one taken from the publication *Acid Rain: The Politics of Pollution*, published by the Acid Rain Information Group (an amalgamation of conservation bodies), and the other from *Acid Rain*, published by the Central Electricity Generating Board.

EXTRACTS FROM 'ACID RAIN: THE POLITICS OF POLLUTION' (BY CONSERVATIONISTS)

Definition and Causes

•THE SCIENCE OF ACIDITY•

To understand how acidification affects the environment, a brief chemistry lesson is required. Scientists have chosen a scale known as the pH scale (pronounced pea-haich) to indicate whether a liquid or soil is acidic or alkaline. It runs from pH14 – very alkaline to pH1 – very acidic, with pH7 being neutral.

With each drop of one-unit acidity increases ten-fold, so pH5 is ten times more acid than pH6, pH4 100 times more acid than pH6.

Rain is naturally slightly acidic with a pH of around 5.6. But scientists are now worried at the extent of rainfall consistently more acidic than this falling on rocky areas with a poor topsoil.

One side-effect of increased acidity is that at different pH readings different toxic metals are leached from the soil and river-beds. Aluminium solubility especially increases as the pH of the river or lake falls. It is more soluble at pH5 when aluminium hydroxide settles out in the gills of the fish killing them.

So acid rain acts in direct and indirect ways.

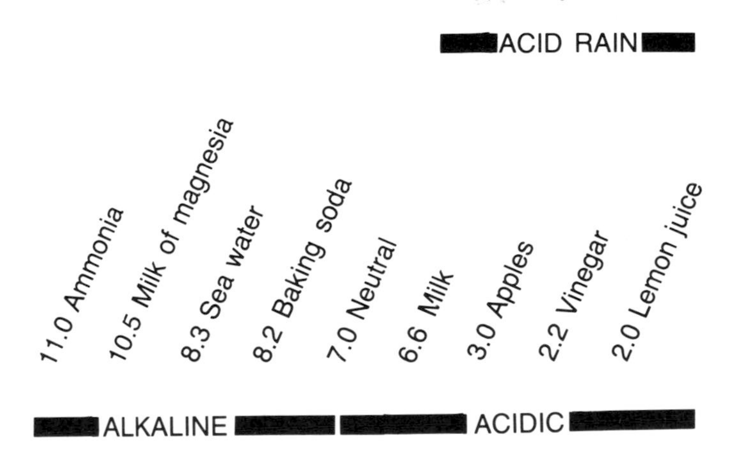

Acid rain is now a major environmental problem causing international concern. Britain, which discharges more of the pollutants causing **acid rain** than any other Western European country, clearly has a lot to answer for. These pollutants not only affect the towns and countryside of Britain – they are causing **acidification** in other countries miles away. The hidden effects of corrosion and loss of life in lakes and rivers is costing millions.

THAT'S THE BAD NEWS. The good news is that it **is** possible to cut down the pollution discharges. This is best done at source – reducing emission from oil refineries, power stations and car exhausts. This too would cost something; but it seems on balance a wise move. It is an essential first step in protecting the countryside and cities from the effects of a silent deadly killer: **ACID RAIN**.

Effects

• Deadly pollutants •

Sulphur dioxide (SO_2) causes **direct damage** when it falls to the earth as **dry deposition**. This includes corrosion of building materials such as metals and stonework, effects on tree and crop growth, and effects on human health – especially when SO_2 levels are high and smoke is present.

Sulphur dioxide and nitrogen oxides cause **indirect effects** when they have reacted with the atmosphere and come down with rain, snow, mist or even fog as **wet deposition**.

This causes a slow acidification of the ground, changing the soil chemistry and interfering with the ability of plants and trees to absorb the nutrients they need. It also releases toxic heavy metals held in compounds in the soil and water. These are poisonous for both plants and fish.

• Forests Dying •

The real possibility of forest loss on a huge scale is causing great concern in West Germany, Austria and Eastern European countries. Here forestry covering an area equivalent to half the size of Belgium is dead or dying. The role of acid rain in tree death is still debated by experts but it seems likely that several factors, including ozone, industrial chemicals and climate all contribute.

It is estimated that every year the equivalent of a large sack of sulphur rains down on a typical acre of West German forest. (30–60 kg per hectare plus 14–30 kg of nitrogen). The effect of dead and dying trees is causing such alarm that new laws have been passed which propose greatly reduced levels of sulphur dioxide emissions.

There are reasonable grounds for concern about trees in upland areas of the UK with thin soils. The effects of acidification on such forestry is being studied but if action to reduce acid rain is delayed it may be that the damage to this enormous national resource will already have been done.

• Crops affected •

Most arable farms are on rich soils capable of neutralising acid rain. Its effect is also slightly masked by use of fertilisers. But pollution damage is well-known on farms near cities, with examples from Canada to Czechoslovakia. Swedish farmers have to apply 50,000 tonnes of lime every year to combat acidification and in Britain agricultural scientists reckon that Scottish farmers and growers lose £25 million a year through the effect on crop yields.

Laboratory tests indicate that sulphur dioxide can reduce crop growth rate. Similar losses on farms throughout Europe might be as high as £100 million. While such calculations must be regarded with caution it is clearly advisable to minimise such direct damage.

• Buildings eaten away •

The sulphur dioxide pollution accelerates corrosion of metals and building materials such as stone, brick, tiles and roofing felts. British towns and cities mainly suffer from **direct** damage caused by local emissions, rather than acidification caused by distant sources.

So, although the Clean Air Act did effectively stop the London smogs and cut out smoke and soot fall-out, the erosion of especially limestone and sandstone buildings continues.

More disturbing is the fact that a priceless monument like the Acropolis in Athens has suffered more severe deterioration in the last twenty years than in the previous 2,000! How much do we value our urban heritage of famous buildings and monuments? Do we care that our children may never see the precious buildings and works of art that we have taken too much for granted? Because the cost of dealing with the effects of this corrosion is enormous:

- Dutch figures for repairs to old buildings put the annual cost at 30m Guilders (£6.6 million).
- Annual repairs to Cologne Cathedral alone cost 6m Deutschmark (£1.5 million).
- Buildings in Cracow, Poland, now need restoration every five years.
- The US Environmental Protection Agency estimates that acid rain causes $2,000 million worth of structural damage every year in the USA alone.

Acid rain doesn't only attack beautiful buildings and statues. It dramatically increases the **maintenance costs** of machinery, fences, transmission towers and shortens the life of parts of buildings like guttering, roofing and so on. Taken all together these enormous widespread costs could swing the economic balance completely in favour of stricter emission controls.

Controls?

• Technical solutions •

Once the problems caused by increasing acidification have been identified the hunt is on for ways to deal with the causes of acid rain. Up to now the main focus has been on **dispersing** sulphur dioxide discharges. The technique used has been:–

- **Dilution**. The so-called 'tall stack' method simply raises pollutants high into the air and spreads them over a wide area. This reduces pollution in the breathing zone near to the source of emission; but it does not affect the total quantity of discharges or prevent long range transport of pollutants. **It is really no solution to the problem**.

One method of **reducing** discharges now adopted in both Sweden and Japan is to pass regulations banning use of fuels with a high sulphur content. There are two ways to approach this:–

- **Low sulphur fuels**. Fossil fuels contain varying amounts of sulphur. North Sea oil, for example has a low sulphur content of around 1.5% compared to the 3.5% content of heavy fuel oils burned in some UK power stations.

Brown coal – lignite – is typically high in sulphur. This is widely used in Eastern European countries where it is most easily available.

The world's supplies of naturally low sulphur fuels are limited so their use is only a temporary solution.

■ **Fuel desulphurization**. Coal can be crushed and washed to halve the sulphur content. It can also be chemically treated to reduce sulphur levels. The sulphur content of oil can be reduced by processes at the oil refinery.

These approaches certainly reduce the amount of sulphur dioxide produced in the furnace. To **control the emission** from the combustion of the fuel requires new machinery or adaptions at the boiler-house:–

■ **Desulphurization of flue gases.** There are a number of systems which reduce sulphur dioxide content of the gases going up the chimney. They are generally called 'scrubbing' and involve water or dry powder spraying. The residues are pretty nasty but can be recycled for other use.

■ **Fluidized bed combustion**. This is a new boiler development which burns fuel more efficiently and at a lower temperature. The latter means a reduction in nitrogen oxides produced. Many small boilers of this design are now in operation and bigger units for large industrial sites and power stations are being developed.

Their main advantage is the ability to add into the furnace small amounts of crushed limestone. This reacts with the sulphur dioxide which then stays in the ash. This reduces gaseous emissions of sulphur dioxide by up to 90%.

• Liming lakes •

It is estimated that 18,000 of Sweden's 85,000 lakes and lakes in a 33,000km² area of S. Norway (equivalent to the area of Holland) are all now affected by acidification. This awesome fact has caused Scandinavian governments to try some **first aid measures** to counter the effect of acid rain.

One of these is to spread **lime** on the lakes and rivers. This temporarily reduces the acidity and restores some of the fish and other life; but it doesn't really solve the problem and Norwegian researchers feel that **'liming is like taking aspirin to cure cancer'.**

• Cleaning up •

Acid rain is affecting building owners, foresters, farmers and the fishing industry. It is costing them a fortune in increased maintenance and loss of produce. It is also costing in terms of irreversible damage to unique landscapes and ancient buildings.

Measures to reduce acid rain will also cost money. An international survey prepared in 1981 by the OECD estimated that halving sulphur dioxide emissions in Europe would increase energy prices around 5%. But even during the preparation of this booklet a new process for 'spray drying' removal of sulphur dioxide **and** nitrogen oxides has come to light.

The Danish company who have developed the technique quote clean-up costs of half the previous estimates. So, while the Central Electricity Generating Board threaten electricity price rises of 10–15% if emission regulations were introduced, there are technical developments in hand which would improve the economics.

EXTRACTS FROM 'ACID RAIN' BY THE C.E.G.B.

Definition & Causes

The fuels we burn are an inheritance from living organisms which existed millions of years ago. Natural processes have converted them into the 'fossil' fuels: oil, natural gas and coal. These contain the hydrogen, carbon, sulphur, nitrogen and other elements that life depends on. They were vital parts of the original living organisms.

1 Atmosphere receives oxides of carbon, sulphur and nitrogen, hydrocarbons etc. from natural and man-made sources on the ground. Some of the emissions come down and are deposited dry.

2 Sunlight stimulates formation of photo-oxidants. These slowly convert sulphur and nitrogen oxides into sulphuric and nitric acid.

3 Sulphur and nitrogen oxides, photo-oxidants, and other gases including ammonia, dissolve in cloud droplets. The products are acids, sulphates and nitrates.

4 Acid rain containing dissolved sulphates and nitrates.

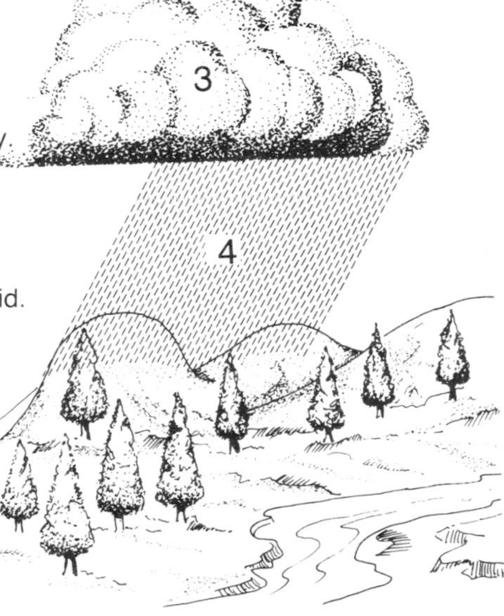

When fuel is burnt the oxygen in the air joins the other elements to form their oxides. Emissions from chimneys and exhaust pipes therefore include these oxides.

One of them is water (which is hydrogen oxide).

The others – oxides of carbon, sulphur and nitrogen – are gases that can travel dry and can partly dissolve in water to form weakly acidic rain.

The higher acidity that is actually found in rain is a result of processes in the atmosphere.

Most of the public anxiety that has been expressed is in relation to the oxides of sulphur. But the oxides of nitrogen are no less important.

Effects

Dry deposition of sulphur dioxide gas is blamed for damage to metals, stonework, trees, crops and human health. Wet deposition is said to sour the soil, interfere with plant growth, and poison vegetation and fish by releasing toxic metals from their compounds in soil and water.

Nobody suggests that acid rain has any direct effect on human health. Very high concentrations of sulphur dioxide in the air do damage health but a review by the World Health Organization concluded that, the concentrations currently found in most of Europe are not harmful.

The UN Economic Commission for Europe has assessed damage to buildings, monuments and materials by atmospheric pollution and has concluded that the main attack in urban areas has come from high local concentrations of pollutants. These arise mainly from a very large number of small individual sources. Large industries make very little contribution.

Out of town, in rural areas, there is the possibility of harm to crops if concentrations are sufficiently high. Attempts have been made to estimate what and how great the damage might be. A study for the European Commission by Environmental Resources Ltd reflects a scientifically held view that current rural concentrations of sulphur dioxide are probably too small to be harmful. Compared with the use of fertilizers and other agrochemicals, sulphur dioxide hardly affects the acidity of soil. But, there is evidence that fairly high concentrations of ozone – a pollutant as well as an oxidant – do visibly injure certain types of crop.

In the UK there are some streams, lakes and other surface waters that are more acid than the rain that falls on them. Acid surface waters seem to occur either where there is heavy rainfall and little acid-neutralizing matter in the soil and bedrock, or where the catchment embraces forest, moorland and peat in its make-up. These conditions usually occur away from modern industrial centres. There are various possible causes of the acidity including geological conditions, drainage, afforestation, vegetation, and farming practices.

Controls?

The EEC seeks reductions of 60% in sulphur dioxide emissions and 40% in nitrogen oxide emissions from large plants by 1995, compared with 1980 levels.

The only way in which the UK electricity industry could comply with the timescale proposed by the EEC is by modifying about 12 existing large coal-fired power stations with flue gas desulphurization equipment (FGD). This equipment would be expensive to buy – up to £160 million per installation and also to operate – up to £35 million a year for each station.

A power station equipped to desulphurize its flue gas would have to be enlarged, as this drawing shows.

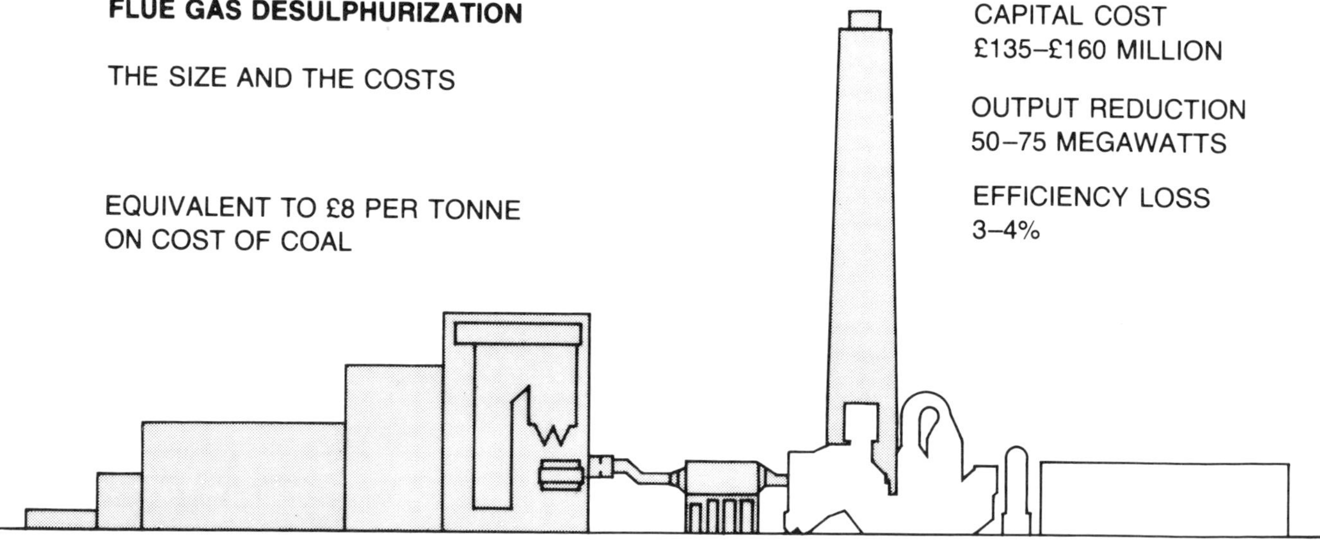

The building of three desulphurization units was announced by the British Prime Minister on her visit to Norway in September 1986.

There would be a great deal of sulphur to dispose of without hurting the environment. Electricity would cost up to 15% more to generate at a modified station. Spread over the generating industry's total costs the effect would be equivalent to raising the price of electricity to the consumer by about five per cent.

FGD equipment would not reduce nitrogen oxide emissions. Every boiler would have to be modified to do that. The substantial reduction in emissions already achieved in the UK will continue as the nuclear contribution to electricity supply increases. Current research into coal preparation and burning technology should also make a significant impact on emissions from future coal-fired plant. Thus the emission levels proposed by the EEC could probably be achieved anyway, given a few more years to do it.

The booklet Acid Rain, *was published by the Electricity Council, 30 Millbank, London, SW1P 4RD.*

The Electricity industry was privatised in the UK between 1990 and 1991. Bodies such as National Power and Power-Gen also issue literature on topics such as this.

38 Make a brief summary of what each organization thinks causes acid rain and what can be done about it.

39 Which argument do you find most convincing and why?

Decision-making Exercise M: An International Environment Conference

An International Environment Conference has been called to discuss the problem of acid rain. Delegates have come from the developing as well as the developed worlds. The meeting is taking place in Copenhagen.

1. (**EITHER**) If you are working by yourself, you should write a report of the conference, outlining the positions taken by people from the countries listed below. Finally, write the text of the resolution that was agreed at the end of the meeting.

(**OR**) If you are working in a group, elect a chairperson to run the meeting and then decide who should play the parts of the delegates from each of the following countries. Where possible, sources of information have been indicated.

The United Kingdom: various information available from Friends of the Earth, 26–28 Underwood Street NI and from the various power generation companies in the UK.

The United States: Acid Rain Foundation Inc., 1630 Blackhawk Hills, St Paul, Minneapolis 55122.

Canada: Information Directorate, Environment Canada, Ottawa.

Norway: The Norwegian Forestry Society, Wergelandsvei 23B, 0167, Oslo 1.

Sweden: National Swedish Environmental Protection Board, Box 1302, S–17125 Solna, or Swedish Forestry Association, Box 273, S–182 52, Djursholm.

The following will supply general information on industry, but not specifically on acid rain:

Japan: Japanese Embassy (Information Section), 9 Grosvenor Street, London, W1.

Australia: Australian High Commission, Australia House, Strand, London, WC2.

New Zealand: New Zealand High Commission, New Zealand House, Haymarket, London, SW1.

Brazil: Brazilian Embassy, 32 Green Street, London, W1.

Other addresses can be obtained from the relevant pages of the London Telephone Directories at your local library.

2. The meeting should be chaired by the delegate from Denmark, who should decide on the order in which prepared statements should be received from each of the delegates. After the initial speeches, delegates should have a 'coffee break' during which they should question each other on their positions, form alliances and decide on the resolution they wish to present to the meeting.

3. One delegate should be allowed by the chairperson to put a resolution to the re-convened meeting. This resolution will need a seconder. Anyone is now free to propose an amendment to the resolution, as long as it is seconded. After all the amendments have been tabled, they should be voted on in reverse order. Any amendment which achieves a two thirds majority becomes the resolution that is adopted by the meeting.

4. Each delegate should now write a summary of events for the government concerned and comment on the resolution passed, if any.

Fig. 4.60 The Ferrybridge power station

Methods of dealing with qualitative information; questionnaires and bi-polar analysis

Questionnaires

It is quite possible to get accurate quantitative results from asking questions. These can be dealt with effectively by working out percentages, illustrating responses with histograms or pie charts or by employing statistical techniques. The qualitative results are not as straightforward to deal with.

If you design a questionnaire that requires some comments, you may need to generalize on those comments in order to make sense of the responses, particularly if they are large in number. Professional market researchers use a device known as coding to deal with this. It is best illustrated with an example.

Let us suppose that you are researching the reaction of people to a planned new development. You carry out a questionnaire among one hundred people in which one of the questions is, 'What are your reactions to the planned use of King's Meadow for a new A and B Superstore?' You now want to be able to say what the majority feeling is. Start looking through the answers you have received and make a note of the different responses thus:

1. Couldn't care less
2. It will spoil the beautiful countryside
3. It would be better on the industrial estate.

As you read on, you will find other answers which are very similar to the answers you already have. Thus:

'It won't affect me at all', and
'I don't know where King's Meadow is'

would in effect be the same as 1. Carry on making your list of very different responses as you go through the questionnaires and allocate them a different number. Sometimes, a reply will need to be coded with more than one number. Thus, 'It will spoil the view from James's Hill and I don't see why it is not built on the industrial estate', would be coded 2 and 3. Mark on the questionnaires themselves the numbers you have given to them on your coding list so that you can check back if necessary. You now have a way of quantifying the results. You may find in some cases that you get a few responses that stand alone. In this case, have a coding for 'others'.

Bi-polar analysis

If you give two extreme statements that you could apply to something, these are poles apart, or bi-polar views. They can be written at either side of a grid, inviting someone to place a tick somewhere between them to judge that phenomenon. An example might be:

The landscape is. . . attractive 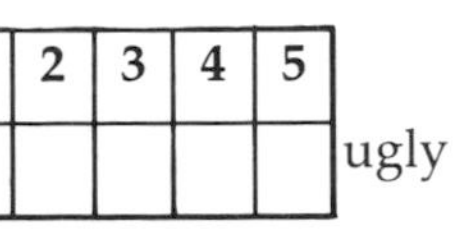 ugly

Values can be attributed to the boxes, as shown above. If there were ten responses to the survey, the maximum it could score would be 50. To ascertain its actual score, sum the numbers it received, which might be 42. Since a neutral reaction to the question would give a total of 30, this represents a tendency for people to think of the landscape as ugly. People have very different reactions to things, however, and it might be sensible to make a note of the variation of responses you have received. In this case, it might have been 2 to 5.

If you are assessing the reaction of groups of people to the same thing, you could use the following technique. Make up your grid, ask your first group for their reactions to the points and work out the average score for each item. Points have been located with crosses on Fig. T4.1 to show the average response of teenagers to a town. The same exercise can then be done with adults in the same town and their reaction shown (circles in Fig. T4.1). The lines which have been drawn to join the points plotted for both teenagers and adults show **perception profiles** for the two age groups.

Another type of bi-polar analysis has been developed using a **weighting factor**. This has been devised in conjunction with research in psychology and perception. It is useful when many people are being asked for their reactions to the same view. The procedure is as follows:

(i) Individuals must look in the same direction and mark their responses on a grid (Fig. T4.2) to show their perception. This can be most easily arranged if photographs have been taken of the required view. When completing this exercise, you must use the exact layout shown in Fig. T4.2, as the weighting factors have been devised for this format.
(ii) When a suitable sample has been taken, the number of responses in each category must be totalled and an average (mean) value calculated. This

Fig. T4.1 Views on Reading

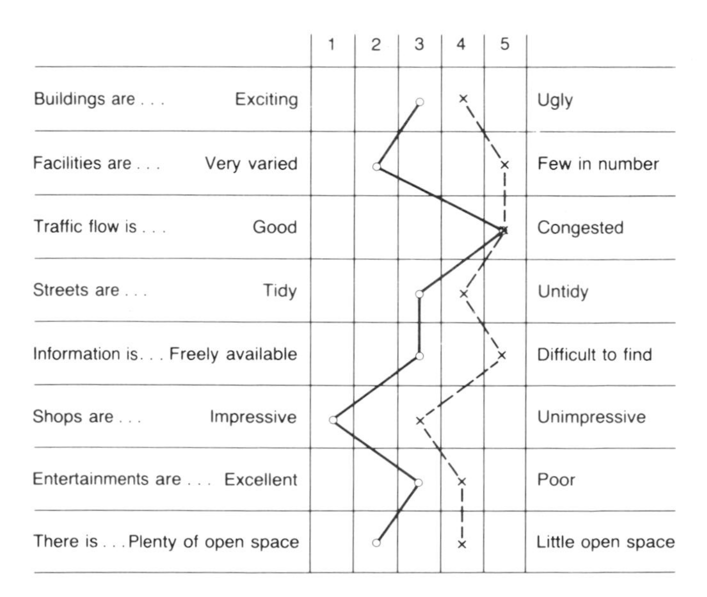

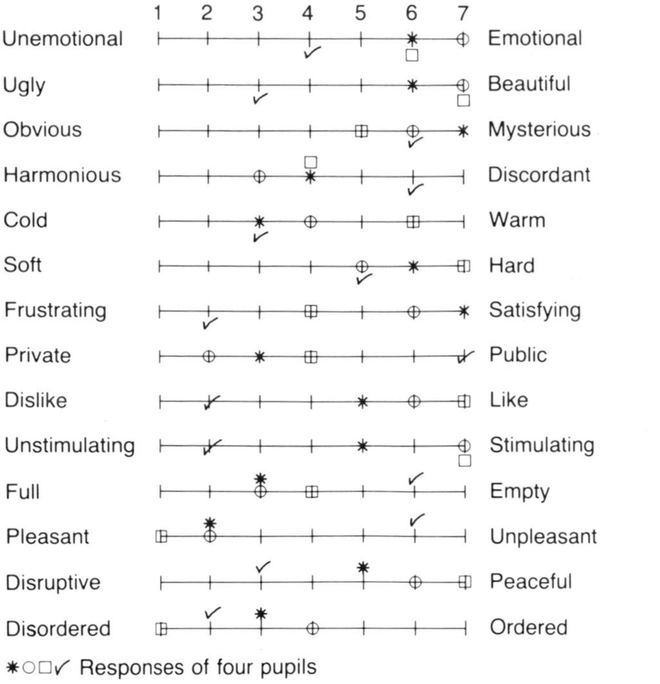

	1	2	3	4	5	6	7	
Unemotional								Emotional
Ugly								Beautiful
Obvious								Mysterious
Harmonious								Discordant
Cold								Warm
Soft								Hard
Frustrating								Satisfying
Private								Public
Dislike								Like
Unstimulating								Stimulating
Full								Empty
Pleasant								Unpleasant
Disruptive								Peaceful
Disordered								Ordered

✱○□✓ Responses of four pupils

Fig. T4.2 Responses in bi-polar analysis using a weighting factor

is calculated by multiplying the total in each box by the scale value and dividing by the total number of responses. The example shown in Fig. T4.2 would be:

$$\frac{(1\times4) + (2\times6) + (1\times7)}{4} = 5.75$$

(iii) The average value is then substituted into the appropriate place in the weighting factor column and the score calculated.
(iv) The scores are then added together, divided by 343 and multiplied by 100 to get a percentage. The value of 343 has, along with the weighting factors, been calculated by psychologists as being the score for a totally 'natural' landscape.

Try this technique by working on two different pictures in this book, one showing an industrial landscape and one showing a more natural region. Calculate the scores from a number of people, for instance a class group, and see if they give you the sort of result you expect, with higher values showing greater natural dominance.

Bi-polar scale valves	1	2	3	4	5	6	7	Mean	Weighting factor	Score
Unemotional/emotional				I		II	I	5.75	(8–5.75)	2.25
Ugly/beautiful			I			I	II	5.75	5 × 5.75	28.75
Obvious/mysterious					I	II	I	6	3 × 6	18
Harmonious/discordant			I	II		I		4.25	5 (8 − 4.25)	18.75
Cold/warm			II	I		I		4	4 × 4	16
Soft/hard					II	I	I	5.75	3(8 − 5.75)	6.75
Frustrating/satisfying		I		I		I	I	4.75	5 × 4.75	23.75
Private/public		I	I	I			I	4	2 (8 − 4)	8
Dislike/like		I			I	I	I	5	5 × 5	25
Unstimulating/ stimulating		I			I		II	5.25	3 × 5.25	15.75
Full/empty			II	I		I		4	8 − 4	4
Pleasant/unpleasant	I	II				I		2.75	5 (8 − 2.75)	26.25
Disruptive/peaceful			I		I	I	I	5.25	4 × 5.25	21
Disordered/ordered	I	I	I	I				2.5	3 × 2.5	7.5

(Sum of scores ÷ 343) × 100 = 64.7% Σ score 221.75

The final value does not really tell you much by itself, but is useful in assessing changes before and after a development or in assessing landscape differences from place to place. You can extend the exercise by comparing the values achieved by different groups of people to see if perception changes with age group or sex. The variations can be endless. These techniques are used as part of **environmental impact assessments**.

Lorenz curves

The **Lorenz curve** is a graph which shows how well distributed something is. The line of perfect distribution runs diagonally across the graph and the greater the distance between the Lorenz curve and this line, the more uneven is the distribution of the thing you are plotting.

Fig. T4.3 Lorenz curves

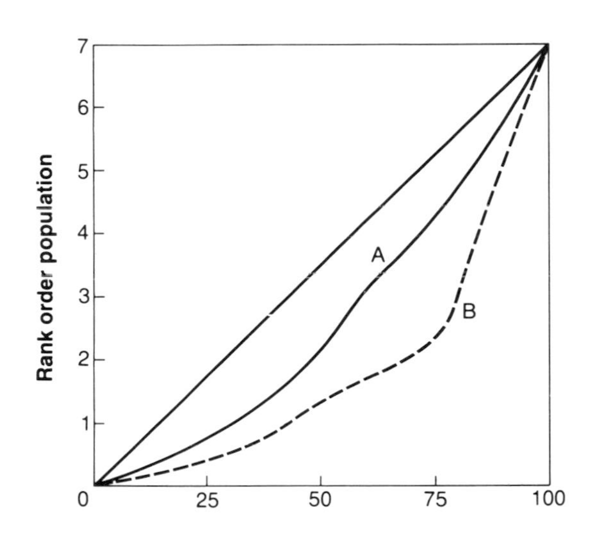

The phenomenon plotted by line A is more evenly distributed than that shown by line B.

How to plot the curve

In order to show how the curve can be drawn, the example of the number of people working in primary industry in five towns in one area will be used.

(i) List the towns concerned in Column 1 in descending order of size.
(ii) This gives the rank of the town. Rank 1 is the largest town (Col. 2).
(iii) List the number of people employed in primary industry in these towns (Col. 3).
(iv) Work out the percentage of people employed in primary industry by the formula:

$$\frac{\text{Number of people employed in primary industry in town A}}{\text{Total number of people employed in primary industry}} \times 100$$

Write the percentages in Column 4.
(v) In Column 5, write down the percentages of people employed in primary industry in A. Then add the percentage of people employed in primary industry in B to that of A and write it down in the row for B. Add the percentage of people working in primary industry in C to this total and so on.
(vi) Using a square graph, choose suitable scales for the x and y axes. Now plot the rank of the town against the cumulative percentage of people employed in primary industry. The shape of the curve tells you how evenly the distribution of people employed in primary industry is.

In a nearby area of similar size there are five more large towns. Table T4.2 shows their population statistics and their employment figures for primary industry. Is the distribution more or less even than in the area shown in Fig. T4.2? Can you offer an explanation for your answer?

Table T4.1 Statistics for drawing a Lorenz curve

Column 1	Column 2	Column 3	Column 4	Column 5
Town (Population)	Rank (size)	People employed in primary industry	Percentage	Cumulative percentage
A – 8017	1	660	34.73	34.73
B – 8000	2	570	30.00	64.73
C – 6321	3	430	22.63	87.36
D – 3571	4	215	11.32	98.68
E – 3023	5	25	1.32	100.00
Total		1900	100.00	

Fig. T4.4 Lorenz curve to show the distribution of employment in primary industry

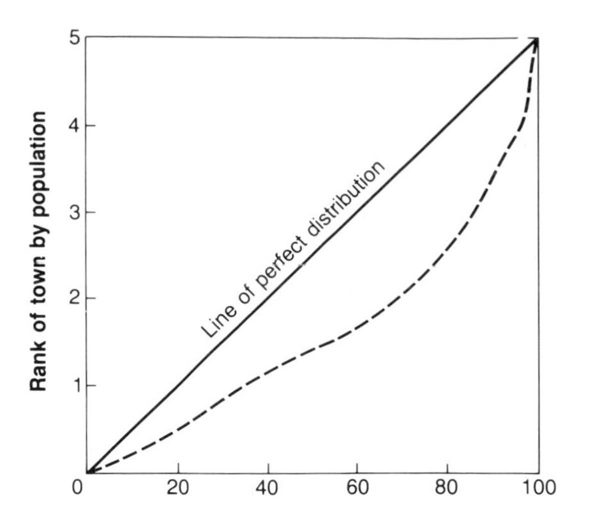

Table T4.2 Statistics for towns in another area

Column 1	Column 2	Column 3	Column 4	Column 5
Town (population)	Rank (size)	People employed in primary industry	Percentage	Cumulative percentage
10 300	1	3800		
10 000	2	3600		
9840	3	2800		
9735	4	2300		
9120	5	2000 14 500		

Choropleth maps

A **choropleth map** shows the distribution of a phenomenon by the use of colour or the density of shading. Such a map can reveal a pattern of distribution at a glance if it is constructed carefully.

It is important that the classes of values to be shown on the map should be meaningful. For this reason, a **dispersion diagram** of the values to be plotted should first be made. An example of a dispersion diagram is shown in Fig. T4.5.

There are twenty-seven values plotted on the diagram. The middle or median value is therefore the fourteenth point reading from either extreme of the diagram. In this case, it has a value of 32.

The **upper quartile** is the point midway between the median value and the maximum. The **lower quartile** is the point midway between the median value and the minimum. The **inter-quartile range** is therefore the central 50% of the readings.

Fig. T4.5 A dispersion diagram

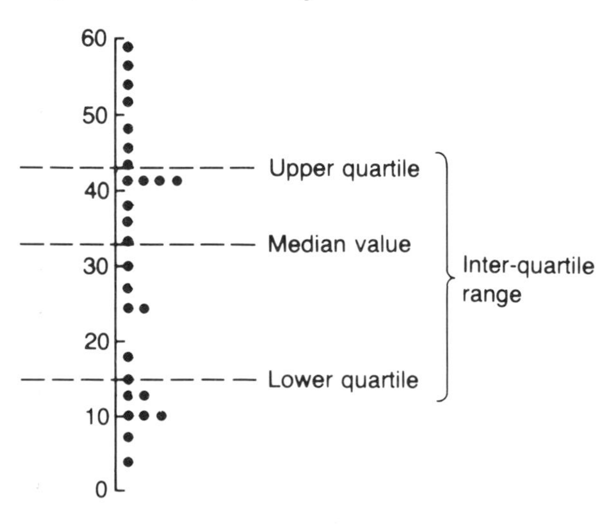

If you wished to show four classes on a choropleth map of the data shown in Fig. T4.5, you could therefore choose to show classes well above average (upper quartile and above), just above average (median to upper quartile), just below average (median to lower quartile) and well below average (lower quartile and below).

Alternatively, you could divide the figures into equal sized classes. In this case, you might use six classes with the ranges:

0– 9.9
10–19.9
20–29.9
30–39.9
40–49.9
50–59.9

Next you will have to select the colours or shading to represent each class you have chosen. If you wish to highlight a particular thing, you should choose the darkest colour or highest density shading to represent that. Thus in a choropleth map showing unemployment rates, the highest rates of unemployment should be shown with the darkest colour or highest shading density. Successively lower figures in this case should be shown by successively lighter colours or lower densities of shading.

1 Look at the map of France which shows French government aid to industry using a choropleth technique (Fig. T4.6).
   (a) Describe the distribution of those areas receiving the maximum level of aid.
   (b) Why should the area in the centre north of the country not receive any government aid to industry?

2 Draw a choropleth map using the data on employees in industry as a percentage of the total work force in the regions of Italy, 1981, from Table T4.3

Fig. T4.6 French government aid to industry

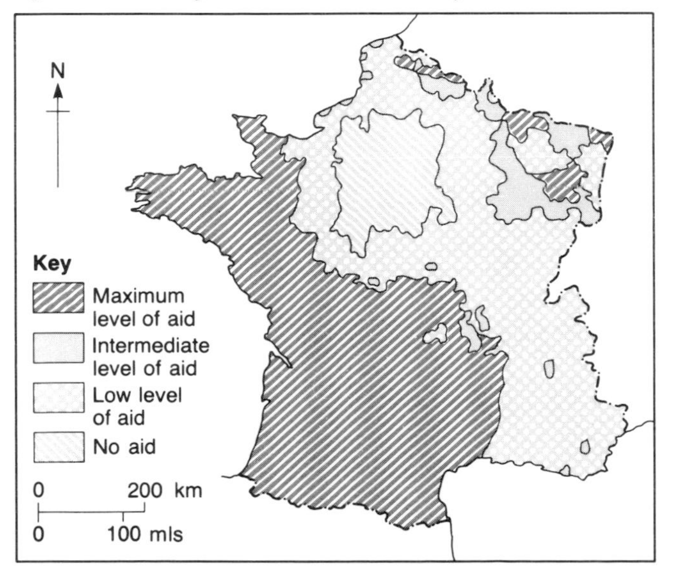

Table T4.3 Industrial employees in Italy as a percentage of total workforce

Region	%	Region	%
Piemonte	36	Marche	22
Valle d'Aosta	26	Lazio	14
Liguria	18	Campagnia	17
Lombardy	38	Abruzzi	16
Trentino-Alto Adige	18	Molise	12
Veneto	27	Puglia	15
Friuli-Venetia Giulia	27	Basilicata	10
Emilia-Romagna	24	Calabria	7
Toscana	26	Sicilia	12
Umbria	25	Sardegna	11

Industrial location quotients

The extent to which an industry is concentrated in an area can be measured precisely by using the **industrial location quotient**. It is calculated by using the formula:–

$$\text{ILQ} = \frac{\dfrac{\text{number of people employed in industry P in region Z}}{\text{number of people employed in industry P in the whole country}}}{\dfrac{\text{number of people employed in all industrial types in region Z}}{\text{number of people employed in all industrial types in the whole country}}}$$

The significance of the figure reached by working out this equation can be found by comparing it with the following:

ILQ greater than 1	=	the industry being investigated has a higher concentration than the national average in that region.
ILQ of 1	=	the industry is of average concentration in that region.
ILQ of below 1	=	there are fewer people working in that industry in that region than the national average.

The location quotient can be used for distributions other than industrial ones.

For the figures given on employment in France, 1981 (Table T4.4), the industrial location quotient for chemicals and man-made fibres in the Île de France is worked out thus:

$$\text{ILQ} = \frac{\left(\dfrac{89.8}{327.0}\right)}{\left(\dfrac{1407.0}{6852.1}\right)} = \frac{0.275}{0.205} = 1.34$$

Table T4.4 *Employment figures by region of certain industrial groups in France*

Region Key for Fig. T4.11	Total Industrial Employ-ment	1 Chemicals + man-made fibres	2 Metal articles	3 Motor vehicles	Industrial location quotients		
					1	2	3
IdF Île de France	1407.0	89.8	107.3	147.8	1.34		
BP Bassin Parisien	1370.7	69.0	156.5	92.8			
NP Nord-Pas De Calais	543.0	20.1	51.0	28.9			
E Est	766.4	25.8	81.8	83.2			
O Ouest	791.3	15.4	47.8	43.9			
SO Sud-Ouest	565.2	24.8	37.6	12.4			
CE Centre-Est	903.3	53.5	112.7	50.1			
M Méditerranée	505.2	28.6	24.3	2.3			
France (except Corsica)	6852.1	327.0	619.0	461.4			

1 Complete the calculations in Table T4.4.

2 Use the figures to complete choropleth maps of France showing location quotients near average, above average and below average for each of the industrial types shown.

3 What patterns do your maps reveal? How do they compare with the choropleth map of regional aid (Fig. T4.6)?

Fig. T4.7 *Base map of regions of France for choropleth maps*

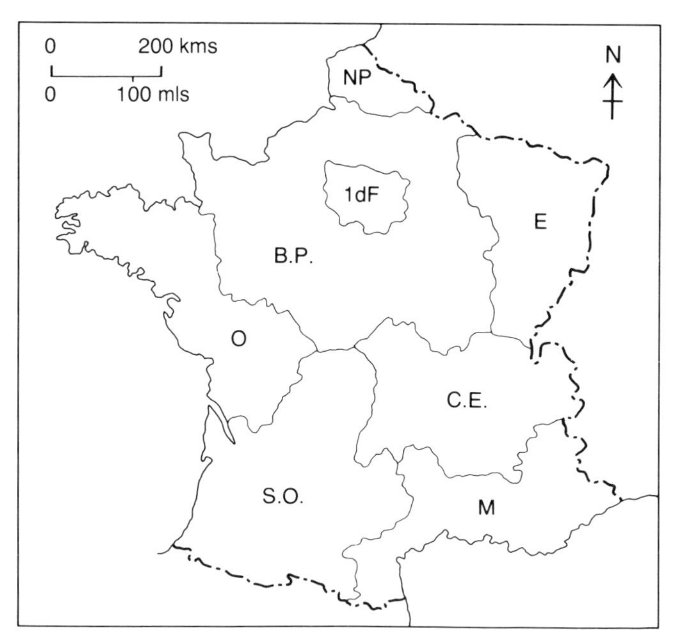

Theme 5 *Energy*

Energy Consumption and Levels of Development

Energy is fundamental to all forms of economic and social activity. It is essential in all manufacturing industry, in agriculture and in the home. It powers machines in the factory and on the farm. It is used in cooking, heating, lighting and cooling. It is also used for the collection of raw materials and the distribution of finished or semi-finished products. It is therefore one of the most important indicators in deciding upon the level of development in a country. The source of energy can vary from the use of animal and human muscle in the most primitive society to the highly sophisticated nuclear reactors generating electricity in the most advanced society.

Hence the energy problem is a different problem in different parts of the world. In a developed country such as Britain, the problem may be in overcoming the inertia in moving from one source of energy to another. This change may lead to changing mining locations, sites for power generation and the distribution of energy, all of which bring about impacts on the physical landscape and the social environment of services and employment. In a developing country, such as Brazil, the problem may be one of reducing dependency on imported fuels and changing to more indigenous sources of power, while the demand for energy continues to rise with the growing population. For the poorest people of the world the problem may be simply the availability and the accessibility to sources of power and their attempts to break free from the vicious cycle of poverty reinforced by and maintained by a low level of energy availability. At this level, energy is required for survival and this may apply to as many as 25% of the world's population.

The relationship of energy consumption and levels of development is illustrated in Table 5.1.

Table 5.1 The relationship of energy use, economy and development

Development scale	**Low ←**		**→ High**
Economic activity	High level of subsistence agriculture. Low level of manufacturing. Export of primary or semi-finished products	Cash crop economy with subsistence food agriculture. Development of manufacturing industry and export orientation	Mechanized agriculture – high inputs. Developed manufacturing and service industries
Energy production and consumption	Mainly fuelwood, crop residue, cow dung. Small commercial production of power based on imported fuels	Rural areas still use fuelwood. Large scale development of indigenous resources. Import substitution	Demand < supply. Wide range of sources. High demand in domestic sector. Consideration of alternatives.
Distribution systems	Isolated communities. Small areas of commercial power	Interconnected urban communities linked to power grid. Rural areas still isolated	Wide choice – fully connected network. Very few isolated communities
Environmental impact	Deforestation – soil erosion – rivers silting up	Large scale environmental damage from large schemes – deforestation	Unemployment and social considerations. Agricultural disruption. High public awareness.
Examples	Bangladesh	Sri Lanka, India, Brazil	UK

1 The following table shows the gross national product per capita in US $ and the energy consumption per capita in units of coal equivalent. The countries have been chosen randomly from information in a World Bank report.

Country	Gross National Product	Rank of GNP	Energy use	Rank of energy use	d	d^2
Ghana	420		258			
Mauritania	440		196			
Syria	1340		925			
Netherlands	11 470		6597			
Nicaragua	740		446			
Congo	900		195			
Kuwait	19 830		6159			
Zimbabwe	630		783			
Central African Republic	300		46			
Sri Lanka	270		135			
Norway	12 650		11 749			
Dominican Republic	1160		490			
Paraguay	1300		234			
Mexico	2090		1535			
Yugoslavia	2620		2415			
Indonesia	430		225			
Finland	9720		6001			
					Σd^2	

(a) Make a copy of the table and use it to test the hypothesis that energy consumption increases as the Gross National Product increases (or wealthier countries use more energy), by

(i) Calculating Spearman's Rank Correlation Coefficient (see Techniques p. 192)

(ii) drawing a scattergraph of the information and a best fit line, if appropriate (see Techniques p. 13).

(b) Comment on the results obtained.

2 Draw scaled pie charts for the developed and developing world using the information in the following table. (See Techniques, p. 193 for the method of drawing scaled pie charts.)

World Energy Consumption–1978 ($\times 10^9$ gigajoules)

	Oil	Coal	Nat. gas	Hydro Nuclear	Biomass	Total
Developed world	93.6	58.24	47.84	6.24	2.08	
Developing world	22.08	23.92	3.68	1.38	40.94	
Total						

3 Describe the differences between the developed and developing world's energy consumption.

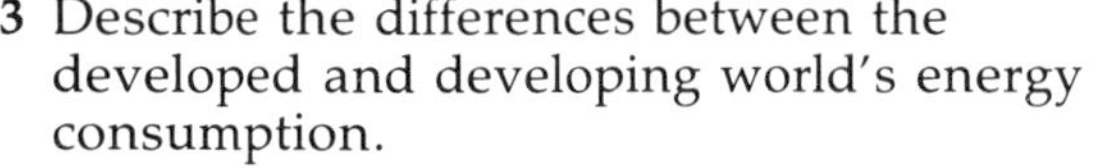

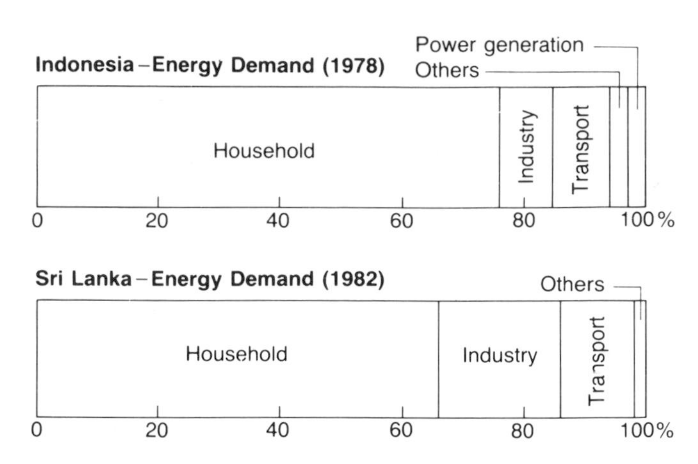

Fig. 5.1 Energy consumption in Indonesia and Sri Lanka

The bar graphs in Fig. 5.1 show the energy consumption in sectors of the economy for two of the world's poorest nations (see table in Question 1). As they develop, the demand for energy will grow in all sectors of the economy, but it will be most pronounced in the industrial and transport sectors. There will be a need to provide basic infrastructure such as roads, which will in turn require considerable amounts of energy before vehicles use the roads and consume even more energy. Hence the demand for energy in these developing countries will grow faster than the increases in wealth of Gross National Product of the country. The need to produce energy affects the rate of development of the country and so more of the income of countries such as Indonesia and Sri Lanka will go into increasing energy production.

The future demand for energy is uncertain, and inexact, but at present the global demand for energy is rising relatively slowly as shown by Fig. 5.2. Demand from 1973 to 1984 has gone back to the pre-1945 level after a near doubling between 1945 and 1973. However, these trends of slow growth have only been brought about by a decrease in demand in the developed world. Consumption rates are still high in the developing world; for example, electricity consumption went up by 8% between 1973 and 1978 and a recent World Bank survey put the rise at over 6% through the 1980s.

Fig. 5.2 World energy use, 1800–2035

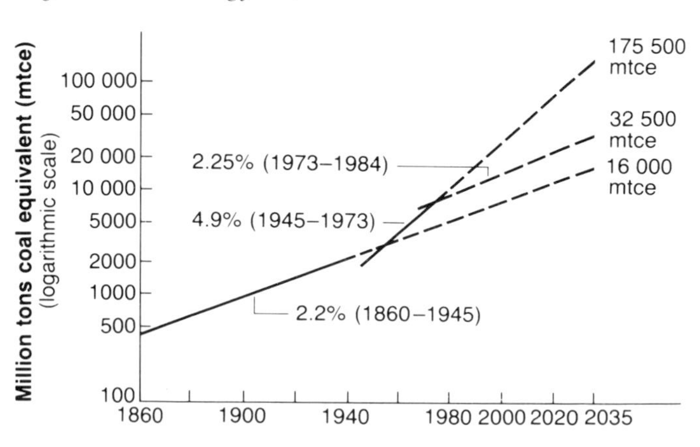

> **4** Write a list of factors which have contributed to a fall in the rate of increase of energy consumption in the developed world. Compare and discuss your list with other members of your class.

The developing world
Many energy demand surveys, however, are only concerned with commercial supplies of energy. The energy crisis in many developing countries is the provision of energy for survival. Many countries have a high dependency on wood as a main source of fuel. Table 5.2 shows estimates for some developing countries.

> **5** Find a map of world vegetation types in your atlas and locate the countries shown in Table 5.2. What types of natural vegetation do these countries have?
>
> **6** What problems would each of the types of natural vegetation you have identified pose for the collection of firewood? It may be useful to refer to Theme 2.

Furthermore the UN Food and Agriculture Organization (FAO) examined the present position and future prospects for fuelwood supplies throughout the developing world. Countries and sub-regions were classified into zones according to their wood-supply position:

- Acute scarcity situations were defined as those 'with a very negative balance where the fuelwood supply level is so notoriously inadequate that even overcutting of the resources does not provide the people with a sufficient supply, and fuelwood consumption is therefore clearly below minimum requirement'.
- Deficit situations are those where the people are still able to meet their minimum fuelwood requirements, but only by overcutting the existing resources and jeopardizing future supplies.
- Other areas were classified as being either in a satisfactory or a prospective deficit situation.

The study found that more than 100 million people, in some twenty-six countries, already faced acute firewood scarcity. The most serious situations were identified in the following zones:

- the arid and semi-arid zones south of the Sahara
- the eastern and south-eastern parts of Africa
- the mountain areas and islands of Africa
- the Himalayan region of Asia
- the Andean plateau
- the densely populated areas of Central America and the Caribbean.

The study also concluded that nearly 1.3 billion people – 39% of the total population of the developing countries – live in 'deficit' areas and face a looming crisis. By the year 2000, some 2.4 billion people living in rural areas will be consuming wood faster than it can be grown. The situation in Sri Lanka is typical of many areas.

Table 5.2 Fuelwood consumption estimates in developing countries

Country	Wood as % of total energy consumption
Angola	74
Benin	86
Brazil	33
Burundi	89
Cameroon	82
Central African Republic	91
Chad	94
Chile	16
El Salvador	37
Ethiopia	93
Ghana	74
Guinea	74
Honduras	45
India	36
Ivory Coast	46
Kenya	70
Liberia	53
Madagascar	80
Malaysia	8
Mali	97
Morocco	19
Mozambique	74
Nepal	98
Nicaragua	25
Niger	87
Nigeria	82
Pakistan	37
Papua New Guinea	39
Rwanda	96
Senegal	63
Sierra Leone	76
Somalia	90
Sri Lanka	55
Sudan	81
Tanzania	94
Thailand	63
Tunisia	42
Upper Volta	94
Zambia	35
Zimbabwe	28

(Sources: Hall, Barnard and Moss, *Biomass, Energy in the Developing Countries*, (1982) Pergamon Press; Leach *et al.*, *A Comparative Analysis of Energy Demand Structures of Countries at Different Levels of Economic Development* (1983) IIED, London)

Case-study: Sri Lanka
In Sri Lanka about 80% of the population lives in rural areas and 90% of the population uses firewood as the basic fuel. They use it for heating water and cooking and the wood is collected by the women and children. On the tea plantations in the interior, the supply of wood has been reduced due to the dominance of the cash crop, tea. The women are allowed to take old tea bushes for fuel but cutting

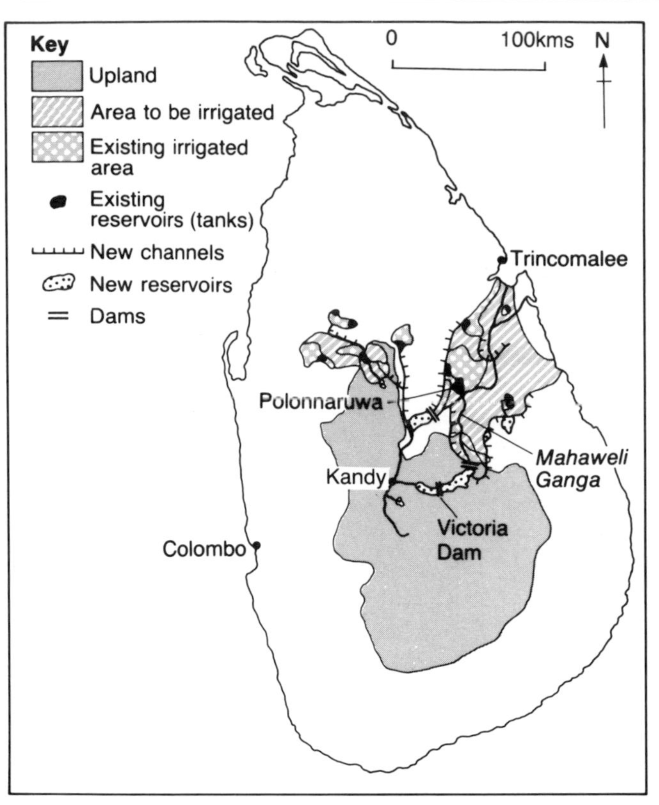

Fig. 5.3 (a) Sri Lanka – the Mahaweli scheme
(b) Soil erosion on a wood-cleared hillside

wood from the forest is prohibited by law. The plantation owners are required to plant trees to provide fuelwood, but few do. As a consequence, wood is stolen from the natural forest with women travelling increasing distances to get fuelwood, some travelling over three miles from the home. Firewood is also on sale from trees felled in other parts of Sri Lanka. These trees have been felled on land which is coming under agriculture for the first time and will be using the irrigation water from the Victoria Dam on the Mahaweli Ganga. Trees have also been felled in the valleys which will be flooded by the lake created by the Victoria Dam (see Fig. 5.3). One day's firewood costs about 25% of the earnings of a worker on the plantation.

Alternative sources of fuel in Sri Lanka include the production of bio-gas and charcoal. A typical bio-gas plant is shown in Fig. 5.4, and there are 250 of these plants installed on farms in Sri Lanka, and in particular in the region around Kandy.

Charcoal is produced by burning wood in an oxygen-poor environment. However, 10 tonnes of wood are required to produce 1 tonne of charcoal. It is also twice as expensive to cook an average meal with charcoal. Much of the charcoal is used by the cement industry.

The government of Sri Lanka has invested heavily in hydro-electricity, and the Victoria Dam on the Mahaweli Ganga is the centrepiece of the development. Much of the money required for the project was provided by Britain in the form of development aid. Most of the power generated by this scheme will be consumed in the capital, Colombo, and in the industrial Free Trade zone situated to the east of it. About 10% of the population will benefit from this power, and it is seen as essential for the development of the tourist industry. The Victoria Dam has been built to reduce Sri Lankan dependency upon imported oil and ironically has produced a local and temporary increase in firewood for the peasant workers. Even if electricity were available to more of the population of Sri Lanka, then the cost would be prohibitive. The average meal would cost 3.5 times as much to cook than if the power had been provided by firewood.

For most people in Sri Lanka all methods of providing fuel are too expensive. Even the biogas plant costs the equivalent of £750 and is beyond the budgets of most families.

7 Consider the advantages and disadvantages of the hydro-electricity option and the bio-gas option. Draw up columns for advantages and disadvantages.

8 Write two paragraphs, one giving the views of the government regarding the provision of energy for the future and one for the peasant farmer.

Fig. 5.4 A bio-gas 'boiler'

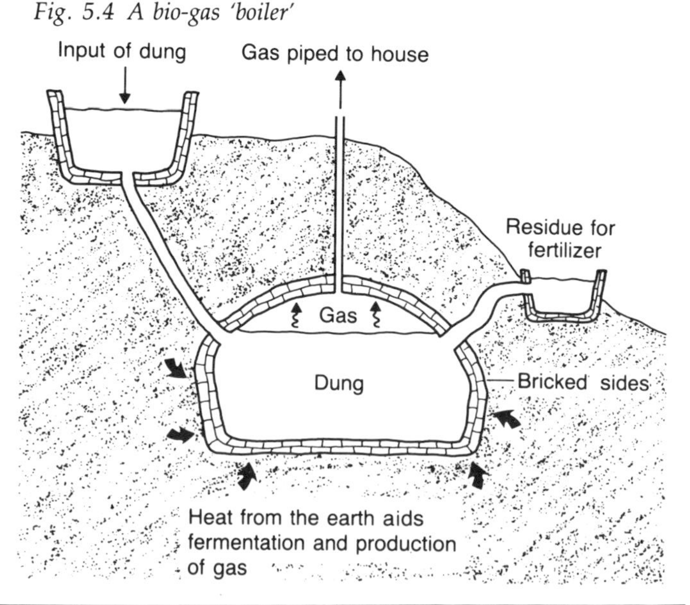

Case-study: Bangladesh

In Ulipur, a small village in Bangladesh, the land is intensively farmed, and the natural woodland has been cleared. The agricultural system of the area is based upon rice, and the farms are owned by Muslims. Poor Muslims and Hindus are fishermen. The fuel used in the village has been surveyed and Table 5.3 summarizes the results:

Table 5.3 Fuels used in a Bangladesh village

Fuel	%
Crop residues from inside the village	54.0
Crop residues from outside the village	5.1
Cow dung	2.7
Firewood trees within the village	10.8
– from the river (bhaza lakri)	4.4
– purchased from the bazaar	5.2
Doinshah (see text)	4.9
Bamboo	3.6
Water hyacinth	1.6
Other dried vegetation	7.7

However, access to the fuel is strictly regulated and controlled by the wealthier groups. The rich peasant farmers make up 16% of the families in the village, and they own 80% of the fruit trees and the firewood trees. However, they get most of their fuel from crop residues and only in the months leading up to the monsoon do they use firewood. They allow the poorest Muslims to use the firewood from their trees. The Muslim and Hindu fishermen will depend upon the bhaza lakri which they hope to find in the river, especially after the monsoon floods. In the period from mid-June to mid-September, i.e. before the monsoon, they will be forced to buy firewood from the bazaar. Doinshah is a tall legume which grows on the high ridges between the paddy fields. It is useful to the farmers in that it fixes nitrogen, but it is also dried for fuel. However, it is difficult to protect and is stolen from the farms by all groups. The access to fuel is shown in Table 5.3 and Fig. 5.5.

9 Construct divided rectangles for the five groups of people and shade the various areas to represent the type of fuel used. Comment on the relationship of accessibility to fuel and social grouping.

One of the major problems is storage of fuel and the space in which to dry it out. This is particularly so for the poorer groups who do not own land.

The supply of fuel is therefore tied to the agricultural system and to chance factors, such as having the good luck to find enough bhaza lakri in the river. The situation could be eased by providing more efficient means of using the fuel, particularly in the cooking of food, and stoves with efficiencies greater than the current 15% would be an improvement. The provision of solar driers to assist the drying of vegetable matter might also help.

Fig. 5.5 Sources of fuel and access by social class in a Bangladesh village

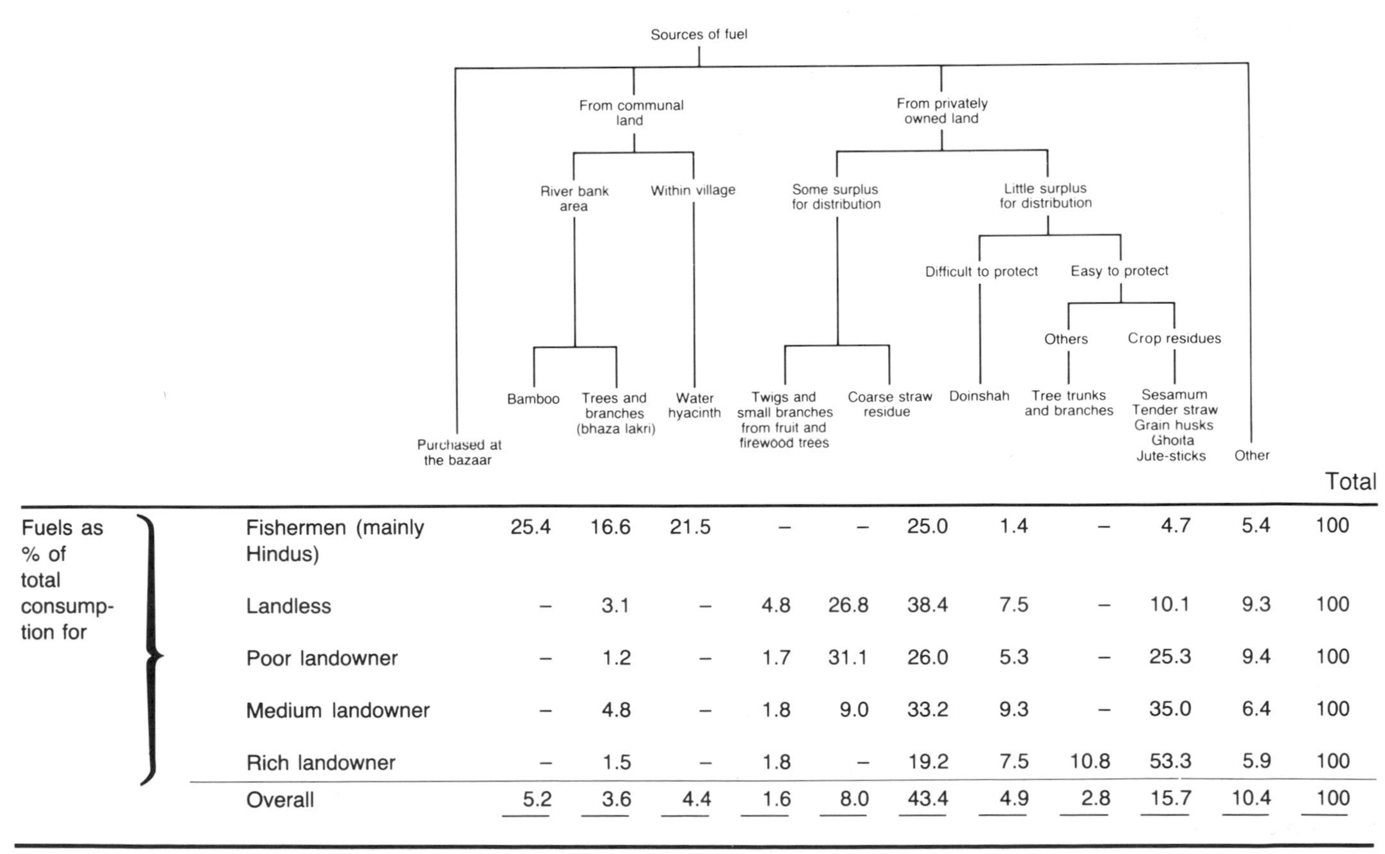

	Purchased at the bazaar	Bamboo	Trees and branches (bhaza lakri)	Water hyacinth	Twigs and small branches from fruit and firewood trees	Coarse straw residue	Doinshah	Tree trunks and branches	Sesamum Tender straw Grain husks Ghoita Jute-sticks	Other	Total
Fishermen (mainly Hindus)	25.4	16.6	21.5	–	–	25.0	1.4	–	4.7	5.4	100
Landless	–	3.1	–	4.8	26.8	38.4	7.5	–	10.1	9.3	100
Poor landowner	–	1.2	–	1.7	31.1	26.0	5.3	–	25.3	9.4	100
Medium landowner	–	4.8	–	1.8	9.0	33.2	9.3	–	35.0	6.4	100
Rich landowner	–	1.5	–	1.8	–	19.2	7.5	10.8	53.3	5.9	100
Overall	5.2	3.6	4.4	1.6	8.0	43.4	4.9	2.8	15.7	10.4	100

Decision-making Exercise N: Predicting Energy Demands for a Tropical Country

N

Controlling the amount of energy required by a country is one of the most difficult areas of management. Too much energy produced will be inefficient and wasteful, but not having enough will lead to dissatisfaction at best and economic breakdown at worst.

In a developing country it is vital to maintain the supply of energy at a fast enough rate so that industry and communications can keep developing and the country can move towards its targets. To do this the energy planners in the country need accurate information on the future size of energy demand. This will include information on the population growth rate, the growth of income and industry. This will be essential to plan the necessary investment in power stations and other power generation equipment.

However, as a country begins to develop, the demands for energy increase at a much faster rate than many of the other factors including the population increase. Fig. 5.6 shows some of these critical rates for Kenya. Fig. 5.6a shows electricity consumption growing at about 8% per annum, with domestic use, industrial use and commercial use showing substantial increases. To this must be added other forms of fuel such as oil products including petrol and aviation spirit.

At the same time that fuel demand is growing by 8% the population of the country is growing by about 6% per annum (Fig. 5.6b) and the Gross National Product of the country growing very fast at just under 13% between 1980 and 1985 (Fig. 5.6c). In the period 1960–80 the GNP per capita has been growing at the slower rate of 2.7%.

The demand for fuel tends to outpace the growth of population as the country develops due to the multiplication of demands for fuel. Development will give some members of the population access to a higher standard of living and all the modern gadgets and energy 'slaves'. These appliances such as washing machines and refrigerators will increase the domestic demand for energy. A new road network will open up new areas for farming. To get the products to market and to serve that market, cars, trucks and lorries will use considerable amounts of petrol and diesel. New manufacturing processes will tend to be capital intensive and therefore dependent upon energy.

Resources such as petroleum will be in higher demand as a feed stock for manufacturing industry. They will be a raw material as well as a source of power. Fertilisers may be manufactured from natural gas. Use of such fertilisers may increase agricultural production but at the cost of using a valuable power source. At present Kenya imports petroleum. This makes up about 32% of the value of imports to Kenya. The cost of the petroleum is about 8% of the Gross National Product.

The Kenyan government has decided to pursue a policy of vigorous growth. It intends to enter a new phase of development which will increase further the Gross National Product of the country. It is the aim of the government of Kenya that this plan will increase the wealth of the country and this will be spread to all the people. This economic growth will lead to improvements in basic living standards – food, shelter and clothing for all Kenyans. It is going to do this by:

(i) job creation to absorb the larger population
(ii) increased productivity in agriculture to feed the population and provide products for export
(iii) increase widespread rural non-farm activity to provide jobs outside the towns to stop rural-urban migration
(iv) encouraging a large informal job sector in which people will set up small businesses
(v) restructuring industry to employ workers in export oriented industries.

To summarise the plan it looks at economic growth being based on agriculture, and the establishment of small rural industrial centres.

These aims all have implications for the use of energy. The current and planned use of energy is shown on Fig. 5.7. Note that this contains information about other forms of energy used in Kenya. Agricultural growth will increase the demand for fuel wood, while urbanisation will increase the demand for charcoal. Increases in rural industries will demand electricity and a grid to deliver it. There will be increases in transportation and therefore an increased demand for imported petroleum.

There will be a planned expansion in woodfuel supplies, by intercropping crops with trees, by a programme of reafforestation and by developing charcoal plantations where prudent management will ensure that wood used for charcoal is replanted.

It is hoped that increased access to electricity will lead to a reduction in the demands on fuelwood and new generating capacity has been installed at Mombassa. The Kiambere and Turkwell hydroelectric schemes will add considerably to the current capacity.

Petroleum will still be needed for transportation but it is hoped that the agricultural centres will develop sufficiently to reduce the need to transport goods. It is the intention to reduce the need to import petroleum.

Your task is to make an analysis of the situation and the future plans.

(i) Try to work out what the future situation will be assuming current trends continue, by extrapolating the lines on the graphs and complete a copy of Table 5.4.

(ii) What changes do you see happening in the critical indicators if the new plan is brought in?

(iii) Read the article (Fig. 5.8) and use it as a basis to help you comment on the implications of the Kenyan plan using the following headings:

- The economy and pattern of trade
- The future industrial structure
- Potential land use conflicts
- The economic basis for future development.

Fig. 5.6 Electricity consumption in Kenya

Fig. 5.7 Population growth in Kenya

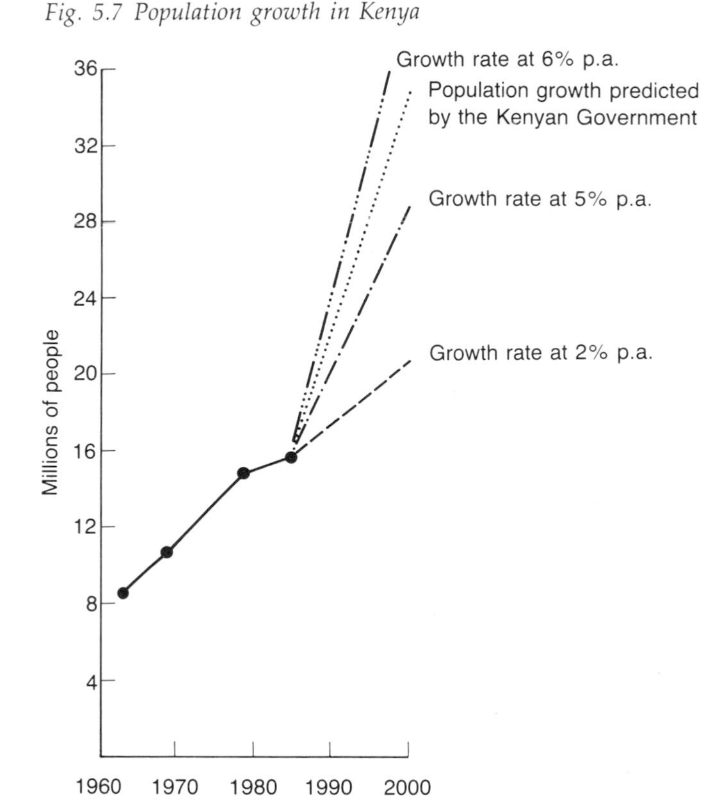

Table 5.4

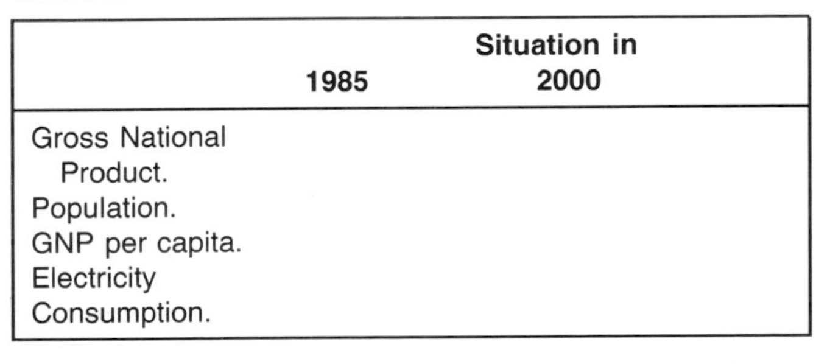

	1985	Situation in 2000
Gross National Product.		
Population.		
GNP per capita.		
Electricity Consumption.		

Fig. 5.7a Growth of Kenyan Gross National Product

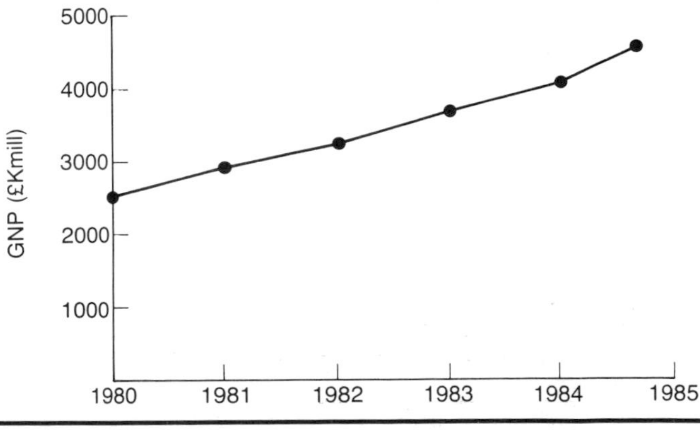

Fig. 5.8 An article from the Economist

Third-world energy

Poor countries have the same energy worries as rich ones – and a few more besides. Yet they have the advantage of flexibility over the rich: since their energy patterns are not fixed by generations of expensive investment, they can escape the Opec trap if they choose (and can afford to take) the right paths. The choices that poor countries make now (and the outside help they receive to make and pay for them) could have a profound effect not only on the lives of their billions of citizens, but on the rich countries' millions, too.

The poor countries' energy worries form a tight circle:

- In the next few years, oil-importing developing countries must cope with last year's dramatic price rises. They must deal with oil import bills that now total $50 billion a year and, on average, swallow up 26% of export earnings. For some the figure is considerably higher – 60% for Turkey in 1980, forecast Morgan Guaranty earlier this year, 40% for Brazil, and 30% for India. The short-term problems of financing these imports are acute. Even countries which manage to avoid severe disruptions to their balance of payments will pay a heavy price in forgone growth and imports of other goods.
- Poor countries are passing through relatively energy-intensive periods of economic development. A World Bank study* published recently forecasts a rise in LDC energy demand of just over 6% a year in the 1980s, lower than in the years before the 1973 oil-price rises, but still dramatically higher than the 2–3% expected in the developed world. In part, this is because poor countries are acquiring the energy-intensive heavy industries and transport systems that the rich countries built up decades ago. It also reflects a shift from traditional, 'non-commercial' energy forms (eg, wood, waste crops and dung) to ones, such as coal and oil, which statisticians can more easily measure. This shift threatens living standards as traditional fuels are exhausted or diverted for commercial use.
- Poor countries can help ease these pressures by increasing indigenous energy supplies. The costs of domestic supplies, of both conventional and less orthodox fuels, may now be much less than the cost of imported oil. The World Bank estimates, for instance, that domestically produced oil is unlikely to cost more than $15 a barrel (and could be less than half that). Imported oil costs over $30 a barrel. Domestic coal might cost between $4.50 and $15 a barrel of oil equivalent – including the cost of building, eg, the railway to take it to its final destination. Imported coal, at $14 a barrel of oil equivalent, is considerably cheaper than imported oil, but more expensive than much locally produced coal.

None the less, the investment needed to meet a significant part of developing countries' energy needs from domestic or lower-cost sources of supply is huge – and the short-term oil-price squeeze is starving them of the foreign exchange that such an investment programme demands. The World Bank calculates that between 1966 and 1975 the developing countries invested about 1.3% of GNP in energy production and conversion (mostly in electricity). By 1980, the likely level of investment is put at 2.3% of gnp; by 1990 it could be 3.2%. Even this figure, says the bank, does not take account of 'the additional investment that would be needed if the developing countries made a maximum effort to develop their energy resources', a course it is eager to encourage and, in part, finance.

Such a course would imply a tripling of the share of GNP going to energy investment in the course of little more than a decade. Roughly one investment dollar in every seven that the developing world spends would have to go on energy resources. These projects would absorb much of the third world's skilled manpower.

*Energy in the developing countries, Washington, World Bank, 1980.

Case-study: Brazil

Brazil is an example of a country from the middle range of development as shown on Fig. 5.1. It experienced rapid growth through the 1960s which was based largely upon oil, but this fuel was imported and the subsequent price increases have highlighted a crippling dependence on oil. The price increases, the result of the reduction of supply by the OPEC countries in 1974, meant that some 40% of Brazil's export earnings were to pay the bill for importing oil. The truck was the main form of transport and most of the infrastructure investment had gone into road development. Agriculture had also been mechanized and farmers had been encouraged to use a high proportion of oil related inputs. Fertilizer, based on oil, was encouraged with a government subsidy. The country was caught in a trap. How could it escape?

The Brazilian response was to look at three solutions – they looked for new technology in the form of the nuclear programme; they looked at their own indigenous resources of conventional power, and stepped up the prospecting for oil within Brazil; and they looked to alternatives in the form of hydro-electricity (something which was highly developed in Brazil anyway) and production of energy by biomass conversion.

(a) The nuclear programme

In 1971 Brazil invested in its first reactor at Angra dos Reis, buying the technology from the Westinghouse Corporation, USA. In 1975, Brazil signed an agreement with the West German government to buy the complete nuclear technology, including reactors and reprocessing plants. In addition to four reactors with an option to buy four more, Brazil owned enough technology to be recognized as a full nuclear power.

In fact the first reactor has been beset with problems of a technical nature and has only been operating at half capacity, while the high cost of the investment, 25% of the national debt, has slowed the building of the West German reactors and it is doubtful if they will be completed.

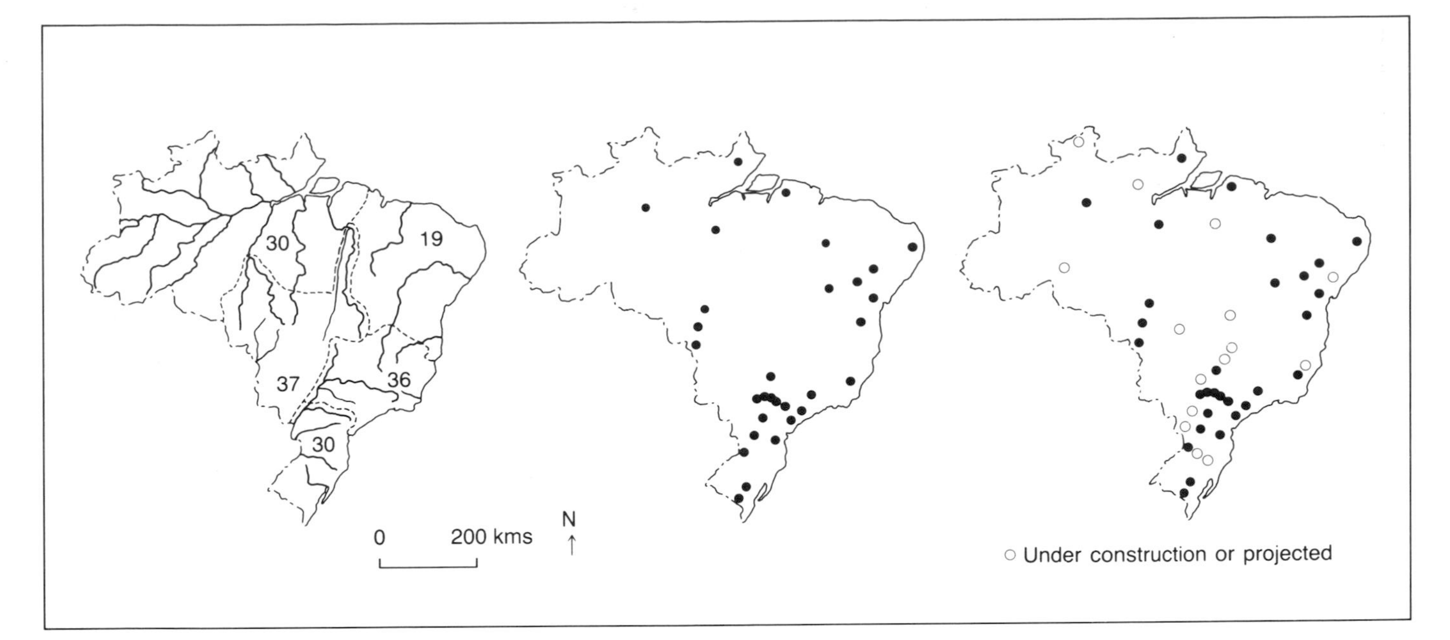

Fig. 5.9 (a) Hydro-electric power potential (b) Hydro-electric power stations, 1979 (c) Hydro-electric power stations, 1985

(b) Hydro-electric power
Brazil has a vast potential for hydro-electricity as shown on the map (Figure 5.9a).

Brazil has taken advantage of this and Fig. 5.9(b) gives an indication of the number of hydro-electric power stations in 1979. Brazil has developed considerable expertise in planning and building such large scale schemes and it is to Brazil's advantage that energy potential can be built up using indigenous supplies without the need to acquire foreign technology. However, the impacts, particularly to the physical environment, are vast, as indicated in Theme 2.

10 Using an atlas map of Brazil, try to identify the principal rivers involved in hydro-electricity production.

11 What types of natural vegetation are threatened by the expansion of the hydro-electricity programme?

(c) Production of energy by biomass conversion
This is a method of producing energy by converting the natural energy derived from photosynthesis into fuels such as ethynol and burning them in the same way that petrol is now used. This can be done with many crops but is particularly successful with sugar cane and cassava. Sugar cane can yield 3500 litres of ethynol per hectare per year while cassava can yield 2160 litres of ethynol per hectare per year.

Brazil has been producing sugar cane since the sixteenth century, but a collapse in the world market in the mid-1970s coincided with the need to look for alternatives for oil. Ethynol had been used for years before 1975 for blending with petrol (gasoline) to produce gasohol, but the main development came with the establishment of proalcool. This development took place when the price of oil rose from US $3 per barrel to US $12 in 1974. The method used by the Brazilian government was to offer incentives to private enterprise to accept government approved schemes to increase the crop area for the growing of **energy crops**. In the initial stages, distilleries were constructed next to existing sugar mills. The production in 1974 was 625 million litres of ethynol which has risen to 4000 million litres in 1980. Oil prices per barrel rose in 1979 from US $12 per barrel to US $32 and new targets for ethynol were introduced of 10 700 million litres by 1985 rising to 14 000 million litres by 1987. To meet the 1985 target needed Brazil needed 325 new distilleries and an additional 1.6 to 1.8 million hectares of land under sugar cane cultivation. By 1980 they were well on their way to this target.

A distribution network of 6000 service stations has been built. There are 2.5 million hectares of land under sugar cane at present, with 100 million hectares potentially suitable.

The programme has had its critics. The system is based upon large plantations rather than giving encouragement to the small farmer. Indeed, one property in Mato Grosso covers 1700 km^2. It is also suggested that food production will be taken over by sugar growing. Stillage, known as vinhoto, has been reported to be dumped into rivers, killing fish. Complaints have been voiced against the wisdom of clearing land in the Amazon to grow manioc (cassava) for biomass conversion.

There are advantages as well, however. Employment can be stimulated in areas of Brazil where conditions are hard, such as in the north-east where manioc can be grown for distillation. Employment is stimulated not only in agriculture but in distilleries and associated industries. It is estimated that 600 jobs are created by every distillery and a further 200 jobs in related industry. It is also held that using alcohol instead of petrol has a beneficial effect on the atmosphere, as alcohol does not have lead additives.

12 Analyse the impacts of increasing the area growing sugar cane in Brazil. What impact will it have on the natural vegetation and on the existing farmland of Brazil? To do this, make a tracing of the map showing the potential areas for this development, Fig. 5.10(a) and place it over the maps showing agricultural regions (b) and natural vegetation regions (c). Mark on the tracing areas of natural vegetation change and agriculture change.

13 What impacts will follow the development of sugar cane estates? Put your tracing over the map showing population distribution (d) and mark areas where you see urban areas growing to supply the estates with labour and managerial expertise.

14 Describe the developments and the impacts you have predicted.

15 Growing sugar cane will bring advantages too. Read the newspaper article Fig. 5.11 'Sugar: Fuel for the Future' and list the advantages.

The growing of crops such as sugar cane for fuel may produce a dilemma for developing countries. Should land be used for growing food crops or fuel crops? Who will get the benefit, the urban rich or the rural poor?

(d) Oil exploration

The search for indigenous fossil fuel has been important to Brazil. Even before the oil price increases of 1972, Brazil was the world's eighth largest oil importer. The oil price increases made her import bill increase from US $606 million in 1973 to US $ 3150 million in 1974. To reduce this massive bill, Petroras, the nationalized oil company of Brazil, intensified its efforts to find indigenous supplies.

Oil had been discovered in Alagoas, Sergipe, and Bahia (see Fig. 5.12) but these supplies were beginning to decline. Attention has been turned to exploration in the offshore continental shelf basins, an area of 830 000 km² and more recently to the onshore sedimentary basins measuring some 3 150 000 km². These have been moderately successful and they provide about 20% of the Brazilian demand for oil. The rest is still imported from the Middle East.

An important impact is the valuable employment provided by the petrochemical industry which uses this oil. This also acts as an import substitution, an important element in a developing economy, which means that less of the country's wealth is required to pay import bills and it can be redirected towards investment at home. The negative effects of such a development, however, make Cubatão, for example, one of the most polluted places on earth.

For Brazil, the oil crisis was a challenge which the government would have to respond to if it was to

Fig. 5.10 Brazil (a) Potential area for sugar cane growing

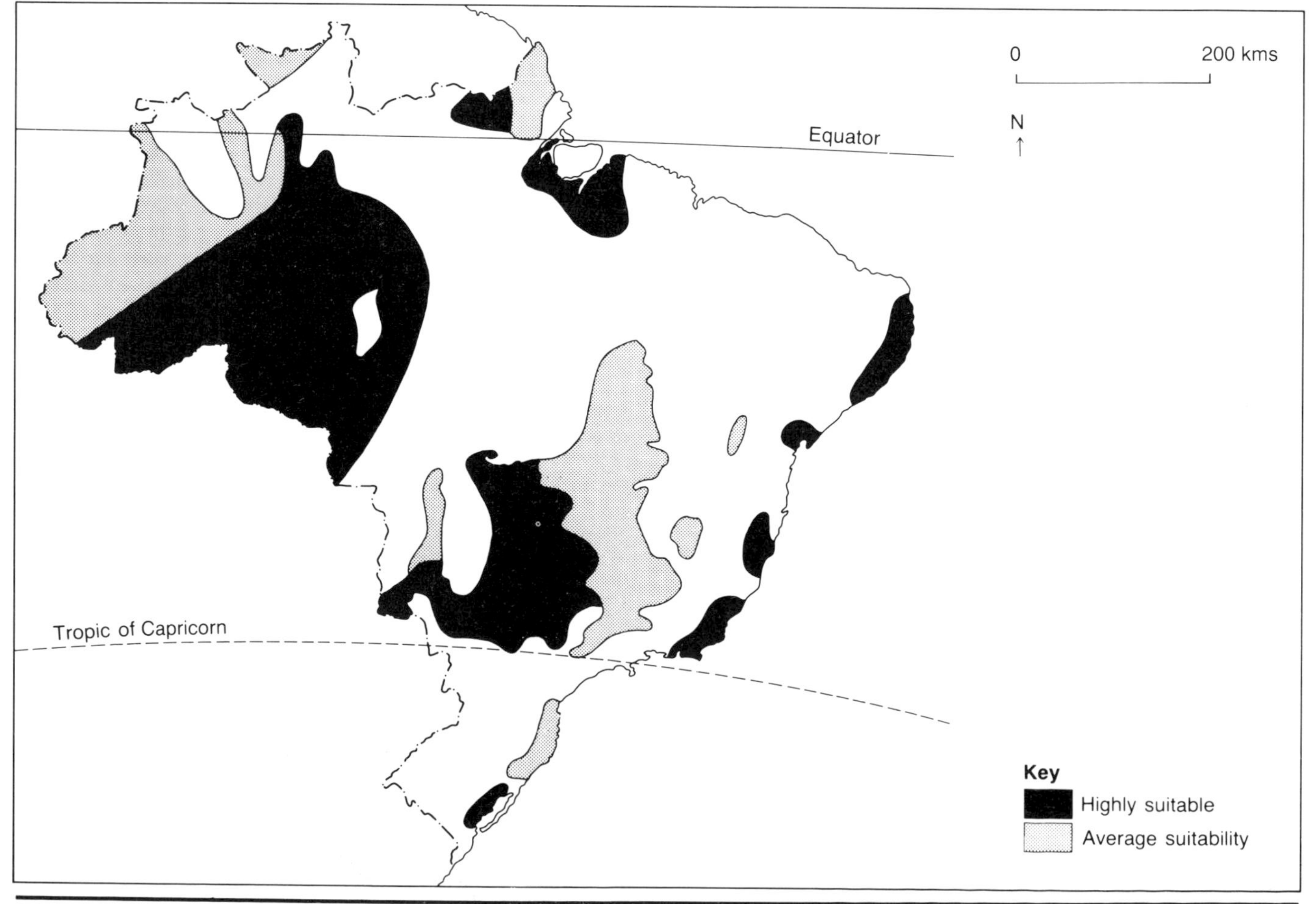

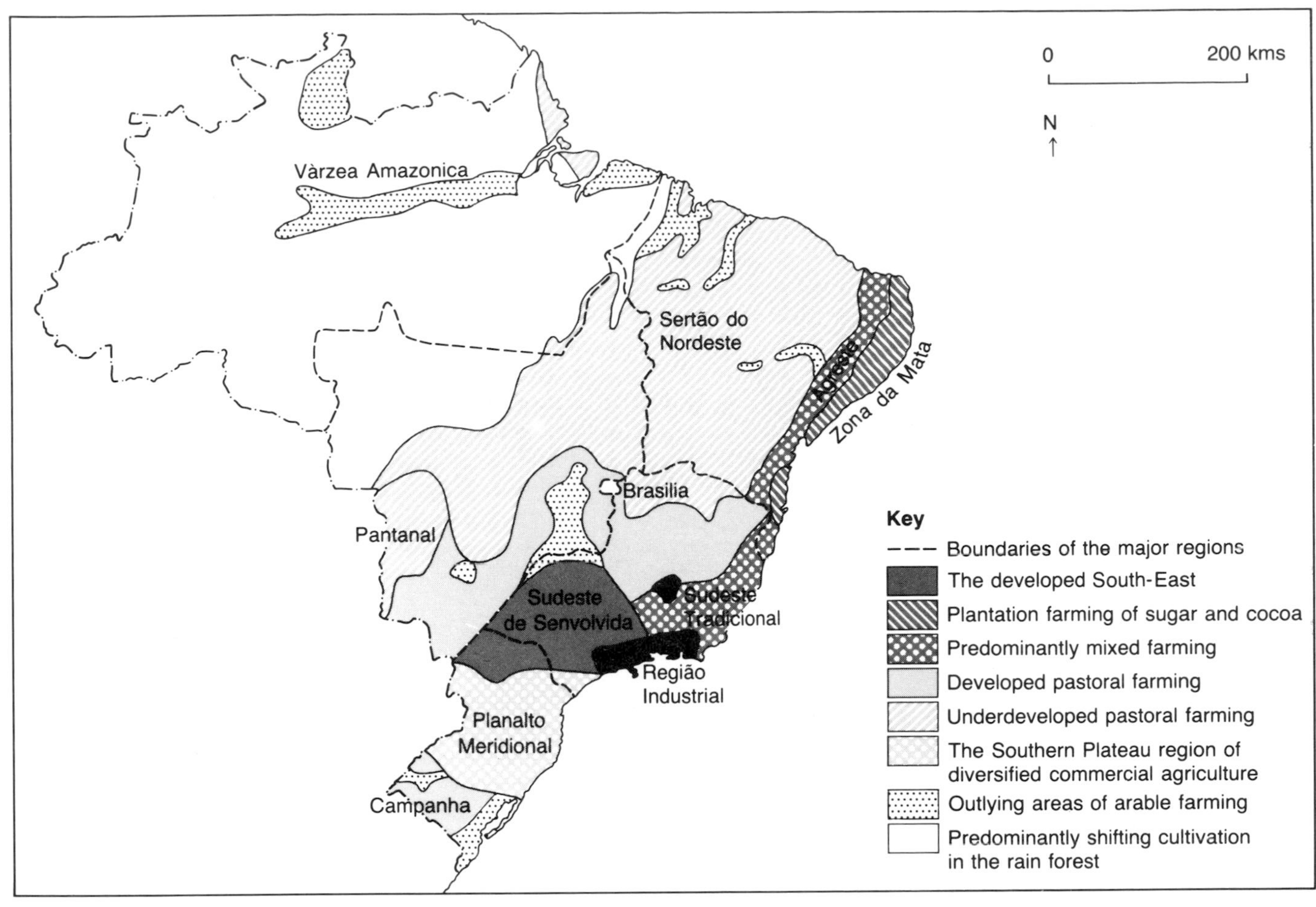

(b) *Major agricultural regions*

(c) *Natural vegetation regions*

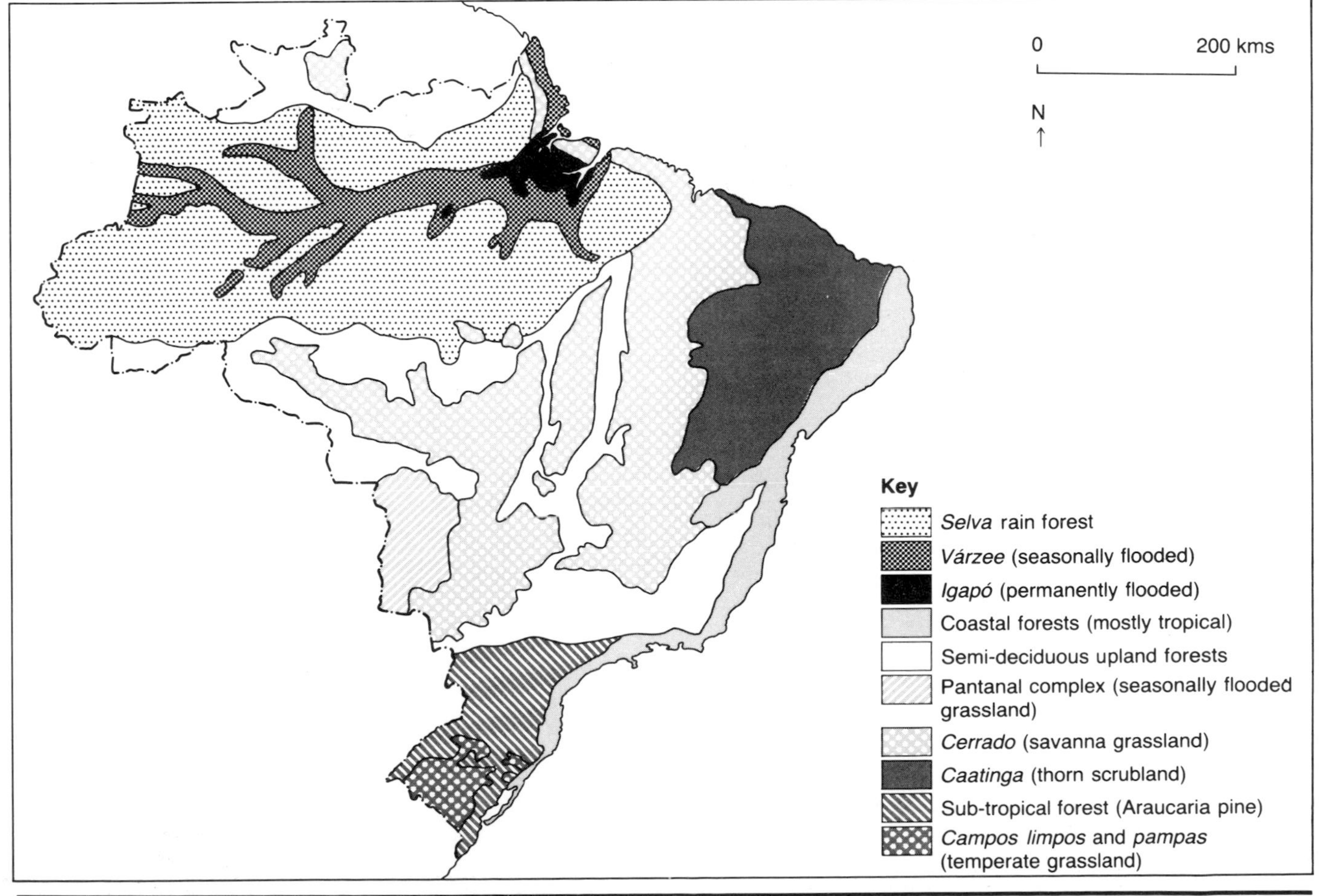

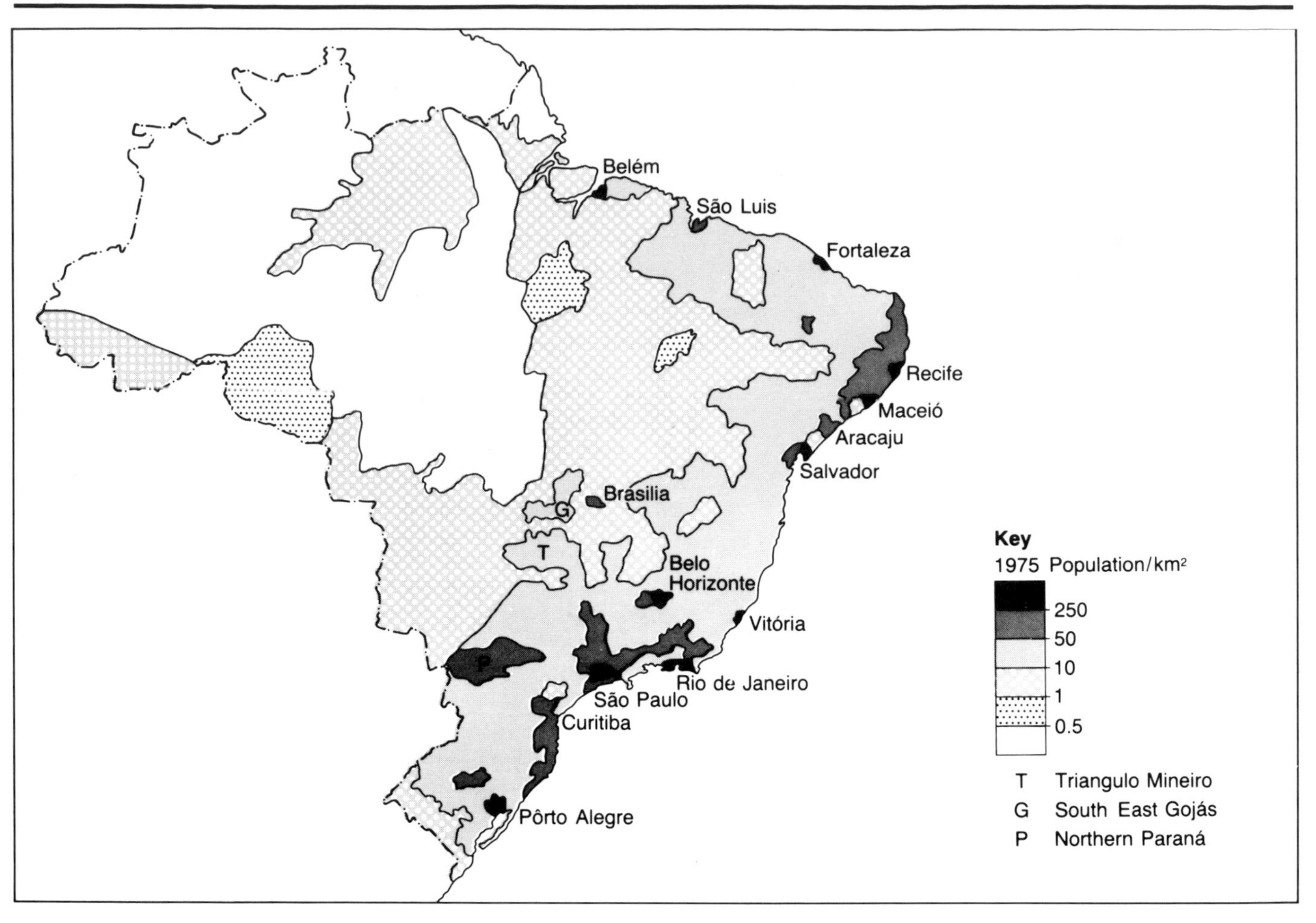

(d) Population density, by micro-regions, 1975

Fig. 5.11 An article from the Observer

Sugar for energy : Cane and its products have a potential finally being realised for a

Sugar: Fuel for the

RISING oil stocks and falling prices make it unfashionable at present to talk of an energy crisis. But for many developing countries suffering from high inflation, the cost of oil has escalated dramatically, soaking up hard-earned foreign exchange. Consequently, the search for alternative energy supplies is on in earnest in many parts of the world.

One of the largest and best known programmes is in Brazil, which now has around two million vehicles powered entirely by alcohol made from the juice of sugar cane. The country's remaining eight million or so vehicles are run on a 20 per cent alcohol/petrol blend. The alcohol (ethanol) is made by fermenting the cane juice using yeast, in a process that mimics that of the home brew beer or wine kit. The alcohol is then extracted, purified and has most of the water removed.

The production of biomass energy (the ugly term used to describe energy derived from plant and animal matter) has become big business in Brazil, which last year produced 10.7 billion litres of fuel alcohol.

It is estimated that by the turn of the century the country will have between 11 and 14 million vehicles fuelled by agriculturally-produced alcohol. But for sugar cane, its new role as a source of energy is only just beginning.

The most abundant source of 'free' energy is the sun; and plants, by means of photosynthesis (made possible by the green pigment, chlorophyll, found in their leaves), are the best converters of solar radiation into energy-rich materials.

The tropical grasses, a group that includes sugar cane, are among the most efficient photosynthesizers.

Modern sugar cane varieties are the result of the wilful selection of those plants producing the most sucrose, since even primitive man and woman apparently enjoyed a sweet chew. So important was this need to satisfy our sweet tooth that around 150 years ago 'King Sugar,' then the most lucrative commodity on the world market, determined the lives of many thousands of labourers and made the fortunes of a few. But these days with the increase in use of artificial sweeteners and the dumping by the West of large surpluses of home-produced sugar on to the world market, the price of sugar has dropped so low that it often does not cover the cost of manufacture. Many sugar-producing countries, therefore, are on the verge of bankruptcy.

It was to rid itself of exc sugar that Brazil embarked its programme of converti sugar into alcohol. It was r until the spectacular oil pr increases of the early Sevent that the replacement of imp ted oil became a priority. Oth countries have followed. Zi babwe, for example, produ 40 million litres of alcohol fr home-produced sugar ca which is blended in a 12 p cent mixture with petrol a saves some $15 million e year on the fuel import bill.

It may come as a surprise learn that sugar production not what sugar cane does be Until now selection has co centrated on those varieties t store the largest amount 'free sugar' (sugar not bou to or incorporated into t plant tissue) which can extracted by simply crushi

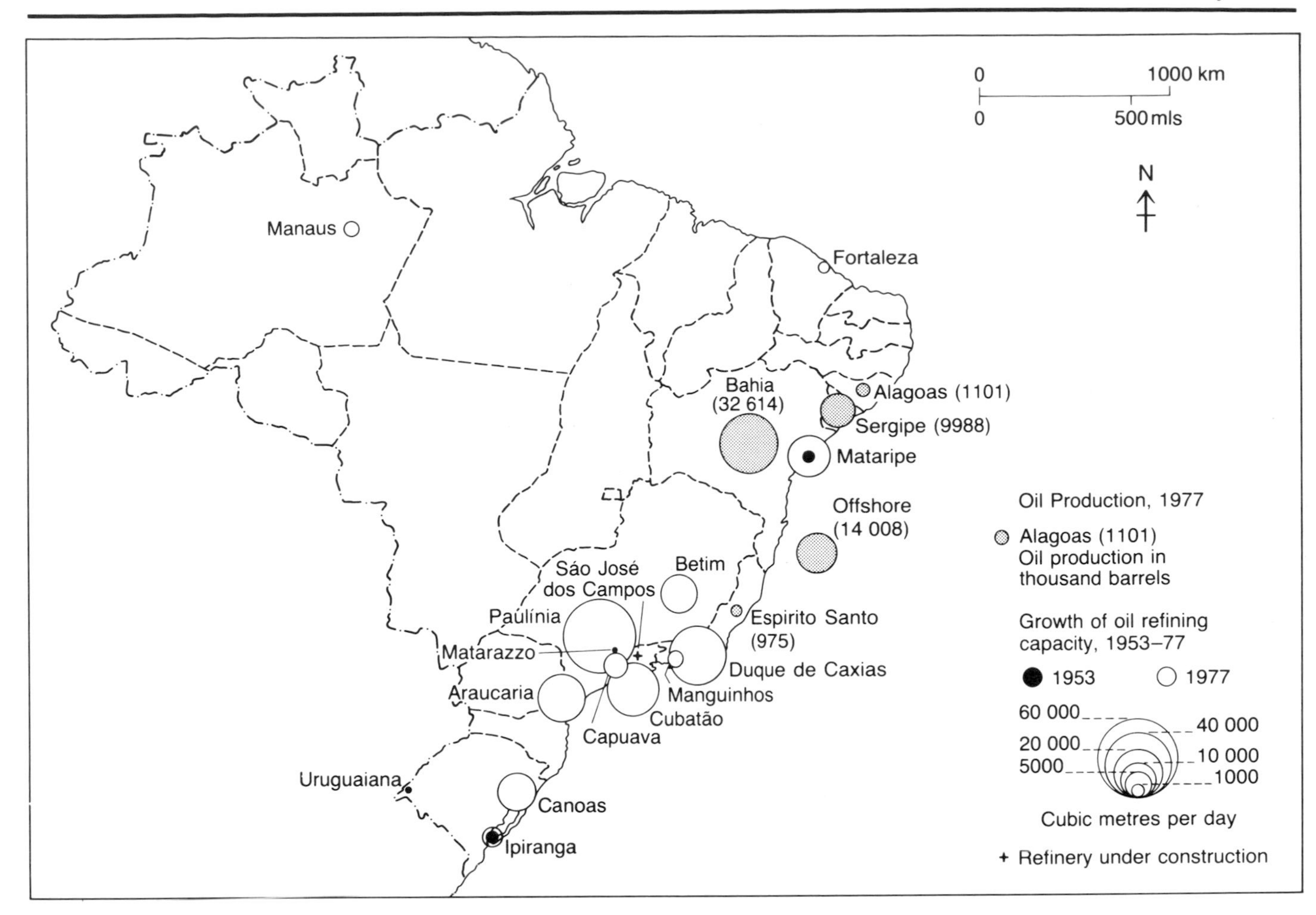

Fig. 5.12 Crude oil production 1977 and refining capacity 1953 and 1977

maintain its high rate of economic progress. An energy programme was launched as part of the Second National Development Programme which aims at reducing energy imports and maximizing the use of indigenous supplies. The oil crisis has compelled the country to make a careful and detailed assessment of its own energy potential and set in motion the most appropriate solutions.

❛King Sugar may be dead, but citizen cane is alive and well and has a rosy future❜

e stem.

The extracted juice is the ısis for sugar, and now cohol, production. The ɔrous waste that is left after e milling process is tradition- ly kept to a minimum.

These practices have, how- ver, prevented sugar cane om realising its full potential an energy crop. The fibrous aste (known as bagasse) is ɪmposed of cellulose, hemi- llose and lignin. They are the ɪiversal building blocks of the ant world and have been the ɪurce of food, shelter, clothing ɪd building material from our ɪrliest history. They are ɪown collectively as lig- ɔcellulose, and making lignocellulose is what sugar cane does best.

Bagasse has around 70 per cent of the heat value of wood and can be burnt to provide energy to powerthe sugar mill. Many sugar factories through- out the world now produce all their own energy from burning sugar cane bagasse. In some countries, for example Maur- itius, Hawaii and South Africa, the sugar industry 'exports' energy in the form of electricity to the national grid. Mauritius derives around 10 per cent of its electricity from sugar factories.

Several research groups, not- ably in Puerto Rico and at the Louisiana State University, are working on ways of capitalising on this potential to obtain energy from the plant tissue itself — or biomass. The results have been staggering. The selective breeding of vigorously growing varieties has produced new types of sugar cane which have several stems, produce a more luxu- rious leaf canopy and grow to more than 3.6 metres. These distinctive new varieties have been dubbed 'energy cane.'

Ordinary sugar cane will yield 60 – 100 tonnes of green biomass per hectare each year, but energy cane will give an average of 315 tonnes per hectare per year—equal to the highest yield of any plant recorded. Furthermore, it is predicted that it will be possible to reach annual yields of up to 380 tonnes per hectare.

Energy cane yields less sugar on a weight basis than tradi- tional sugar cane (8 per cent, against sugar cane's 14 per cent). But the phenomenal growth rate of the energy cane means that the amount of sugar that can be obtained from a hectare of energy cane is at least as much as that from sugar cane. The important point for the growers is that because of the three-fold increase in bag- asse, energy cane produces in total around three times the energy of a sugar cane crop.

In the short term, energy cane bagasse will continue to be used as a boiler fuel. Its heating value increases as it is dried, and once the technology for compression into pellets or briquettes (for easier storage and transportation) is perfec- ted, it will become a substitute fuel for wood and charcoal. King Sugar may be dead, but citizen cane is alive and well and has a rosy future.

PETER DE GROOT

Nuclear Power in Developing Countries

Nuclear power, as an alternative to burning fossil fuels, has been introduced in relatively few developing countries. There are 15 reactors operating in 6 countries which represents 5% of the total number of reactors in the world. However, the number in developing countries is increasing with an additional 21 reactors under construction and an additional 49 planned. If they are all completed, sixteen countries will be involved in this technology. In addition, Syria and Indonesia have completed feasibility studies and Iran had 4 reactors half completed in 1979 before the change of government halted further development. Fig. 5.13 shows the extent of nuclear technology.

In India, the nuclear power programme aims to produce 10% of the demand for electricity by the year 2000. At present it produces about 3% of the supply. The location of the nuclear power industry in India is shown in Fig. 5.13 and it is clear that India is developing a self-sufficient nuclear industry with uranium mines, fuel reprocessing plants and nuclear stations. The first reactor was the station at Tarapur near Bombay in 1969. Four reactors are now producing power, two at Tarapur and two at Kota in Rajasthan. Four larger plants are under construction at Kalpakkam and Narora and two more are planned for Kakrapar, with four more projected for the 1990s.

Nuclear technology has not been made available to all countries for a number of reasons. The by-product of the reprocessed uranium is plutonium, the raw material for some nuclear weapons. Hence many developed countries do not want to sell the entire technology to other countries in case the technology is put to weapons research.

Another consideration for the widespread use of nuclear power is the fact that many of the technologies sold to the developing world are now, in some cases, rather dated. The reactors sold to the Brazilian government were of a technology redundant in western Europe. It may develop an economic dependence on the donor country if the local people or local industry is not sufficiently trained or advanced to cope with repairs and day-to-day working of the plant. Concern is expressed about the levels of safety applied in some countries, made more relevant by the fire and release of radioactivity at Chernobyl in the USSR. (See page 189). The worry persists that safety concerns might be waived by an unscrupulous country anxious to install nuclear power.

Concerns about the more widespread use of nuclear power come from a popular growing concern about global issues. Green politics play a part in West Germany and in the European Community. Pressure groups such as Friends of the Earth and Greenpeace maintain their monitoring, lobbying and information role in UK. The concerns have focused around the issues of health and the possible links between certain cancers and proximity to nuclear reprocessing and handling plants. Dumping and linkage of radioactive materials can enable damaging radioactivity to enter the food chain and water supply. There are still problems with the disposal of high level nuclear waste after the reprocessing of spent fuel rods. At present this is cased in glass, concrete and steel containers. Even if these are safe, where should they be stored while they breakdown over thousands of years?

16 Using information from Fig. 5.14 redraw the flow diagram (Fig. 5.15) and include actual placenames from the example.

17 (i) Write the arguments a campaigner for one of the Environmental agencies might make to a government of a developing country thinking of investing in the nuclear power technology.

(ii) What arguments could be used by the government of the developing country?

Fig. 5.13 Map of countries with nuclear power technology

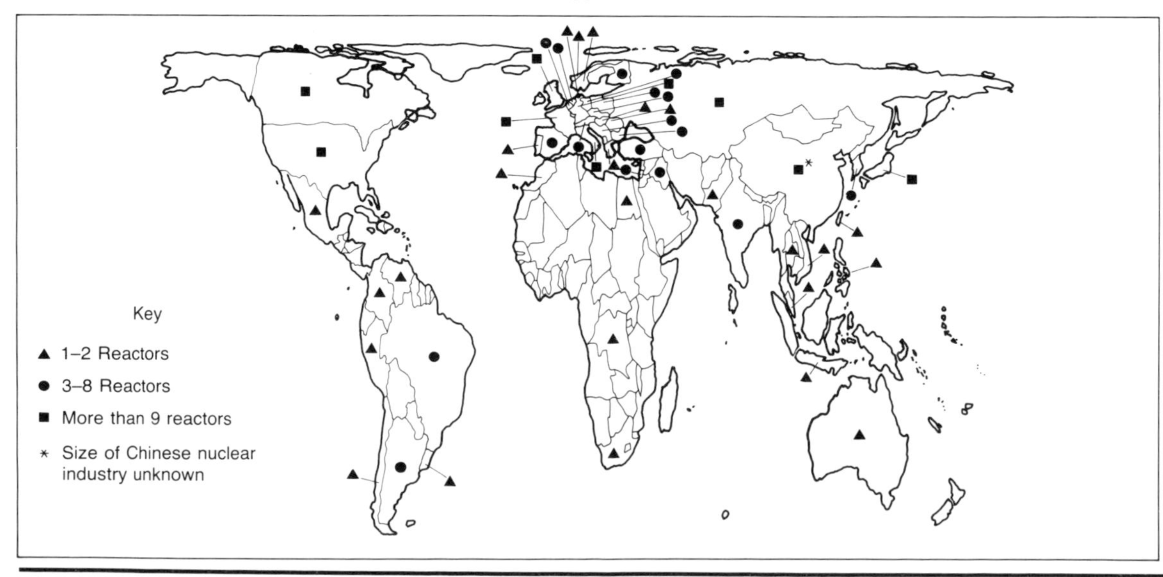

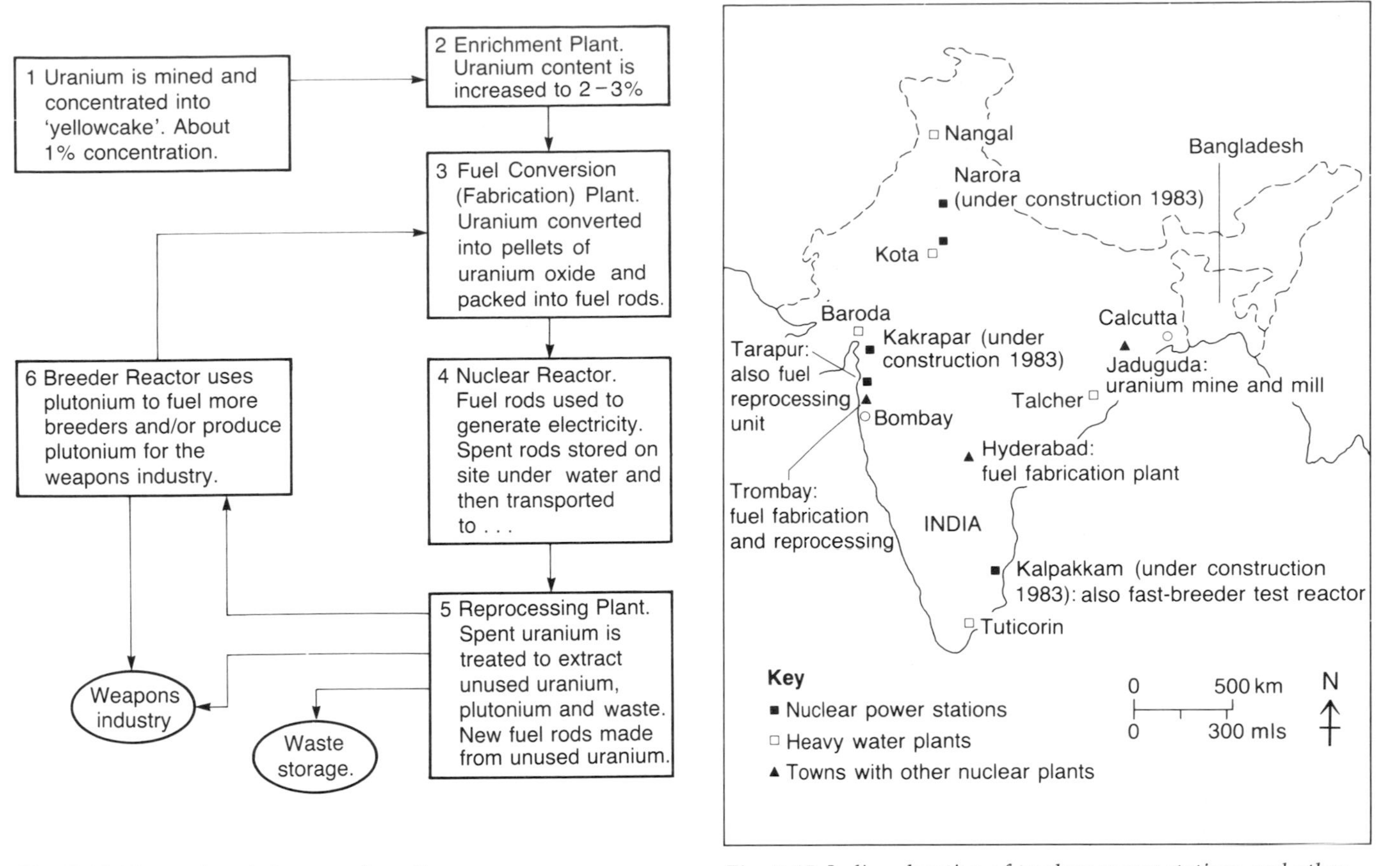

Fig. 5.14 The nuclear industry – flow diagram

Fig. 5.15 India – location of nuclear power stations and other nuclear plants

Decision-making Exercise O: Evaluating Energy Development Schemes

Evaluate the impacts of three energy development schemes, nuclear power, providing incentives to peasant farmers to grow crops suitable for conversion to ethynol, and forest management. The last scheme would involve the management of existing natural woodlands in which trees could be thinned to provide firewood, while replanting areas of depletion. They could be managed alongside farms and would use local technology and the local resource base. Nitrogen-fixing trees providing foodcrops could be planted and energy could be provided by utilizing crop residues. The local social and political structure might need to change, e.g. the ownership of woodlands might have to pass into the hands of the village, to operate this system, but it has worked successfully in Korea.

To do this evaluation, make an enlarged copy of Table 5.5 and write notes on the impacts in the appropriate spaces and score the scale of the impact as indicated below the table.

Score the impact using the following scheme:

+2 . . . Positive (beneficial) impact

+1

0

−1

−2 . . . Negative (detrimental) impact

You may wish to weight the factors, making some more important than others. To do this, multiply the score for the particular factor by an agreed constant. Sum the columns. Write a comment on the result.

Table 5.5 Impacts

Local impacts, e.g.	Nuclear power	Biomass	Forest management
Land-use patterns and agricultural systems			
Industrial and employment structure			
Physical systems – atmosphere, water and land			
National impact, e.g. importation of fossil fuels			
Dependency on foreign technology			
Balance of payments			
Other development projects and priorities			
Total impacts			

Summary of Impacts

(a) *Nuclear power:*

(b) *Biomass:*

(c) *Forest management:*

Criticise the method of evaluation.

Write a short report on the problems of energy for the developing world, as if you are a member of the World Bank investigating this particular problem. You may wish to refer to Theme 2 to find out more about the World Bank. Use the following structure in your report:

(i) Setting the scene—introduction to the problem including some indication of the scale of the problem.

(ii) Identifying the possibilities and the scale of operation.

(iii) Evaluating the environmental impact (i.e. social, economic and physical) on the developing countries of each alternative.

(iv) Recommending and concluding.

Remember that you are thinking as an international banker. What sort of bias will you give your report? Whose interests do you represent? How important will local impacts be to you?

The Developed World

Energy use in the developed world has also changed in the last fifty years. The line graph (Fig. 5.16) shows the changes in the total energy consumption from 1937 for France, West Germany, the United Kingdom, Italy and the Netherlands. The bar graphs (Fig. 5.17) show the energy used per capita in these countries.

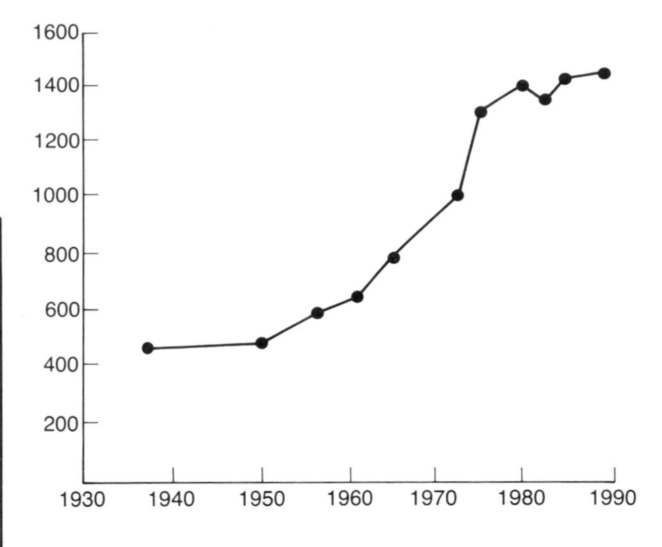

Fig. 5.16 Total energy demand for UK, W. Germany, France, Italy and the Netherlands

18 Why was energy use increasing? Make a list of the reasons for these changes and discuss them with your class group.

19 Use the information given in Table 5.6 showing energy use by fuel type to complete the aggregate percentage line graph in Fig. 5.18 from 1957. Extend the time axis up to the year 2000.

20 Describe the changes in energy use shown on the graph.

21 Extrapolate the lines up to the year 2000 and give an estimate of the contribution of each type of energy to the overall picture.

22 Give reasons why the extrapolation may not be very accurate.

The energy specialists of the early 1970s saw coal being phased out to about 6%, with oil being the dominant fuel at 64%. Gas was thought to be a steady fuel throughout the period and was at 10% in the estimates, and the remainder was made by hydro-electricity and nuclear power. These estimates were made before the oil crisis of the mid-1970s and were based largely upon high levels of cheap imported oil coming into Europe from the Middle East.

The increase in oil prices made the governments of Western Europe reappraise their energy policies. Three main ideas were common to these policies:

(i) Search for new supplies of indigenous fuels—coal, oil and gas

(ii) Increase the nuclear programme

(iii) Reduce the demand for energy by conserving fuel by such things as improving building techniques and by improving insulation.

These methods were used to reduce the level of consumption of energy and also reduce European dependency upon imported fuels, particularly oil.

Fig. 5.17 Energy consumption per head in selected European countries

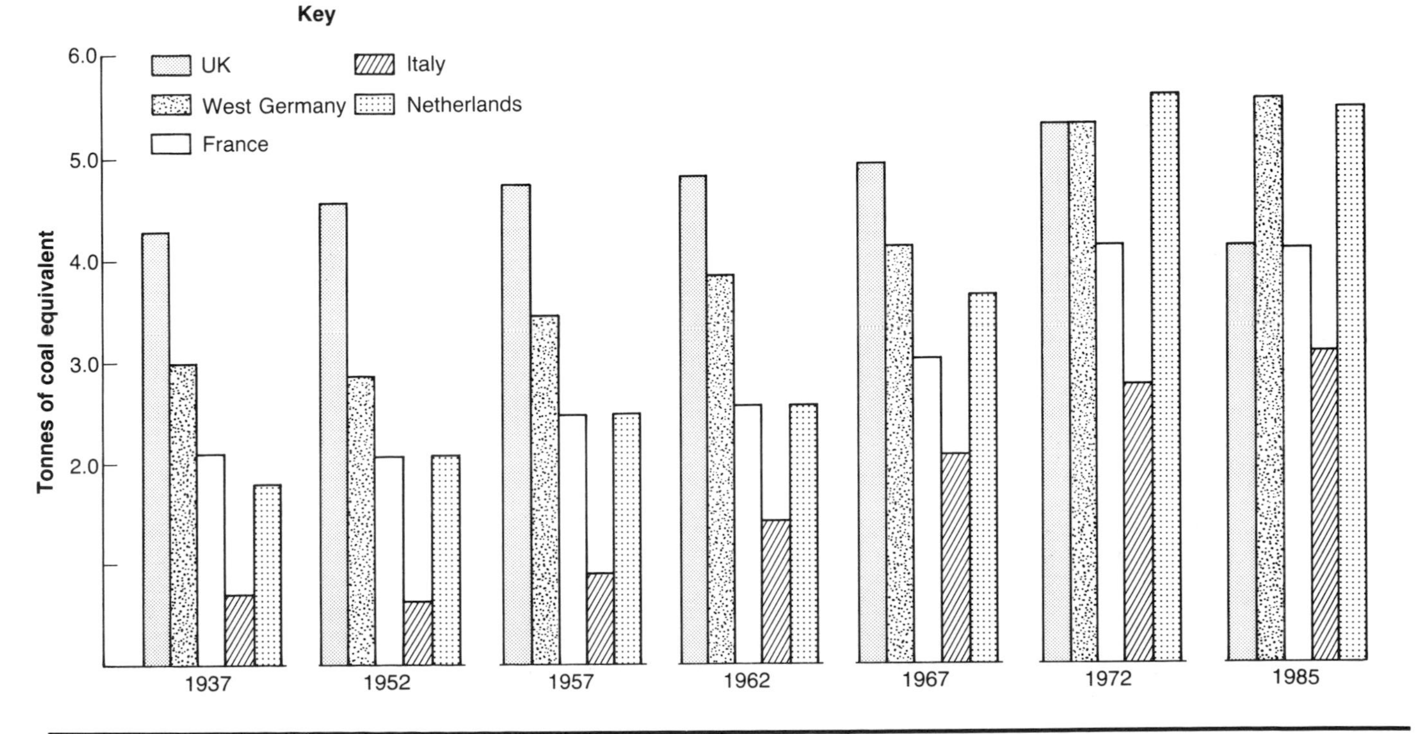

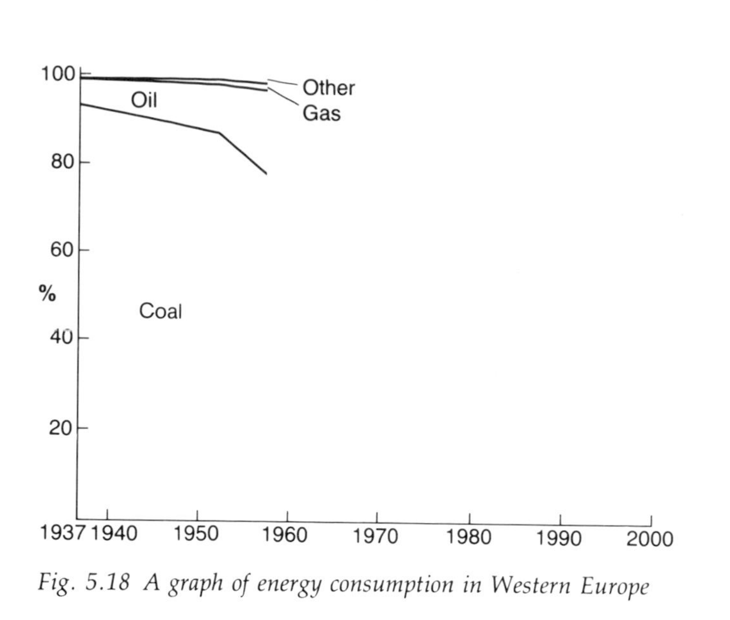

Fig. 5.18 A graph of energy consumption in Western Europe

Table 5.6 Energy consumption in Western Europe, 1937–1986

	1937	1952	1962	1967	1972	1986
Coal	92.77	87.04	62.64	44.95	27.54	21.66
Oil	5.87	11.04	33.02	48.15	55.08	44.86
Gas	0	0.39	2.55	4.72	14.85	18.66
Other	1.36	1.53	1.79	2.18	2.53	14.40

'Other' includes energy generated as hydro-electricity and nuclear power

Fig. 5.19 Graph showing projected energy growth rates for selected European countries

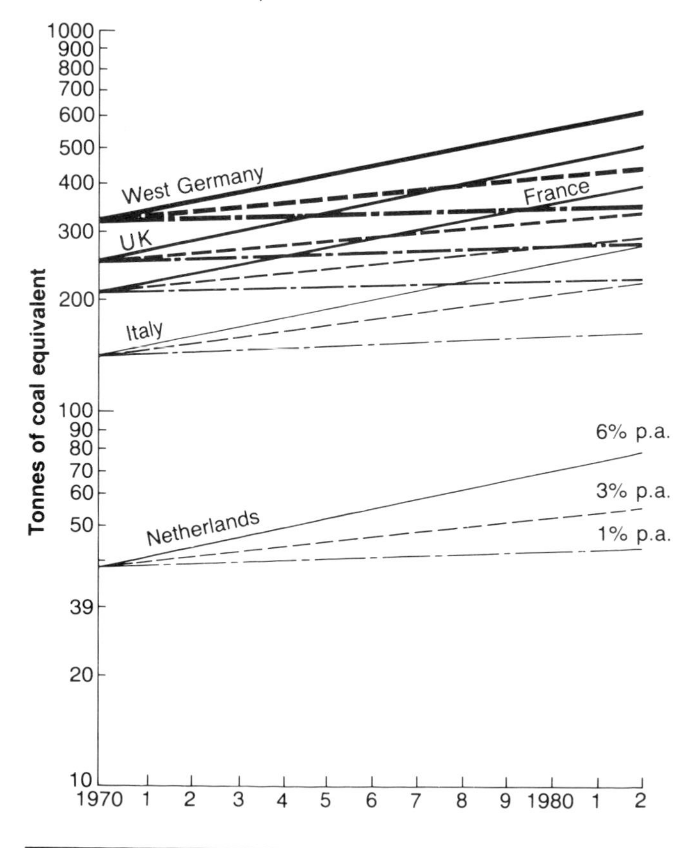

23 The graph in Fig. 5.19 shows the level of energy consumption in France, West Germany, the UK, Italy and the Netherlands in 1970. The lines on the graph show the projected energy growth rates predicted from 1970, with a high level at 6% (which was the popular projection of the time), a middle growth rate at 3%, and a low growth rate at 1%. Using the table showing the actual energy use (Table 5.7), plot the figures for each country on a traced copy of the graph.

24 What conclusion can you draw from these figures on the success of the energy policies outlined above?

In fact the drop in demand for energy and the success in locating new supplies of indigenous supplies of fuel have led to energy experts revising the fuel contributions for the year 2000 so that coal is projected to provide 16.7%, oil 40.7%, gas 33.3% and hydro-electricity and nuclear 9.3%.

Table 5.7 Actual energy consumption for selected European countries (million tonnes coal equivalent)

	1970	1975	1980	1982	1985	1987
UK	351.2	336.0	333.8	323.8	340.2	348.9
West Germany	389.3	402.0	451.4	415.5	444.6	445.4
France	237.1	263.9	308.1	292.6	323.5	336.8
Italy	175.9	203.9	224.4	212.4	221.4	233.6
Netherlands	80.0	98.1	108.6	94.7	102.3	109.4

Case-study: coal in Britain

In Britain the policy was to allow private companies to prospect for oil and gas, both at sea and on land, and also to look at a policy which was based on coal, conservation and nuclear power.

Coal had been declining in output over the last two decades. The output figure for 1972 was only 58% of the output figure for 1952. However, first the 'Plan for Coal' in 1974 and later plan '2000' projected coal output at a stable level of 135 million tonnes per annum (the 1972 output figure was 132 million tonnes per annum). The later plan revised this figure to 170 million tonnes per annum. However, to reach these targets some degree of restructuring was called for, and this meant that older collieries, where coal was being mined 'uneconomically' would close while investment would go into new mines with large reserves of coal. These new mines were to contribute an extra 60 million tonnes per annum to the total from the existing mines.

Fig 5.20 shows the locations of these new mines. Most of them are extensions to existing coalfields, although the Oxfordshire reserve would be a completely new area.

The investment programme reflected the plan with 81% of total investment going to the central area, Yorkshire, East Midlands and South Midlands, with

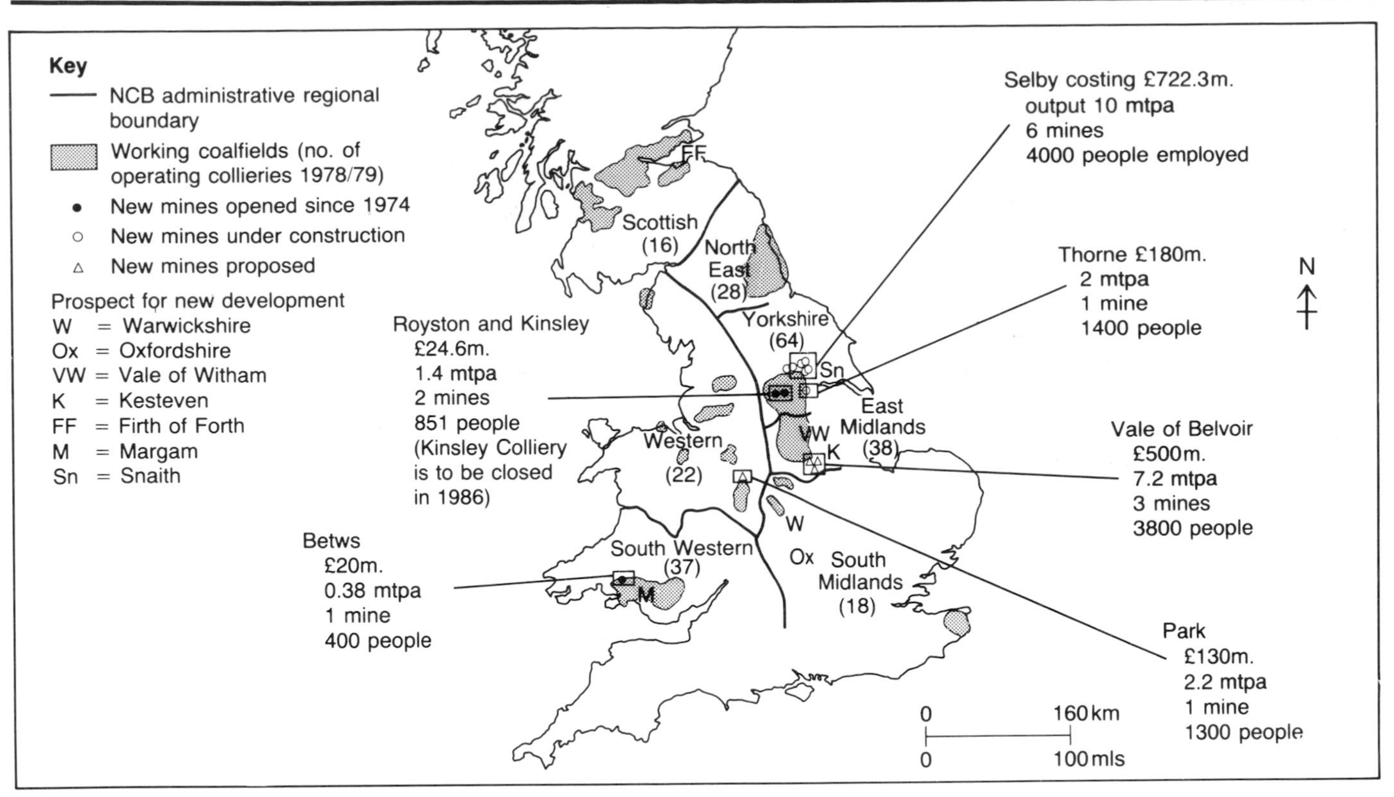

Fig. 5.20 UK coalfields in 1979 – new mines and prospects

the rest of the investment shared in the peripheral regions. Since 1974, 36 collieries have been closed, 31 of them in peripheral areas with 17 800 workers losing their jobs, while there have been 5 closures in the central areas with a net gain of 700 jobs. British Coal is investing in areas such as Selby where it is projected that 10 million tonnes of coal will be produced per annum employing 4000 miners, rather than investing money in an area such as South Wales where the output is 6.9 million tonnes per annum, employing 22 700 miners.

However, these plans have not been free of problems. Some of the new mines are in rural areas either with an agricultural economy or a population of commuters from nearby urban areas or retired people. The Vale of Belvoir (Fig. 5.21) is a good example of such an area and the public reaction is so strong in these areas that a public inquiry is necessary to pacify the local community. These public inquiries can be very lengthy – it took a total of seven years to sink one mine at Ashfordby, in the Vale of Belvoir – and costly to British Coal in reduced

Fig. 5.21 Conflicts in the Vale of Belvoir

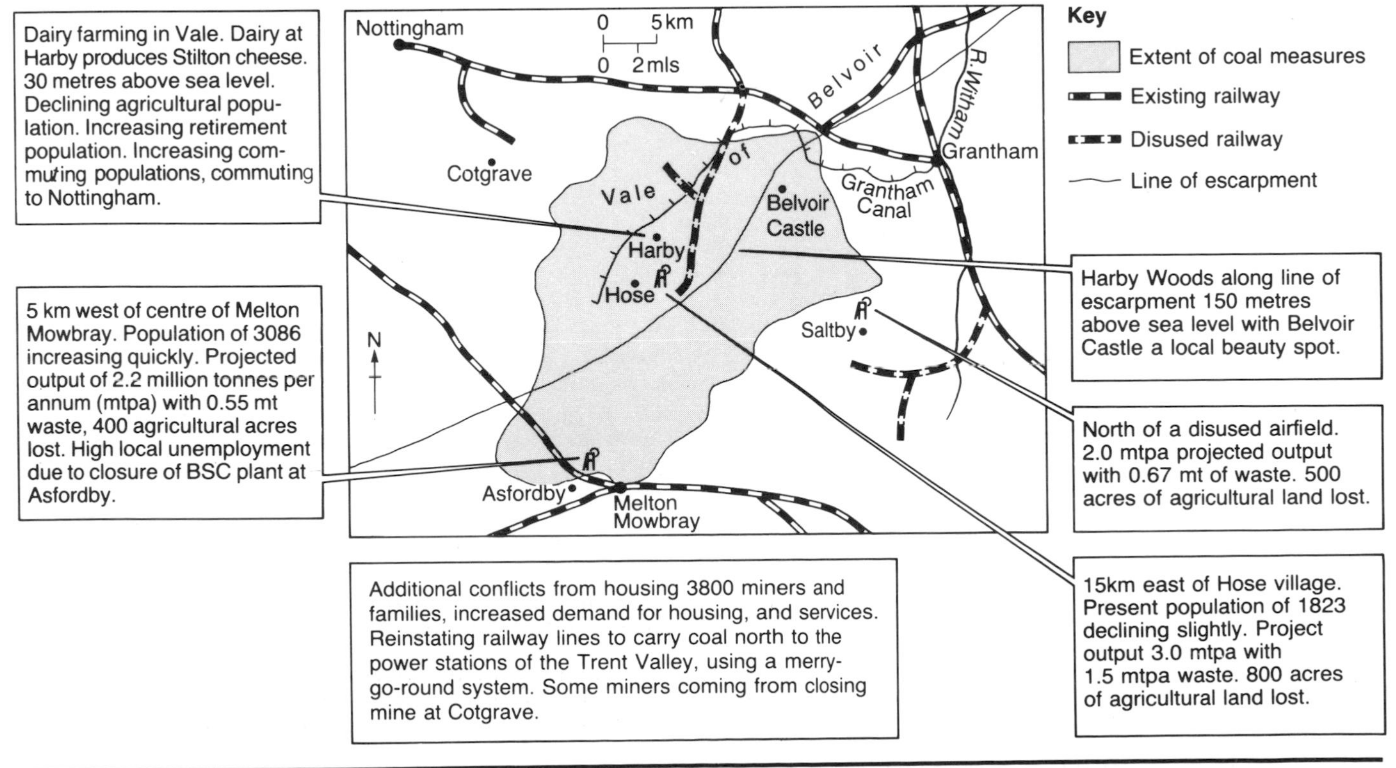

output. These inquiries have made the NCB revise its targets and they have been reduced by 10 million tonnes per annum.

These changes which can take place are similar to those in Fig. 4.1 on p. 104 for Cumulative Causation.

25 Re-draw the model (Fig. 4.1 on p. 104) to show the effect of sinking new mines in the Vale of Belvoir area. The cutting from the *Guardian* (Fig. 5.23) may help you.

26 Make a list of the local impacts on the environment, physically and socially, following the sinking of a mine in a rural area. Try to indicate whether these impacts are positive (enhancing) or negative. You should try to do this from the point of view of the miner moving in with his family from an old coalfield, a farmer who will lose some land in the development, a retired individual who has sought peace and quiet and the local councillor for the District Council. Try to devise your own system of giving positive and negative scores and then summing to get an overall value. Compare your results with your class and criticise the method you have used.

In areas such as South Wales, the model may be seen to work in a negative direction, as collieries are closed down.

27 Re-draw the model (Fig. 4.1) to show how the area may deteriorate following the closure of the colliery.

28 Working in pairs, write a letter each, one as if you are a miner explaining the decline of the community to a national newspaper, and the other as if you were responding for the British Coal explaining the Board's policy.

29 Look at the situation nationally. If the investment programme continues, how will the energy map of Britain look in 20 years' time? What effect will it have on the regions? Draw an annotated sketch map like Fig. 5.21 to summarize your views.

The problem is made more difficult for the peripheral regions as the 'special' coal that some of them produce is not in demand. The more general coals of the central area are more marketable.

British Coal depends on selling coal to the Central Electricity Generating Board (CEGB) and to the British Steel Corporation (BSC). The CEGB bought 71% of the coal produced in 1982–83, but only one new coal-burning power station has been constructed, the Drax 'B' station. And, the BSC, which takes 18% of the coal produced, has also been reducing and rationalizing steel production, particularly in the peripheral regions with the closure of Consett, and reduction at Llanwern and Port Talbot.

Fig. 5.22 A recent newspaper article

Turfing out an unwelcome neighbour

Mrs Margaret Drumbreck's family has been farming mushrooms in Fife for 50 years. Now she fears for her livelihood.

British Coal plans to convert 730 acres of land next to her farm at Dysart, near Kirkcaldy, into an opencast mine. She has assembled evidence from mushroom experts which suggests that her crop will be fatally damaged by the fungal pathogens released into the air by the mining operation.

This will mean the loss of 35 jobs and the destruction of her family business. 'To consider compensation in purely financial terms is gross arrogance,' she said. 'This is our place of belonging, our own hard-won establishment, our quality of life. No amount of money can replace that.'

She, along with Mr Lyle Laird, a tenant farmer who stands to lose 20 per cent of his land, have formed an action group to oppose British Coal's plans, which they say will destroy the environment and harm tourism. There have been angry local meetings about the proposals, due to be considered at a special meeting of Fife regional council's planning committee on July 6.

British Coal says the opencast work will provide about 90 jobs and yield up to 1.5 million tonnes of coal over six years. But, crucially, it links it with longer term plans to develop the huge reserves known to be under the Firth of Forth, promising 1,180 jobs and 45 million tonnes of coal over 25 years.

These undersea reserves used to be extracted via the nearby Frances and Seafield collieries until they closed after fires, the former in 1985 and the latter this year. As well as developing the opencast mine, British Coal wants to sink two new drift mines under the Forth.

The undersea development, however, has been put in jeopardy by the hard bargaining being done by the South of Scotland Electricity Board in the run-up to electricity privatisation. Recent reports have suggested that the kind of short-term agreement likely to be reached would make it difficult for British Coal to justify the £100 million investment.

The action group, which is not opposed to the undersea development, suspect that it is merely bait to help persuade people to accept the more controversial – and more profitable – opencast work, the real aim being to develop more opencast mines.

The group is also alarmed at new evidence on the health effects of opencast mining from Glynneath in South Wales. Doctors there claim to have uncovered alarmingly high rates of asthma, ear infections, rheumatoid arthritis and septic ulcers, as well as enhanced rates of cancer and deformed babies. They blame the dust from an opencast coal mine.

A spokesman for British Coal insisted that the opencast and undersea operations were integral and mutually beneficial parts of the same project. He said that all the steps necessary to minimise dust and other intrusions would be taken. 'There is no evidence to suppose that opencast mining has detrimental effects on health and that is based on sound medical advice,' he said.

NCB plan for £450m pit faces united opposition

House prices and ancient woods find plenty of protectors, writes Paul Hoyland

OPPOSITION groups joined forces to fight the National Coal Board's plans to sink its biggest mine in a prosperous part of the Midlands.

The two sites short-listed for the £450-million mine are on the outskirts of Coventry, where 7,000 residents would be affected, and five miles away in green belt land on the edge of Kenilworth.

The board has commissioned environmental consultants to analyse the impact of the 2,000-job mine. The opposition groups have formed a federation and are preparing detailed research for a public inquiry, whichever site is selected.

Dr Peter Lea, the federation's spokesman, said that whatever the environmental arguments, the opposition groups did not believe the board had proved its case for developing the mine.

"They have not got a market for the coal they are mining at present and there is no guarantee they would have a market for this proposed extra 3 million tonnes a year. Even if there was a market, they could reach the coal from existing sources," Dr Lea said.

The NCB says it can barely meet demand for the Warwickshire seam's high quality coal. By the end of the century some collieries which supplied that market would be exhausted.

To mine new reserves by tunnelling from an existing pit at Daw Mill would increase capital cost by 25 per cent and be less efficient, said an NCB spokesman.

At the proposed site near Coventry at Hawkhurst Moor Farm in the village of Berkswell, the pillars of the local establishment have been organising protest marches.

"This is a beautiful village and the pit would destroy it", said Mr Frank Chambers, deputy director of the Berkswell Society Mine Opposition Group, better known as Smog.

"You won't get a property here for under £100,000. I am shattered and disillusioned that the Government will allow the board to come along and make properties depreciate."

Mr Chambers, a financial management consultant, paid £105,000 for his house 16 months ago and has spent £20,000 on renovation. The proposed pit would back on to his barn.

Leaders of the Allesley and Meriden Mining Opposition Group and Woodlands Against the Mine estimate that up to 7,000 people living within one kilometre of Hawkhurst Moor Farm would be affected by noise and dirt from the mine.

The NCB spokesman countered: "We are perfectly aware of the number of people who live fairly close to this site. However, we reject some of the more panicky statements as to what the effects would be. A large proportion simply wouldn't notice the mine being there, although for those close by it would have quite an impact."

At South Hurst Farm, near Kenilworth, opposition groups hope to protect an area of ancient woods believed to be the site of a medieval village. "As an amenity, this area is unique within the outskirts of both Kenilworth and Coventry," said Mr David McInnes, a supporter of the Kenilworth Mine Opposition Group.

Mr Vernon Watterson, the group's chairman, said the 5,500 students at Warwick University's halls of residence, less than a mile from the site, would be affected.

Kenilworth's tourist industry would also be undermined as the development swamped the area with coal movements by road and rail.

Mr McInnes said the destruction of the green belt would result in Kenilworth being absorbed into Coventry. The destruction of the recreational area would be seen as "a philistine act—an act of vandalism," for which the board would never be forgiven.

The NCB spokesman said: "There is more of an impact on a pretty area but this site is in a valley surrounded by woods. . . . You would have to go up to the fifth floor of the university's halls of residence before the site became visible.

"We are measuring noise and dust at existing collieries so that we can apply the information to the two sites. We will then provide the facts — as opposed to the scare stories that have been based on guesses".

Dr Lea said: "You can't hide a pit of this size. It will cover some 200 acres and will be horrendous. We will make more noise than the campaigners in the Vale of Belvoir, because, as well as the rural communities, we have the town of Kenilworth and much of Coventry behind us".

Fig. 5.23 A recent cutting from the Guardian

30 Plot these locations on the sketch map you produced in answer to Question 29.

The picture is further complicated by the influence of cheap imported coal. Unit costs are cheaper for coal from Poland, Australia, and the USA. In Poland, the stimulation of the coal industry is part of national economic planning strategy, while the other countries mine coal using open cast methods which greatly reduces cost. The future of coal-mining in Britain and the communities that depend on coal may well be determined by decisions made by the CEGB and the BSC as to where they buy their coal.

British Coal are looking more closely at opencast operations as Fig. 5.22 indicates. Is it more disruptive to excavate all the coal with opencast methods and reclaim the land or sink shafts and mine the coal over many years?

Case-study: alternative energy for Britain

Supplying energy from alternative sources is an initially attractive proposal, mainly because of the renewable nature of the resources suggested – from water (waves, tides, hydro-electricity), the sun (solar panels or ponds), and from the earth (geothermal). The main drawbacks are the cost of investment in research and the problem of scaling up from the experimental model. In fact, with the exception of wave power, all the alternatives are in operation somewhere.

Alternative sources are only half the story. Energy can be used more efficiently, as in the combined heat and power station, and design of buildings can be important. Fig. 5.26 shows some of the features which have been incorporated into the design of houses on some estates in the new town of Milton Keynes.

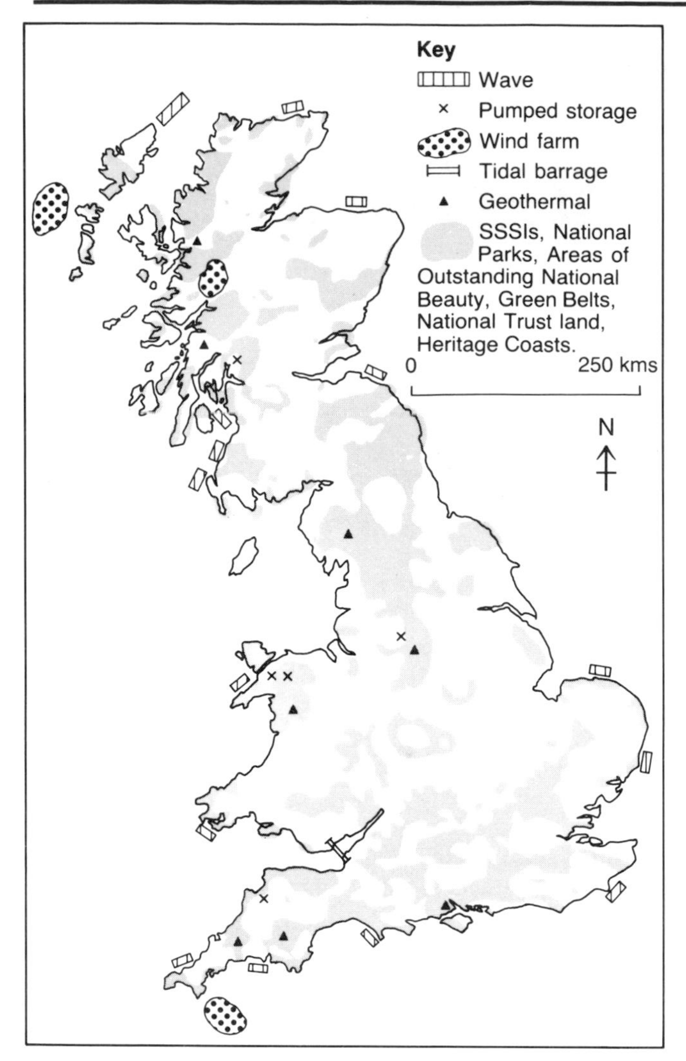

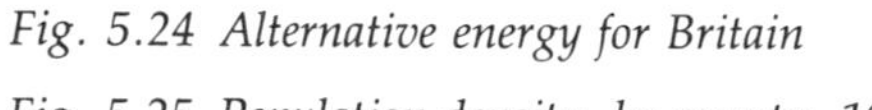
Fig. 5.24 Alternative energy for Britain

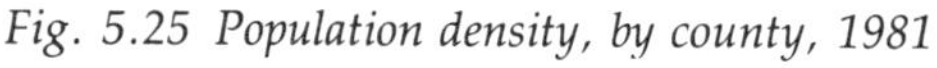
Fig. 5.25 Population density, by county, 1981

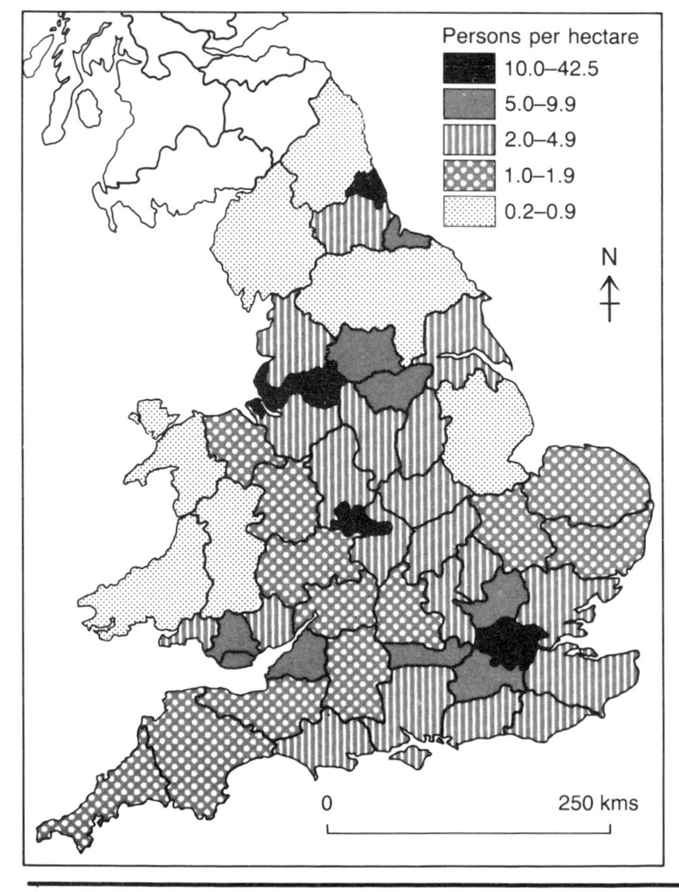

31 Fig. 5.24 shows some possible sites for extra alternative power stations and one of the main limiting factors, conservation areas. Fig. 5.25 gives some indication of where the demand is and where it is likely to be in the future. Make a copy of Table 5.8 and complete it. Do not forget to consider supply of power to the market, i.e. transmission of energy, in your analysis.

In fact Milton Keynes has an energy park in which many builders display energy saving designs. We may be asked to make personal choices to reduce demand for energy and effect environmental improvements. Why not share cars to travel to work and save on petrol? Lanes of the motorway network could be saved for such drivers, or they could pay less road tax. These are the sort of small but significant things we may have to consider, sooner rather than later!

32 (a) Make a sketch of your house or part of your school and try to produce some recommendations of where the design could have been improved to make it more energy efficient.
(b) Try to brainstorm as many ideas as possible which could be used to improve energy efficiency in Britain.

It has been estimated by experts at the Open University that most of the energy demand of the United Kingdom could be supplied by alternative sources, only requiring a small amount of conventional power for peak periods.

Case-study: nuclear power in Britain

The CEGB have been building and using nuclear power stations for a number of years. The locations of these stations are given on the map (Fig. 5.27). This strategy was accelerated in 1979 with an announcement by the government that £15 000 million was to be invested in building new nuclear stations using the Pressurized Water Reactor (PWR). The government wanted ten stations built in the next ten years. However, the public interest was so great that it was decided to submit the first application to a public inquiry.

The procedure for such large-scale planning matters involving a public inquiry is given in Fig. 5.28. The CEGB formally submitted plans on 1 October 1980 to build a PWR on the East Anglian coast at Sizewell, next to an existing Magnox nuclear station on land owned by the CEGB. This new station was to be called Sizewell B and the public inquiry got under way on 11 January 1983. It was held at Snape Maltings and was under the chairmanship of Sir Frank Layfield, an inspector appointed by the Secretary of State for Energy (Fig. 5.30).

Alternative	Description	Impact on (benefits and drawbacks)		Scale (H.M.L.)	Cost (H.M.L.)
		Physical Environment	Social and Economic Env.		
Wave Energy	Floating platforms of ducks, hinged raft, oscillating water cylinders etc. off shore extracting energy from waves. Experimental scale only at present				
Pumped storage	Energy extracted by water falling from upper lake to lower at peak demand time. Pumped back when surplus energy. In use in N. Wales e.g. Dinorwic which has been totally enclosed in mountain.				
Wind farm	Group of wind turbines most probably off shore for maximum wind in wind farms but could be reduced in scale and used over a wide area. CEGB have small turbine in Dyfed. Large ones 150 metres high.				
Tidal Barrage	Barrage which allows water through and generates power from the ebb and flow of the tide. Has been developed in France at La Rance. Could be developed on the Severn Estuary.				
Geothermal	Has been used in New Zealand and Iceland by using heat from the earth along plate boundaries. In UK old crystalline rocks provide potential. Water pumped down heats up and returns to surface. Experimental plant at Cambourne.				
Combined Heat and Power	Small conventional thermal plant built near factories so that waste heat can be utilised in manufacturing process. Some built already, e.g. Hereford with the Bulmer's factory. No large cooling tower as waste heat retained.				

Table 5.8 Alternative energy schemes

Fig. 5.26 An energy-efficient house (Adapted from the Sunday Times, *3.1.85)*

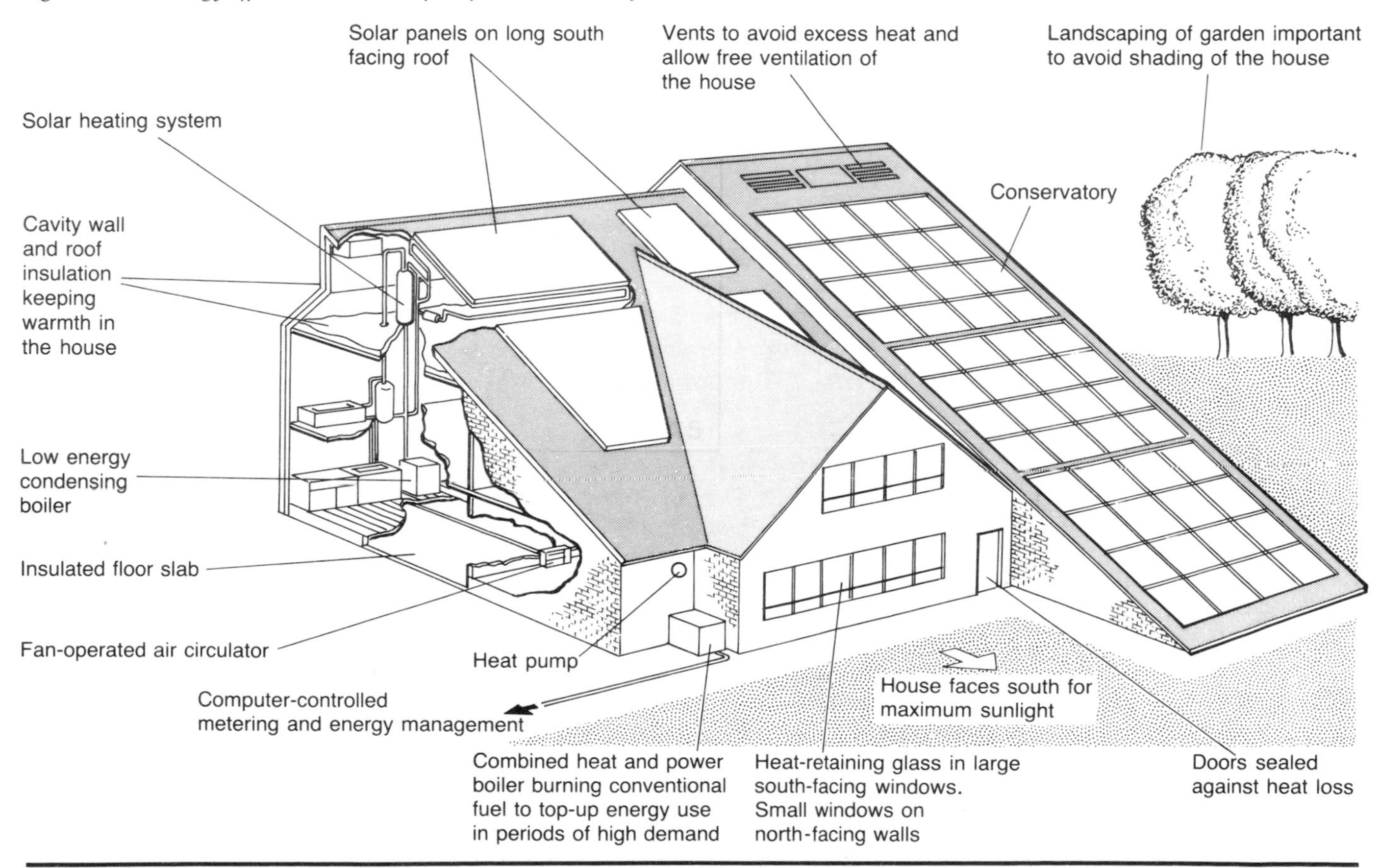

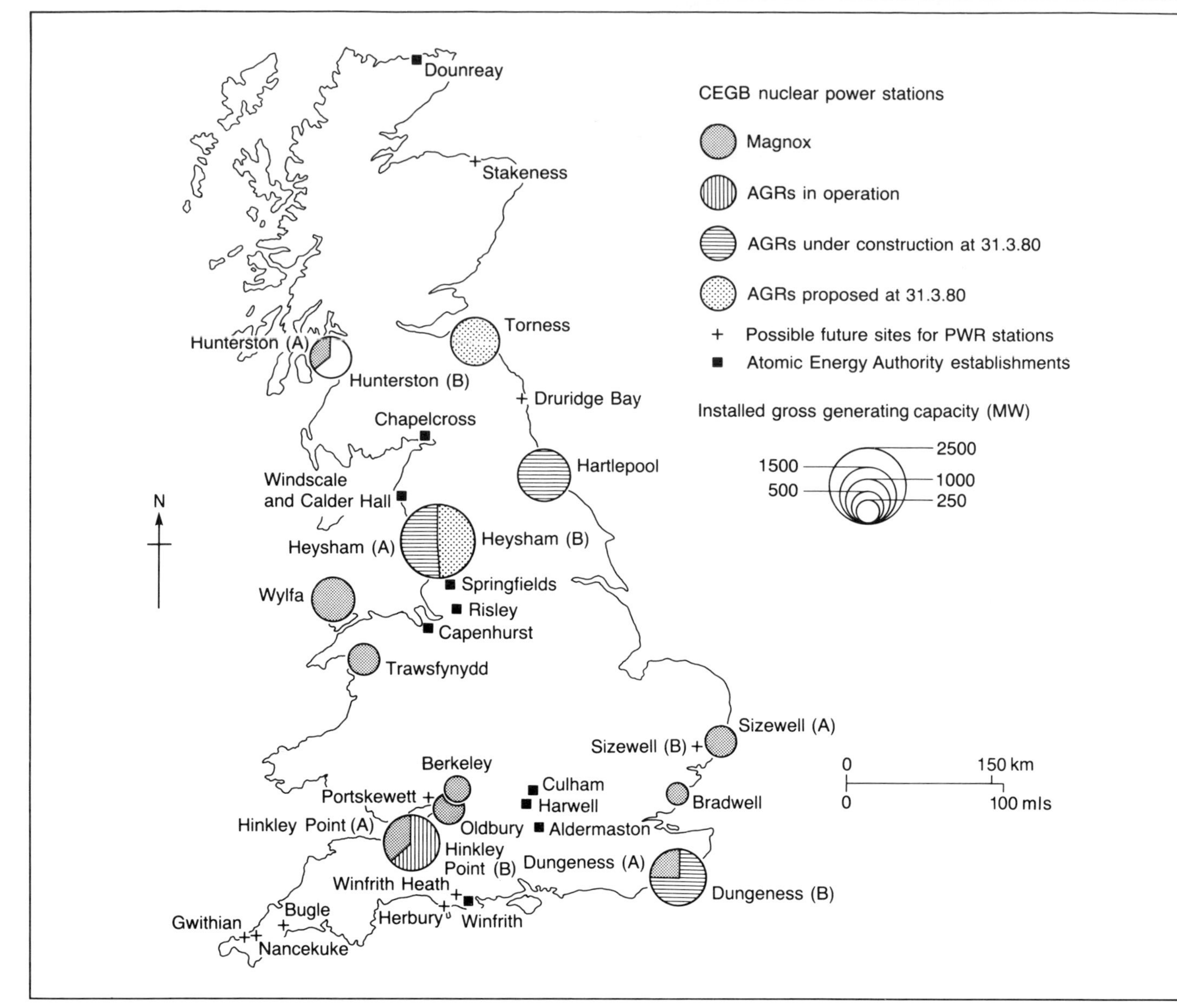

Fig. 5.27 CEGB nuclear power stations and AEA establishments in the UK, 1980

Fig. 5.28 Planning decision-making in the UK

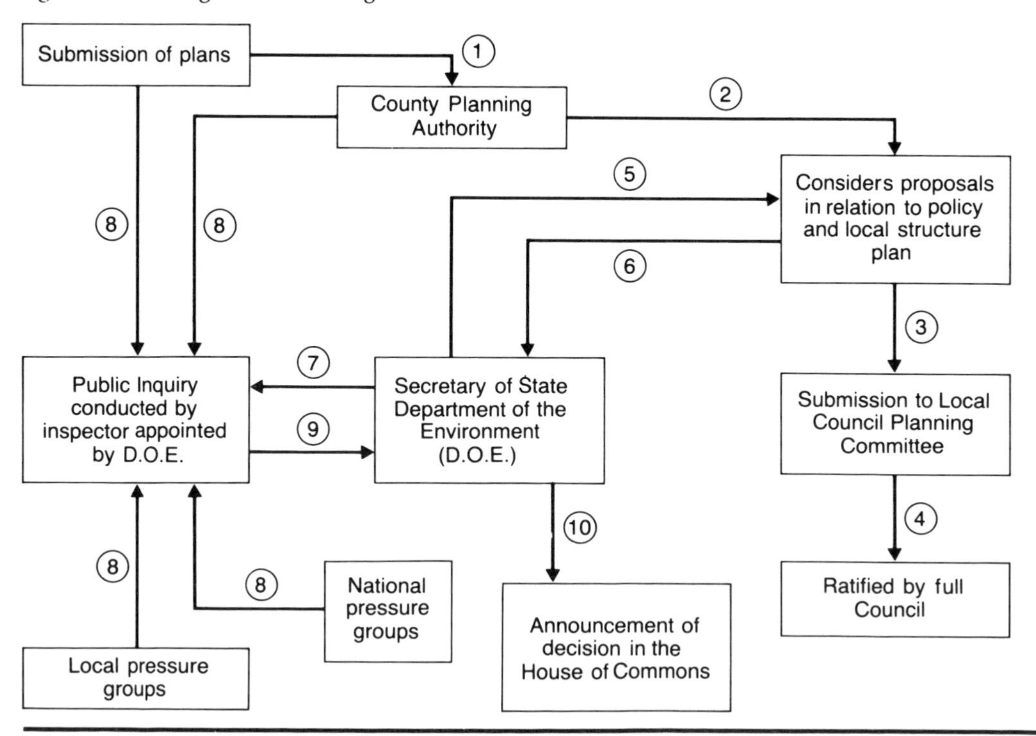

Decision-making Exercise P: Conducting a Public Inquiry

Hold a public inquiry of your own. Appoint a chairperson and allocate the roles given in Fig. 5.29. Read the information given for your role and prepare a speech for the inquiry. The chairperson should ask for contributions, starting with the representatives of the CEGB and their supporters and then going over to the opponents of the scheme. It is up to the chairperson as to when the neutrals are asked to speak. Cross-examination of each speaker should be allowed after each has spoken.

After the inquiry, decide as a group which side presented the best information and won the argument. Try to distinguish between the information which was factual and information which was emotive.

Working in pairs, write two newspaper reports, one for the local newspaper, the *Leiston Gazette*, and one for a national newspaper covering the inquiry. Try to give a factual coverage of the arguments and try to use the style of writing you would expect. When you have finished, compare the reports for bias.

The actual public inquiry cost £25 million and sat for 18 months. The CEGB spent between £5 and £7 million of tax payer's money in presenting its case and in addition paid the expenses, administrative costs and expenses of the inspectors. Before the inquiry started, objectors applied for public funds to help argue their case. This was refused by the Government and hence such organizations as Greenpeace, an environmental group, did not bother to attend. (The attitude of Greenpeace is made clear in Fig. 5.30.) The total time spent on the whole project amounts to many years. Planning the applications will take some 7 years, the formal application including the public inquiry will take another 2 to 3 years, the station will take 7 years to build and will then last between 25 and 35 years with a further 5 years to be decommissioned. All in all, a very expensive enterprise.

Fig. 5.29 The Sizewell B public inquiry

Planning decision-making

Detailed plans are submitted to the County Planning Authority (1) who consider them in the light of the local policies and structure plans for the area (2). These may then go to a sub-committee of the Council for discussion (3) and if passed they go to the full Council for ratification (4). However if the plans affect major areas of land and are the source of public outrage, the Department of the Environment may be notified by the local Planning Authority (5) and the plans may be 'called in' (6). The Secretary of State may decide that the plans are sufficiently controversial to call a Public Inquiry (7) and an inspector is appointed to hear all sides of the problem (8) and to prepare a report. This report is sent back to the Secretary of State (9) who then makes a decision and announces it to the House of Commons (10).

Fig. 5.29 The Sizewell B public inquiry (cont.)

Roles for the public inquiry

P

FOR

CEGB. They represent Westinghouse Corporation and Bechtel, the designers, architects and engineering consultants making the PWR. They argue that Britain needs to build up its capacity for generating power, especially as the costs of fossil fuels increase. It is the responsibility of the CEGB to provide a reliable supply of electricity and the PWR is a proven technology used in the USA and will produce electricity at the cheapest rate. Supplies of uranium are assured as they come from 'friendly' countries, such as Australia and South Africa. They project energy consumption to increase in the future by 1% and see the station lasting for 35 years.

National Nuclear Corporation. They argue that many jobs will be generated by this project and ones like them. They state that 20 000 jobs will be created at CEGB and NNC headquarters and in subsidiary industries and that some 12 000 man years worth of work will be created at Sizewell to act as a boost to local employment opportunities.

British Nuclear Fuels (BNF). These people will be responsible for supplying fuel, and reprocessing the spent fuel rods at Sellafield in Cumbria. They argue that the industry has an excellent safety record and that Britain has the complete nuclear technology so that we will be using all-British resources.

NEUTRAL

Ministry of Agriculture, Forestry and Fishing. Concerned about discharge of nuclear contaminated water into the North Sea affecting fish stocks.

Department of Transport. Carriers of nuclear products to and from Sellafield.

Electricity Consumers Council. Doubtful of CEGB figures but not so extreme as the CPRE. They see fossil fuel prices increasing only marginally and agree with CPRE on the length of time the station will be in operation (25 years), but they extend the construction time by another 18 months longer than the CPRE. Overall they see benefits but they are not as optimistic as the CEGB.

Other neutrals are the **UK Atomic Energy Agency**, **The Department of the Environment** and **Suffolk County Council**.

Although the UK Atomic Energy Agency had a vested interest in the result of the Public Inquiry, they did not give evidence to support the application of the CEGB.

AGAINST

Council for the Protection of Rural England (CPRE). They argue that the demand for energy is not going to rise as fast as that predicted by the CEGB – they favour the growth of −0.4%. They also argue that the building costs have been understated by up to 74%, that the station could take longer than the 90 months to be built given by the board (they say an extra 24 months) and that the station will only last for 25 years. They do not see fossil fuel prices rising as quickly as the CEGB and consider that the decline in demand for electricity will be down as industries have gone out of business in the recession. They favour the combined heat and power station for the future.

Friends of the Earth (FOE). They are concerned with broader issues, such as the safety of nuclear power in general. They cite the examples of the nuclear accident at Harrisburg, USA, when the core of a PWR overheated and pregnant women and children were evacuated. This happened in 1979. They also use the example of a supposed nuclear accident in USSR south-east of Sverdlovsk in 1958 when many people are assumed to have died and many hectares of land laid waste. They favour the development of alternative forms of energy production – solar, wind, wave and tidal.

Town and Country Planning Association. They are similar to CPRE, but also sceptical of CEGB involvement with Westinghouse-Bechtel in a deal to sell PWR technology to China and see Sizewell as being the CEGB's display of technology for any future customers. Other groups who opposed the application were Greater London Council and the Northumberland County Council.

Stop Sizewell B Association. This is a local pressure group concerned with the impact on the local environment and in particular spoiling a stretch of coastline just south of a national bird reserve and on the possible effects of a nuclear spillage on the health of local people.

Campaign for Nuclear Disarmament (CND). Concerned about the possibility of proliferation of nuclear technology and in particular of plutonium, manufactured in the reprocessing of uranium fuel rods, and the availability of this plutonium for the manufacture of nuclear weapons. They cite figures from polls taken in different countries as to the widespread dislike of nuclear technology – percentages of people who would like to disengage from nuclear technology: Norway 83%, Finland 82%, Canada 55%, Switzerland 65%, Spain 50%, West Germany 45%, USA 45%, UK 42% and Brazil 41%.

National Union of Mineworkers (NUM). They are concerned about the affect on coal mining communities due to a reduced demand for coal if the CEGB increases its nuclear capacity. Also they are in conflict with the Conservative Government, as they have leaked cabinet minutes which include the following quotation:

'A nuclear programme would have the advantage of removing a substantial proportion of the electricity production from the dangers of disruption by industrial action by coal miners . . .' The *Observer*, 9.12.79.

Fig. 5.30 A Greenpeace poster

> **33** Consider the stages of the planning application. Is it the best way of making decisions? Is it democratic or could it be improved? Discuss alternatives, in the light of the costs and the length of time each step takes.
>
> **34** Should the Government finance the opposition to such an inquiry?

The decision to go ahead with the construction of a new nuclear power station, Sizewell B, has been given. The government's decision is final but now the construction has started the costs are much larger than estimated.

Case-study: nuclear power and radioactive pollution

The Bhopal disaster was contained within a relatively small area, but radioactive pollution can spread much further. On 26 April 1986, one nuclear reactor hall of a nuclear power station at Chernobyl in the Ukraine, USSR, exploded, leading to the release of a huge cloud of radio-active material. Initially weather conditions led to the deposition of this pollution in the area immediately around Kiev, but later the cloud was driven towards Scandinavia and Eastern Europe, eventually crossing France and the United Kingdom. The highest radiation counts in the United Kingdom were recorded over North Wales and Scotland, where the heaviest rain fell while the cloud was over the country. Later still, the cloud made its way across the Pacific to the west coast of the United States and Canada. This was the greatest nuclear accident recorded, but even so it could have been so much worse. The reactor itself was very near to total 'melt down', where the reaction gets out of hand and melts its way down through the reactor floor, perhaps polluting the groundwater. If this had happened, far greater repercussions would have been felt. Nine people working near the reactor were killed and many others were treated for the effects of radiation. The true death toll will probably never be known because the cancers which will be caused by radiation will be difficult to pin down to this explosion. As it was, the reaction of the British Government was to ban the import of foodstuffs from those countries most affected by the fallout and to advise against visiting those same areas. The other reaction to the disaster was for pressure groups opposed to the nuclear industry to gain credence for what they had been warning and for great political pressure to be brought on the continued use and planning of nuclear power stations in the United Kingdom. There was talk of the outcome of the public inquiry into the Sizewell B station being affected by this disaster. Internationally, the disaster led to a recognition that such industry has an obligation to inform the whole community of any problems and not to attempt to cover them up. In the United Kingdom, it was decided to inform people about any nuclear leak, no matter how small.

35 Under what atmospheric conditions would the cloud from Chernobyl have taken the course shown in Fig. 5.31?

36 In your view, what should be the procedure for a country to announce a nuclear accident to its neighbours?

37 In the light of the major problems associated with nuclear power plants, should they continue to be used? Try to think of arguments for and against their use and come to your own conclusions

38 Should the use of nuclear power be left to one country to decide on for itself, or is it an international issue? In the latter case, how should international decisions on nuclear power be made?

Fig. 5.31 The radiation cloud from Chernobyl

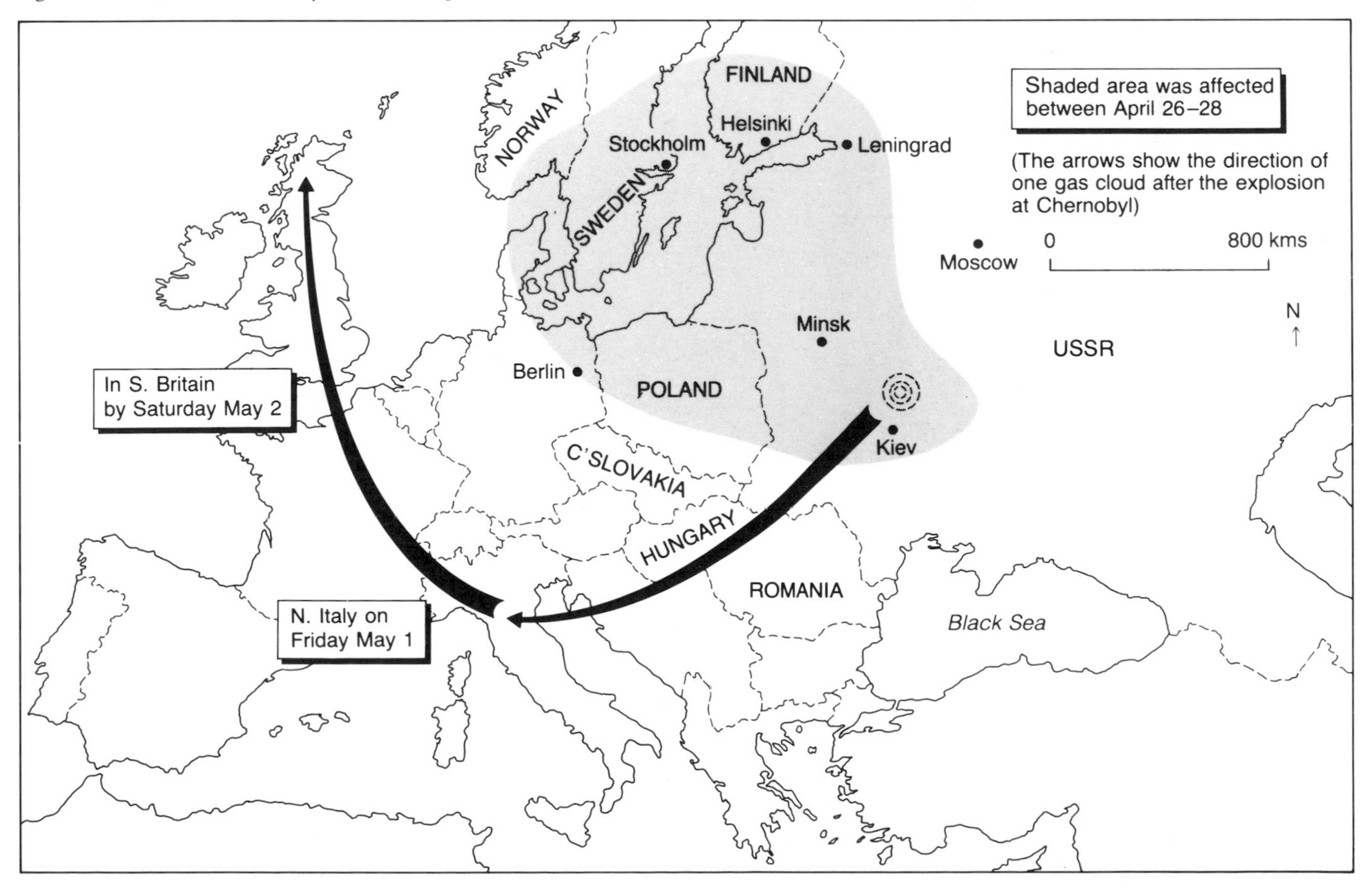

Spearman's rank correlation coefficient

This is a method for showing how well two sets of data relate to one another. As an example, let us look at whether there is a relationship between the size of a factory and the number of people employed.

Factory	Size (00 m²)	Number of Employees
A	500	1000
B	750	1100
C	250	500
D	100	70
E	50	45
F	620	70
G	120	72
H	110	40

In order to calculate Spearman's Rank Correlation Coefficient you need to follow the same steps as with the chi squared calculation shown on p. 79.

a) Formulate the null hypothesis – the biggest factories do not employ the most people.
b) Formulate the alternative hypothesis – the largest factories employ the most people.
c) Agree on the level of acceptance – at what level will I have enough statistical evidence to alternative hypothesis. Usually taken to be 95% i.e. only 5% chance that the relationship calculated has come about because of a chance of set figures.
d) Now draw out the table as shown in Table T5.1 and substitute the appropriate figures into the expression.

Number (n) is 8

Calculate Spearman's Rank Correlation Coefficient by substituting into the formula

$$R_s = 1 - \frac{6 \times \Sigma d^2}{n\,(n^2 - 1)}$$

In this case it is

$$R_s = 1 - \frac{6 \times 5.5}{8(64-1)}$$

$$= 1 - \frac{135}{504}$$

$$= 1 - 0.268$$

$$= 0.732$$

e) You still have to find out if the coefficient is significant and if you can reject the null hypothesis and accept the alternative hypothesis. This is done by using the significance tables shown in Fig. T5.1. Before using these you work out the degrees of freedom (df) which is calculated by taking the number of paired values and subtracting 1. In this case df would be 8−1 = 7. Find df value 7 on the significance table and look at the values across to the right. In this case df value 7 has the value 0.714 beside it. Our correlation coefficient was 0.732 so is larger than the value given in the table. This indicates that our value is significant at 95% which was our rejection level. We have enough evidence to reject the null hypothesis and say that the larger factories do employ more people. Notice that our value is not good enough to be significant at 99%, but 95% is enough for our purpose.

Fig. T5.1 Critical values for Rs
The value of r_s for any given value of N must be equal to or *larger* than that shown for the level of significance required.

N	Levels of significance .05	.01
4	1.000	
5	.900	1.000
6	.829	.943
7	.714	.893
8	.643	.833
9	.600	.783
10	.564	.746
12	.506	.712
14	.456	.645
16	.425	.601
18	.399	.564
20	.377	.534
22	.359	.508
24	.343	.485
26	.329	.465
28	.317	.448
30	.306	.432

Source: Olds, E. G. in *Annals of Mathematical Statistics* (1938), Vol. 9, pp. 133–48 and (1949), Vol. 20, pp. 117–18

Table T5.1 Spearman's rank correlation coefficient

Col 1 Factory	Col 2 Size	Col 3 Rank of Col 2	Col 4 No of employees	Col 5 Rank of Col 4	Col 6 Differences between ranks (Col 3 – Col 5)d	Col 7 Differences between the ranks squared (Col 6)² d²
A	500	3	1000	2	1	1
B	750	1	1100	1	0	0
C	250	4	500	3	1	1
D	100	7	70	5.5	1.5	2.25
E	50	8	45	7	1	1
F	620	2	70	5.5	−3.5	12.25
G	120	5	72	4	1	1
H	110	6	40	8	−2	4
						$\Sigma d^2 = 5.5$

Interpreting your result

If two sets of data correlate exactly, they are said to have a perfect correlation. If one set of values increases as the other set goes up, a perfect positive correlation is found which would have the R_s value of 1.0. A perfect negative correlation (as one set of values goes up, the other decreases) would have a R_s value of −1.0. No correlation would be mid-way between these values, or 0.

The correlation between the factory size and the number of employees is thus a strong positive correlation. This does not mean that the one set of values is the cause of the other set. For instance a large factory might be almost totally automated and have very few workers. It simply means that there appears to be some connection which is worth further investigation. Nobody would suggest that there is a causal relationship, for instance, between the incidence of sunburn and the sales of ice cream. The two are obviously connected by temperature. The fact that there is a strong positive correlation between them should lead you to suggest that they are connected in some way, but not that ice cream causes sunburn!

Drawing pie charts to scale

A pie chart is an effective way of showing the relative sizes of the constituent parts of a total figure. The size of the segments (or slices) of the pie represent the size of the component. Pie charts can also show the relative size of the total in comparison to another total. This can make the graph doubly effective. Care must be exercised in the drawing to eliminate any unintentional bias.

Consider the two sets of figures in Table T5.2 showing electricity consumption in regions of Brazil for 1976 and 1980. The units are in gigawatts.

Table T5.2 Electricity consumption in Brazil, 1976 and 1980

Region	1976	1980
North	1020	1750
North-east	8800	17 320
South-east	51 570	78 860
South	8390	14 420
West central	1730	2690
Total	71 410	115 040

If you scale the two total figures to use the radii for the pie charts then the result would not be accurate. The finished result should show the area of the 1980 pie to be 60% bigger than the 1976 pie. Try drawing circles of radius 3.5 cm for 1976 and 5.5 cm for 1980 (using a scale of 1:20 000) and judge for yourself. How much bigger is the circle for 1980? Clearly simply scaling the totals produces an inaccurate result.

Below is a suggested procedure to produce correctly scaled pie graphs.

(i) Calculate the square root of the total values.
(ii) Scale the square root values and use these values for the radii of the circles; make sure that the scale you use is appropriate for the size of paper.
(iii) Calculate the size of the segments by calculating the percentage of each value and multiplying by 3.6 to get your answer in degrees.
e.g. North region (1976)

$$\frac{1020 \times 100}{(71\ 510)} \times 3.6 = 5.1 \text{ degrees}$$

Draw the two pie graphs to practise the method and remember that you must square root the totals when you are comparing figures by using areas such as circles or squares. You are using an area to represent a total and you can produce biased results very easily if you do not use the square roots.

Moving means

Fig. T5.2 is a graph of annual rainfall for Tombouctou, Mali, from 1951 to 1971. It shows fluctuating totals. On the other hand, the dashed line shows a possible trend which has been superimposed on the actual figures. Is there any more accurate way of suggesting trends in such data?

One way of smoothing fluctuations is to calculate 'moving' means. In figures for three years, the mean is the totals for those years divided by the number of years; what is usually referred to as 'the average'. Thus, for Tombouctou, the mean for 1951 to 1953 is (261+239+262)/3 or 254 mm.

Means can be plotted for any number of years, but the most common numbers used are 3, 5 or 7 years. The mean is plotted on a graph for the central year, i.e. 1952 in this case. The next moving mean is calculated by dropping the first figure and adding the following one, in this case subtracting 261 and adding 380 (for 1954). The mean for these figures is then plotted for the central year, 1953. In this way, three-year moving means can be plotted for the Tombouctou figures to even out major fluctuations.

Now plot five-year moving means for the rainfull figures. Is a trend clearer than it was in Fig. T5.2? Under what circumstances would you use moving means for production figures rather than a normal graph?

Fig. T5.2 Annual rainfall in Tombouctou

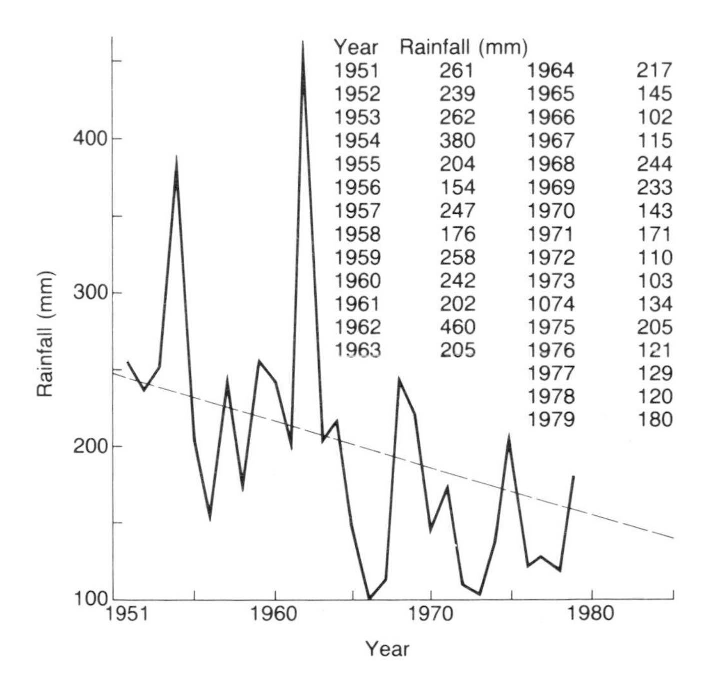

Theme 6 *Urban Issues*

Urbanization

There were no records for Taotou (or Baotou) in China in 1950. By 1980 there were 2 033 000 people living in it. At present rates, Lima-Callao in Peru will have increased in size by 1883% between 1950 and 2000. Nor is this phenomenon restricted to developing nations. It is expected that Los Angeles – Long Beach will add 256% of its 1950 population by the year 2000, a similar figure being expected for Osaka-Kobe. In other words, there is a world-wide shift to the urban areas, a process known as **urbanization** (Table 6.1).

1 Use the data in Table 6.1 to draw a world map of urban populations in the continents for the year 2000 (refer to Techniques p. 230 on mapping techniques).

2 Compare the relevant world map on p. 231 (Fig. T6.1 or T6.2) with the map that you have constructed. What are the main trends in world urbanization that you can see?

The reasons for this rural to urban shift are numerous and only some of them are shown in the migration model (Fig. 4.5 in Theme 4). There are almost as many reasons for it as there are people moving, but they may be summarized by the fact that people are unhappy with their existence in rural areas and perceive greater opportunities for themselves and their families in urban areas.

Although their perception may in many cases turn out to be false hope, once the movement has been made it is very difficult indeed to reverse it. Even the process of **suburbanization**, which has been going on particularly in Western Europe and the United States of America since mass transport became available in the early twentieth century, has only led to an outward expansion of the cities and not the dilution of urban living that the people moving to the suburbs hoped.

Urbanization is a very important theme throughout the world and this chapter deals with some of the consequences of it.

Case-study: Khartoum-Omdurman

Urban models

Geographers have analysed the layout of land-use in urban areas by constructing **models**. These models (Fig. 6.1) explain layout in terms of the growth of the urban area. In the case of the concentric model, growth was supposed to be at equal rates in all directions; in the case of the sector model, it was supposed to be outwards along transport routes, and in the case of the multiple-nuclei model, it was supposed to come about from the merging together of a number of small individual settlements. All of these models were derived from work done in the developed world. Do the models apply to cities in the developing world?

The map of social status in Khartoum-Omdurman (Fig. 6.2) was drawn from one of the infrequent

Table 6.1 World urban population, by region, 1980 (est) and 2000 (est)

World Region	1980 (est.) Urban population (Millions)	(%)	2000 (est.) Urban population (Millions)	(%)
North America	196	79	256	86
Western Europe	268	74	321	83
Oceania	17	73	26	78
Latin America	237	64	464	75
Eastern Europe/Soviet Union	243	62	344	74
North Africa/Middle East	112	48	243	50
East Asia	358	33	591	43
Southeast Asia	90	24	207	34
South Asia	199	22	441	31
Sub-Saharan Africa	80	22	210	37
World	1800	41	3103	50
More developed countries	850	72	1107	77
Less developed countries	950	30	1996	35

(Source: Derived from Table 1 in *Global Review of Human Settlements, Statistical Annex*. United Nations: Department of Economic and Social Affairs, published by Pergamon Press, New York, 1976, pp. 22–49)

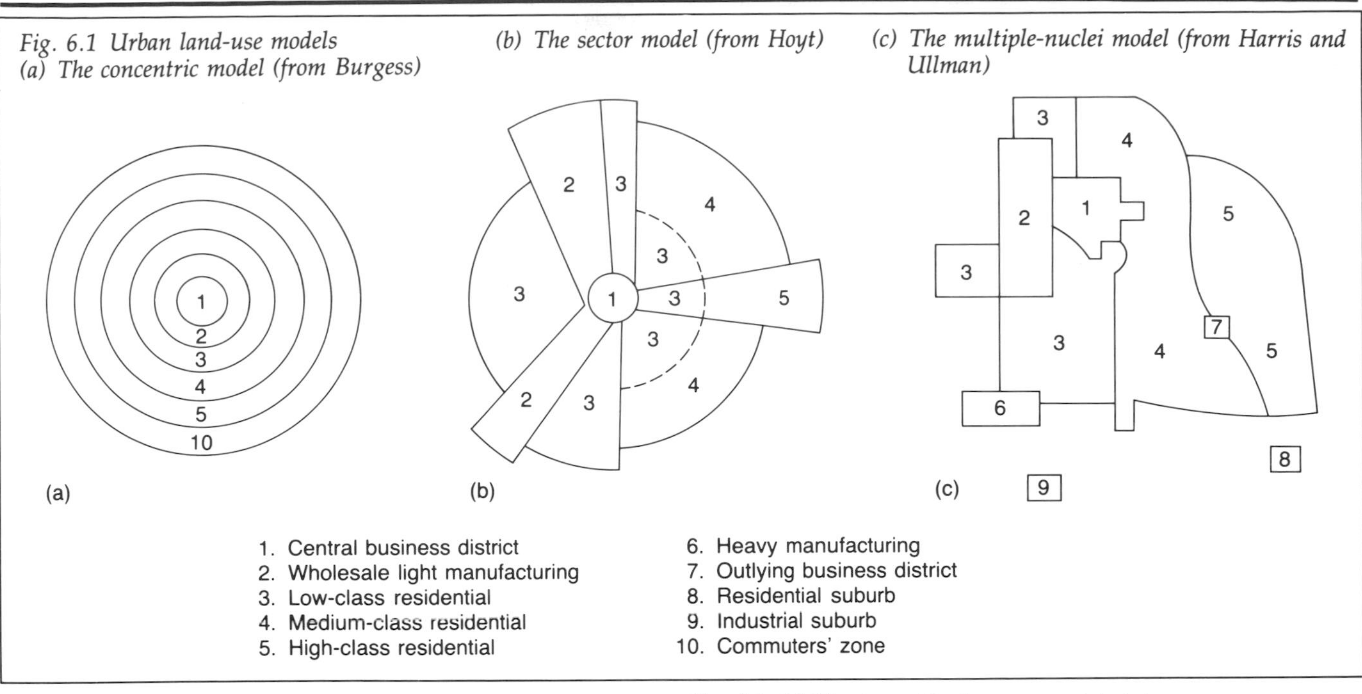

Fig. 6.1 Urban land-use models
(a) The concentric model (from Burgess) *(b) The sector model (from Hoyt)* *(c) The multiple-nuclei model (from Harris and Ullman)*

censuses carried out in the developing world. It relates well to the sorts of land-use shown in the three models from the developed world. At least part of the growth took place under the influence of British imperial rule, so we might expect to see some similarities with the models already mentioned.

3 Which of the following statements appear to be true of Khartoum-Omdurman?

(a) The high class areas are found:
only near the rivers
near the centres of the three settlements
only near the suburbs.

(b) The low status areas are found:
in the low-lying land by the rivers
around the edge of the central business district (CBD)
on the outer edge of the built-up area.

4 From the evidence provided, do you think that any of the models of urban land-use fits Khartoum-Omdurman? Give reasons for your answer.

In fact, it would be more sensible to look for models of urban land-use that have been derived from the African situation in order to explain layouts such as that found in Khartoum-Omdurman. One model (Fig. 6.3) was drawn by Ismail from his work in the northern Islamic nations of Africa, of which Sudan is one. Often, old cities in Islamic areas have walled cities (medinas) which represent the original medieval city. When the colonial period started, the occupying power usually built a fort or garrison next to the wall of the medina so that its influence could be imposed easily. This led to the development of a European quarter near the citadel. Depending on the length of time that the colonial power was involved in the country, this quarter became more or less significant in the layout of the urban area. The Central business district (CBD), which was originally

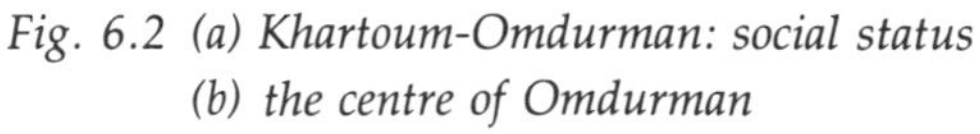
Fig. 6.2 (a) Khartoum-Omdurman: social status
(b) the centre of Omdurman

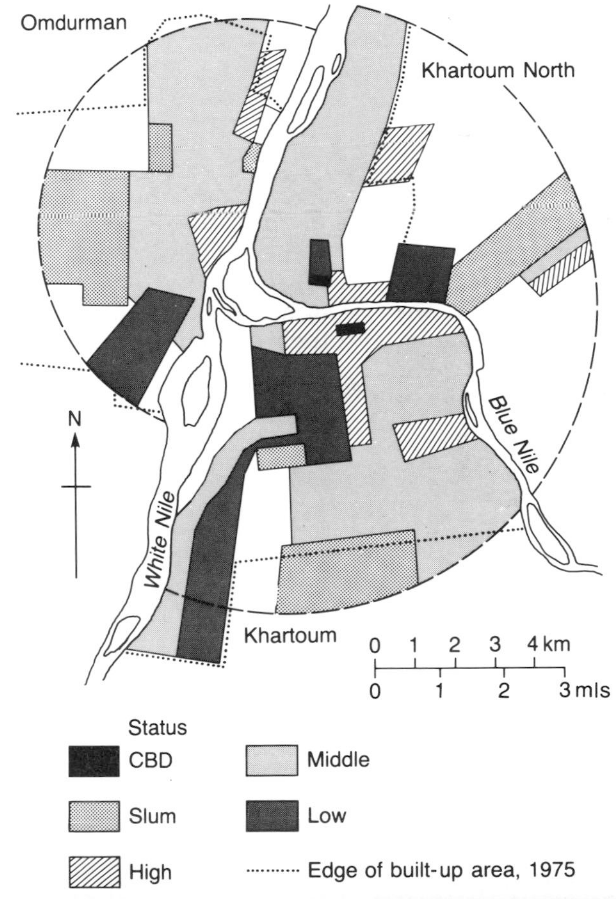

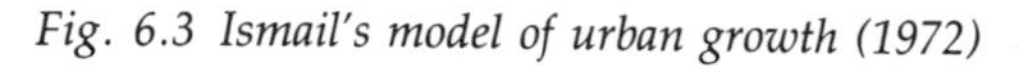

Fig. 6.3 Ismail's model of urban growth (1972)

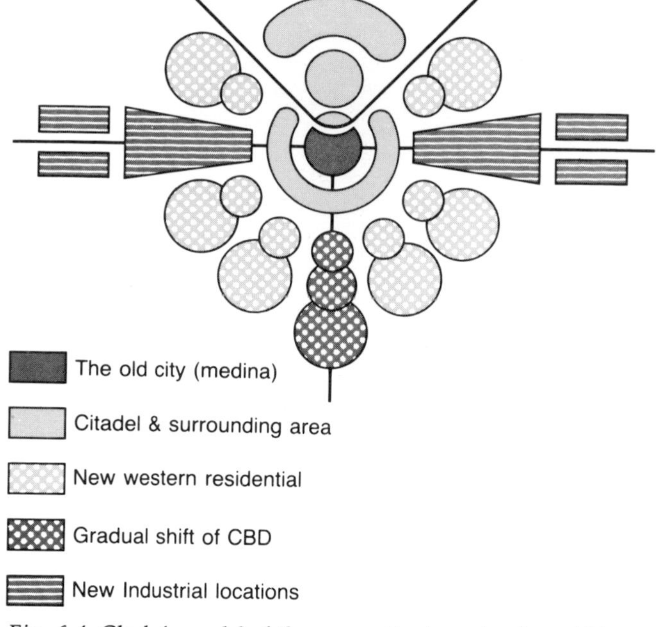

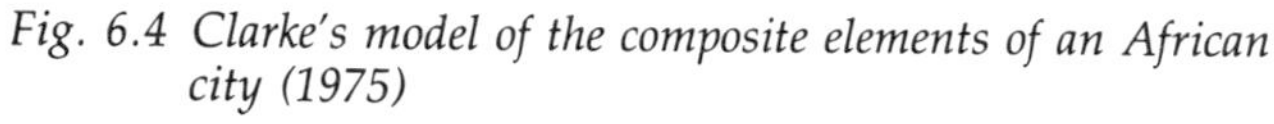

Fig. 6.4 Clarke's model of the composite elements of an African city (1975)

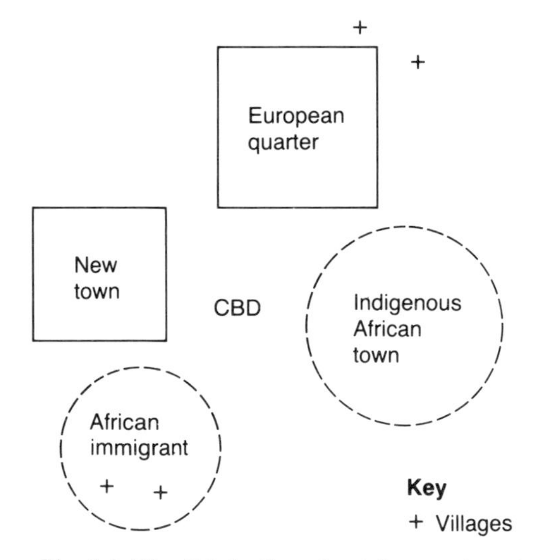

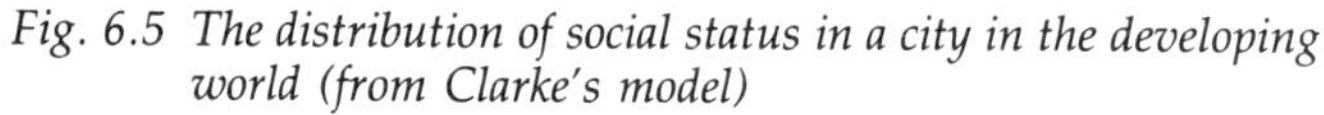

Fig. 6.5 The distribution of social status in a city in the developing world (from Clarke's model)

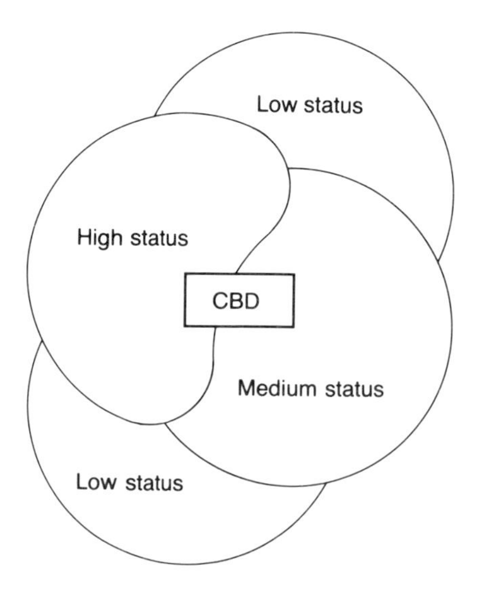

within the medina walls, had to expand when the city grew, and it did so in the direction which gave it the greatest accessibility to the whole urban area. Industrial areas, which again expanded from their original site inside the medina, also grew along transport routes. The resulting pattern was shown by Clarke (1975) in his model of the composite elements of an African city (Fig. 6.4). An interpretation of social status from Clarke's model is shown in Fig. 6.5.

5 Does Clarke's model provide a better explanation of the layout in Khartoum-Omdurman than the previous models? What further information would you need in order to give a more accurate answer?

6 The pattern of social status in a city in the developed world is shown in Fig. 6.6. How does the pattern shown in Fig. 6.5 differ from this pattern?

7 Using Fig. 6.5 to help you, say why cities in the developing world should have such a layout of social status areas.

Fig. 6.6 Social status in a city in a developed area

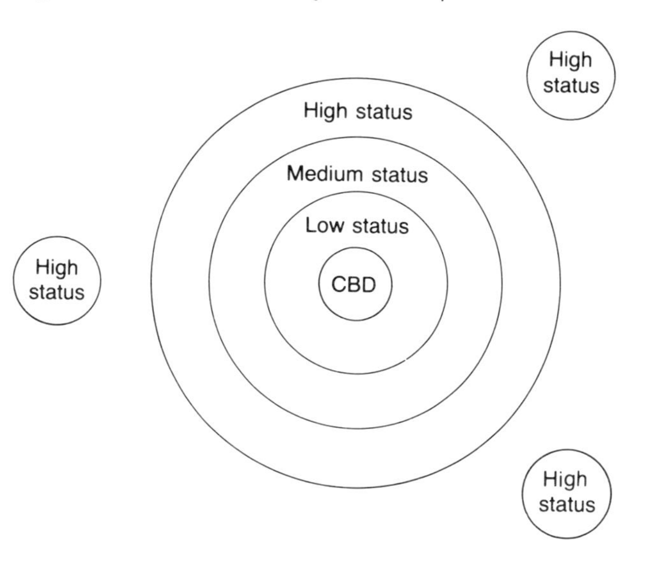

New Towns

The largely unplanned growth of the urban area has had serious consequences both in the developed and the developing worlds. Khartoum-Omdurman, for instance, has serious traffic congestion, especially at the Nile crossing points. The unplanned growth of slums is another problem that it has. In order to overcome problems such as these and to provide a better environment in which to encourage new enterprise, new towns have been built in many countries.

The new towns built in the United Kingdom since 1946 (the New Towns Act) were designed to fulfill certain objectives. They had to provide decent housing for people who had often moved from inner city locations where the standard of housing was abysmal. They were meant to provide decent surroundings for these people, by designing open space to run between housing blocks. Industry was attracted by placing it in separate units away from housing, and communications were speeded up by designing road junctions for the car, the first time this had been done for entire settlements. Designing a town from scratch enabled services to be planned carefully so that they were within the reach of the consumers. Neighbourhood centres provided clinics, halls, and small shops for everyday items. Fig. 6.7 shows the design of one of the early British new towns, Bracknell in Berkshire.

> **8** Make a list of the main aims of the New Towns Act in the United Kingdom.
>
> **9** How have each of these aims been catered for in the case of Bracknell?

New towns have also been built in the developing world. Well known examples include Brasilia, the new capital for Brazil, Ciudad Guayana in Venezuela and Islamabad in Pakistan. Often, new towns in the developing world are seen as the focus for something other than the relieving of urban pressure as they were in the United Kingdom. In the case of Brasilia, a new capital was built which was sited in the interior of the country, thus providing a focus for the nation which did not exist before. The design was futuristic, as befits an emerging country, and acted as a symbol both in Brazil and abroad to show how forward-looking and modern the country was. Some countries have used the building of new towns as an aid to development, which indeed was another of the aims of Brasilia. Saudi Arabia is a country which has set about a development programme using the building of new cities as a focus.

Fig. 6.7 The layout of the new town of Bracknell in 1981

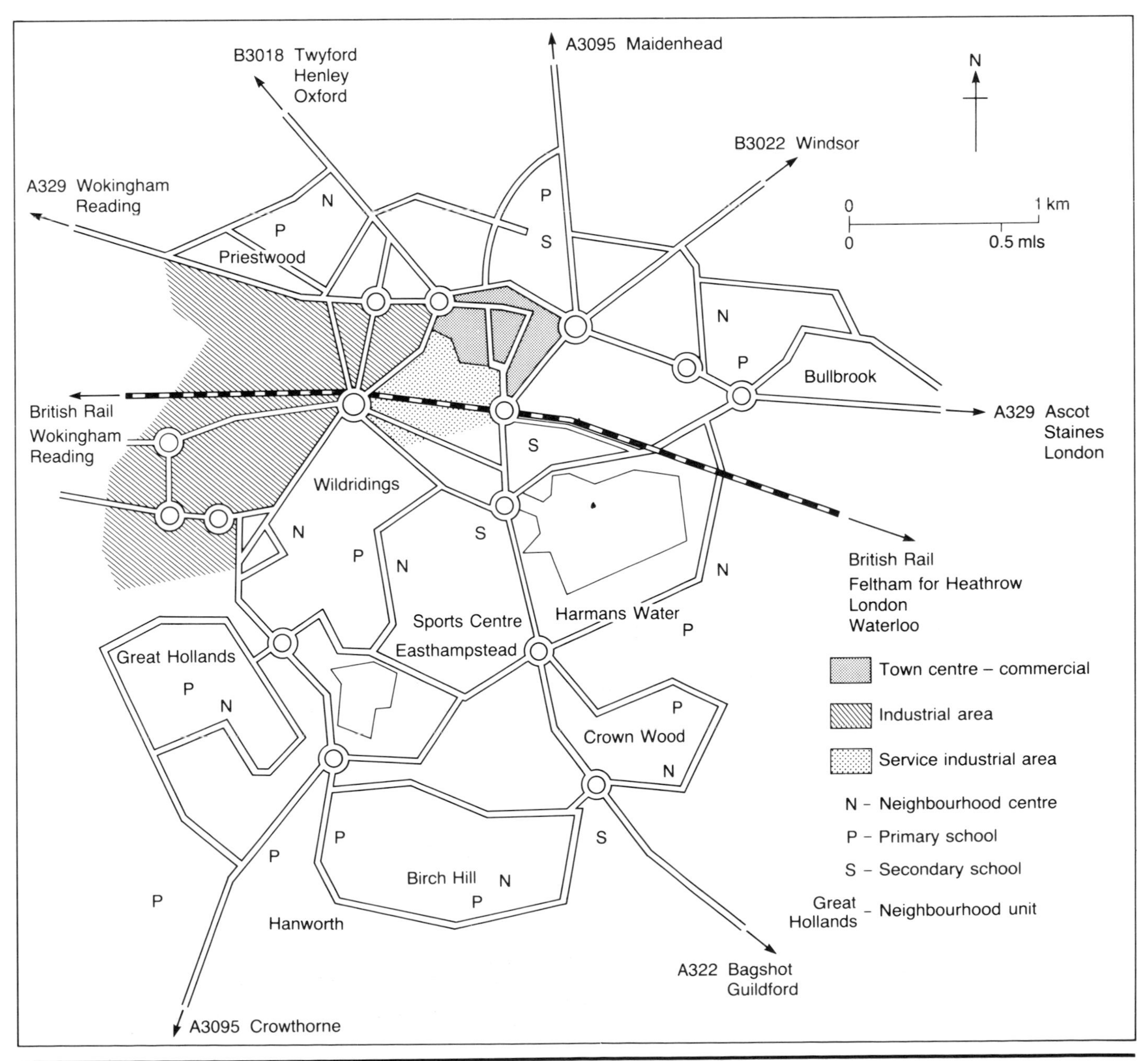

Case-study: Yanbu and Jubail

In 1981, Saudi Arabia had a balance of payments surplus of over 80 billion US dollars. This has been the pattern for a large number of years since the production of petroleum started in the country. With a population of under 10 million, Saudi Arabia is therefore able to 'think big' in terms of spending this money. The big problem in an oil economy is what will happen when the oil runs out (which is predicted for Saudi Arabia in 40 to 70 years' time). Unless planning takes heed of this possibility in time, the danger is that in the future the country will have to live off the capital gained from its oil production days. Saudi Arabia is planning to avoid this possibility by developing several new urban areas into which oil revenue can be ploughed as an investment against a future without oil.

The Royal Commission for Jubail and Yanbu (1981) set out the plans for developing two new cities. Yanbu is to be on the Red Sea coast and is to be a city of some 150 000 people by the year 2000. It will cover 188 km^2 and will be a major crude oil and gas processing complex, able to supply both the home and foreign markets. It will have a major deep water port, and will become the focus for the development of the region (a growth pole). If this were the end of the story, then the city would be short lived, but the plans reveal that oil is only being used as the spur to get the development started. Also included in the plan is the proposal to base another industrial area on Saudi Arabia's other mineral deposits, including copper, lead, zinc, gold, silver, iron, phosphates, gypsum, lime, marble and salt. The encouragement of service industries for the surrounding area is also a principal aim. Plans are laid to utilize solar power, wind power and even nuclear power in the future development of the complex. Food for the city will be at least partly provided by the Controlled Environment Food Programme (CEFP) which is at present providing considerable amounts of food for Abu Dhabi. CEFP entails the use of greenhouses to give adequate humidity for the growing of fruit and vegetables in desert conditions. Also planned is the use of **aquaculture**, where freshwater fish are grown in artificial ponds, and **mariculture**, where seafood is raised in salt water bays or in cages suspended in the sea. It is planned that Yanbu should develop health, education, commerce, trading and tourism functions and thus become a major regional centre as well as an international trade centre. The drawback to all of this is that Saudi Arabia has a desperate shortage of skilled manpower. As a result, only industries with a clear labour-saving characteristic were used in the planning. Houses were built and then sold by the government at a price which reflected the ability of the people to pay rather than the cost of each house. In this way, the government was able to encourage the skills it wanted into Yanbu. Firms which are building in the city take on many unskilled workers, and train them on the job. This is a great benefit to Saudi Arabia, which has a scheme to pay fully for such training on condition that the trainees spend as much time working in the city as they spend in training.

Fig. 6.8 shows the master plan for Yanbu. The heavy, oil-related industries are situated to the

Fig. 6.8 Yanbu master plan

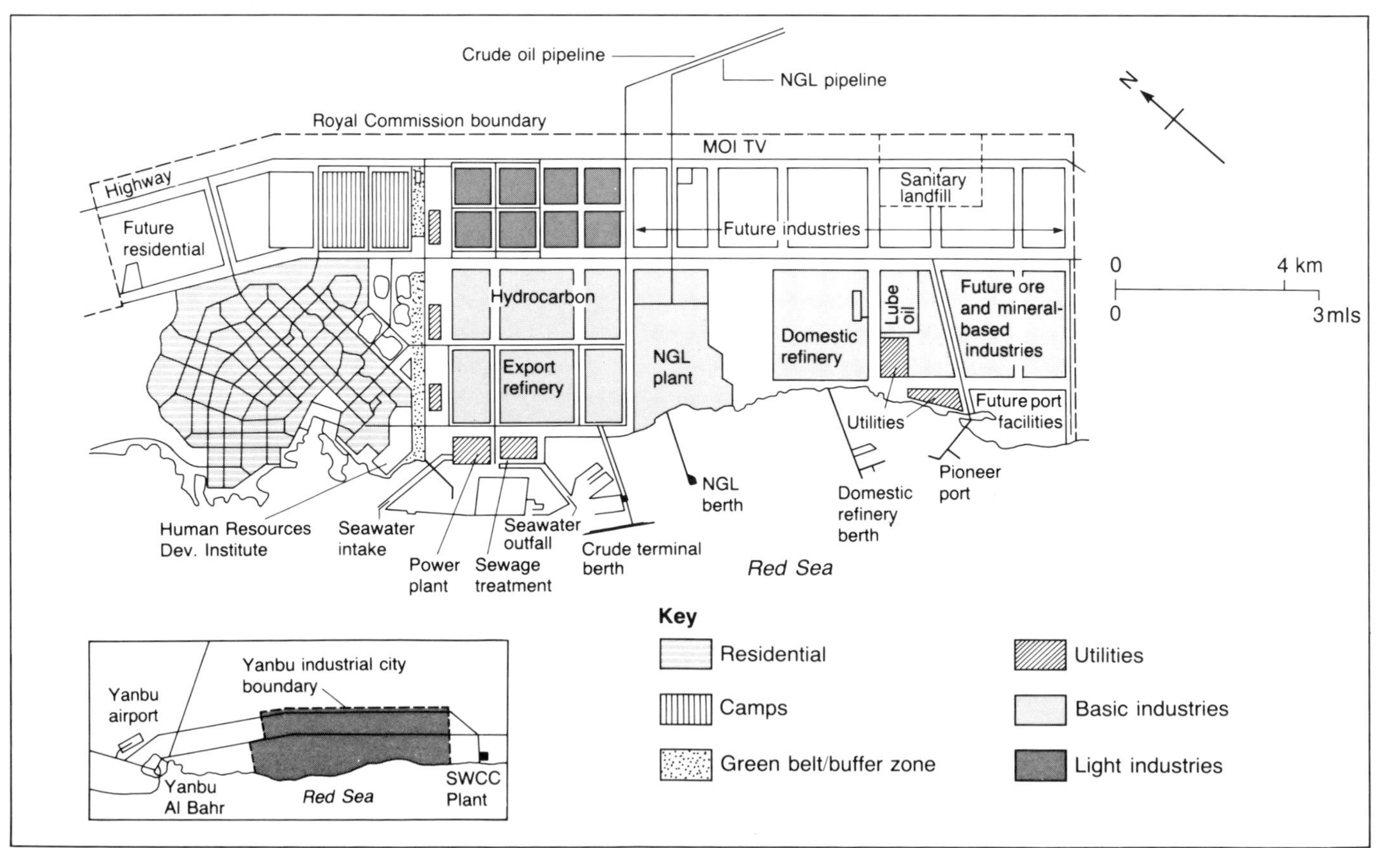

south-east of the residential area, though both of them have a sea frontage. The long-term plan is for both uses to expand inland, with the new, non-oil industries occupying an area inland from the pioneer port.

> **10** What aims does the plan for Yanbu appear to share with those for the British new towns?
>
> **11** If you were planning Yanbu, what 'utilities' would you build in the city?

Jubail is to be a city of some 270 000 by the year 2000. It is to be on the opposite side of Saudi Arabia from Yanbu, that is, on the Persian Gulf coast. Its industrial harbour will be one of the largest in the world, handling 26 million tons of liquid products, mainly hydrocarbons, per year. There will be petroleum plants, oil refineries, a steel mill and a dozen other industries not directly associated with oil. This will allow the old port of Jubail, some distance to the north of the planned new town, to return to its former important functions of being a pilgrimage port and a fishing port. At present, the functions are intermixed somewhat unhappily. The Royal Commission for Jubail and Yanbu said, in 1978, that 'the fact that the two cities are located on opposing coasts of the country makes it more likely that economic and social integration will be encouraged between geographically diverse regions.' The aim then is to strengthen the economic base of the country through urban development while encouraging greater national coherence (Fig. 6.9).

> **12** What aspects of planning do the master plans for Jubail and Yanbu have in common?
>
> **13** The Royal Commission hopes that at the end of the development of Jubail and Yanbu, the following will be achieved:
>
> (a) Saudi Arabia will become a better educated, technologically stronger and more independent country
>
> (b) The country's infrastructure will be much more adequate
>
> (c) The country will continue to realize a sound financial return on its capital investment.
>
> Do you think the building of the new industrial cities will achieve these goals? Explain your answer.

Fig. 6.9 Jubail master plan

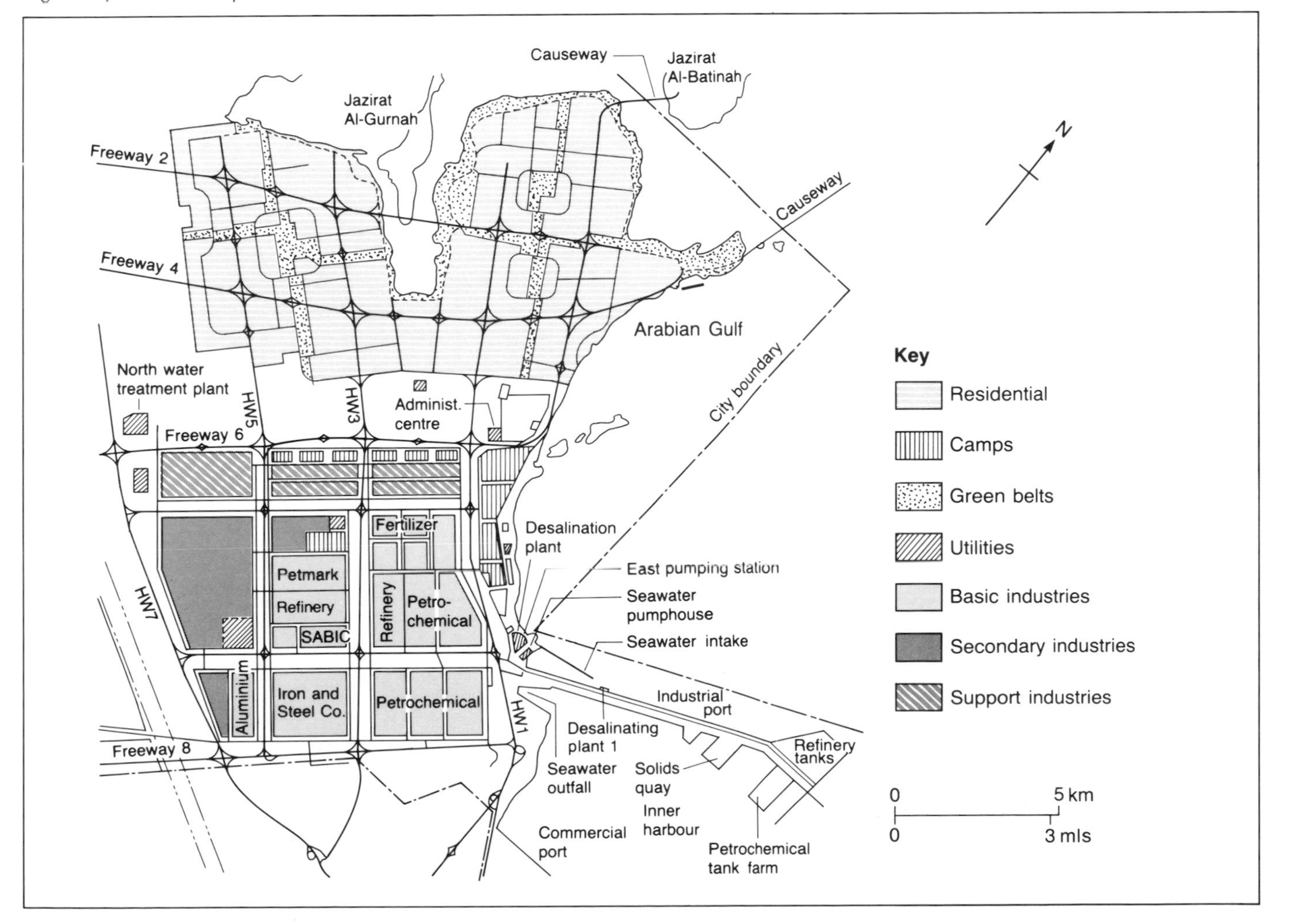

Decision-making Exercise Q: Building in the Desert

Q

The information provided below concerns an area of the Californian Desert around the northern end of the Coachella Valley in the United States of America (Fig. 6.10). The information relates to the year 1930, before there was much urban development in the area.

1. Prepare a report for the State Governor to indicate the potential of the area for urban development. Your report should contain reference to:

(a) the present use of the area (1930)

(b) resources and potential of the area

(c) problems of urban development

(d) your recommendation to develop or otherwise, quoting reasons for your decision. Include a diagram to show the decision-making sequence needed for the decision to be taken. (This should include reference to detailed consultation with the local people, but also take into account the wider interests of the country.)

*2. Complete the sketch map of a suggested town to house 100 000 people. This should be drawn on the base map (Fig. 6.12) and should include the following:

(i) housing areas

(ii) employment areas, showing types of industry to be encouraged

(iii) utilities, showing the major public services required

(iv) main lines of communication.

3. Write a short justification for the layout you have shown in your map (not more than 200 words).

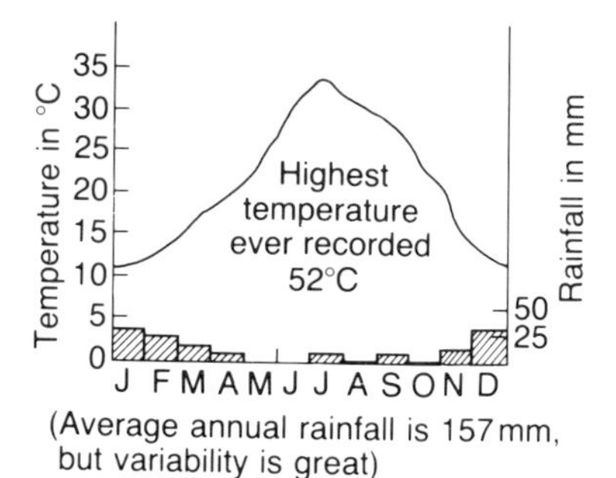

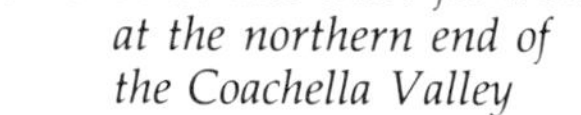

Fig. 6.10 A climate chart for a site at the northern end of the Coachella Valley

Fig. 6.11 A transect diagram for the Coachella Valley, 1930

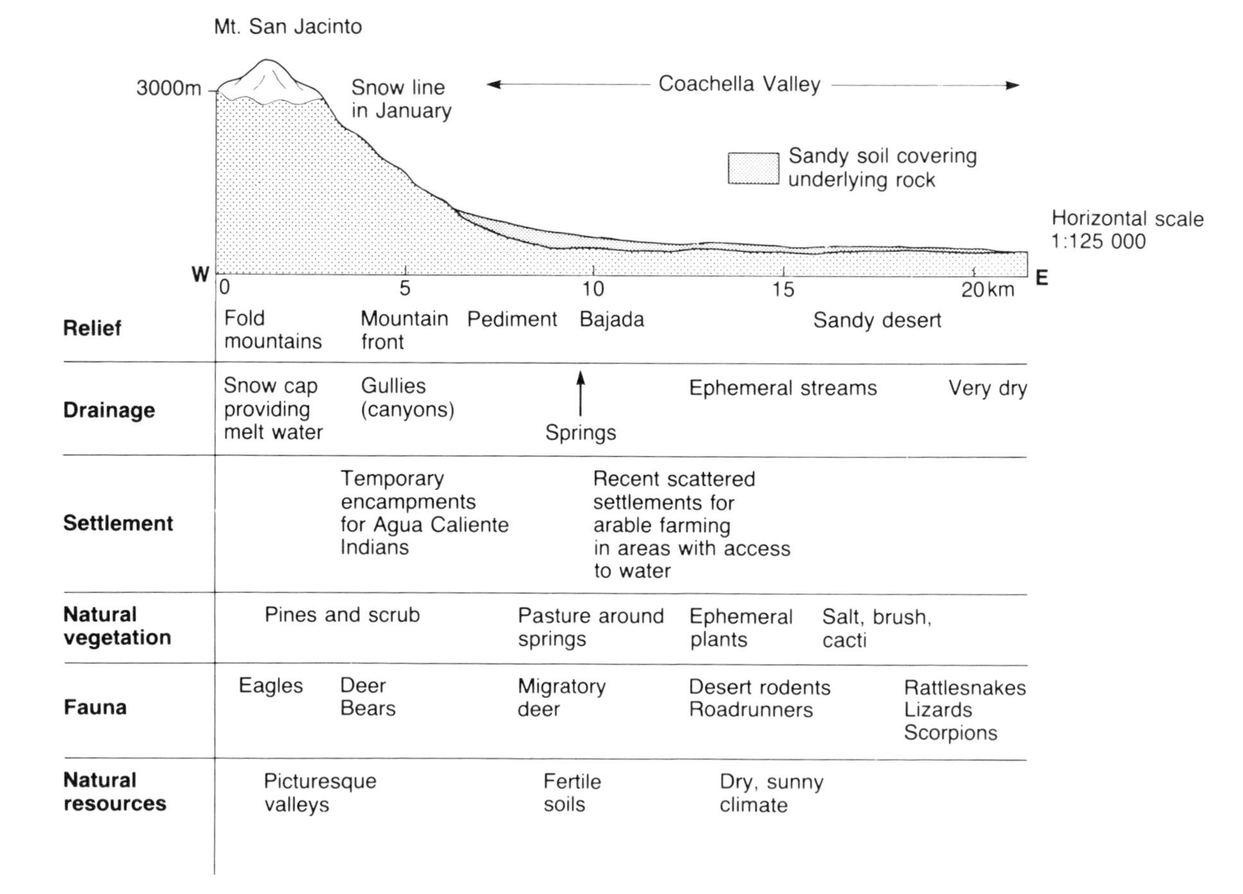

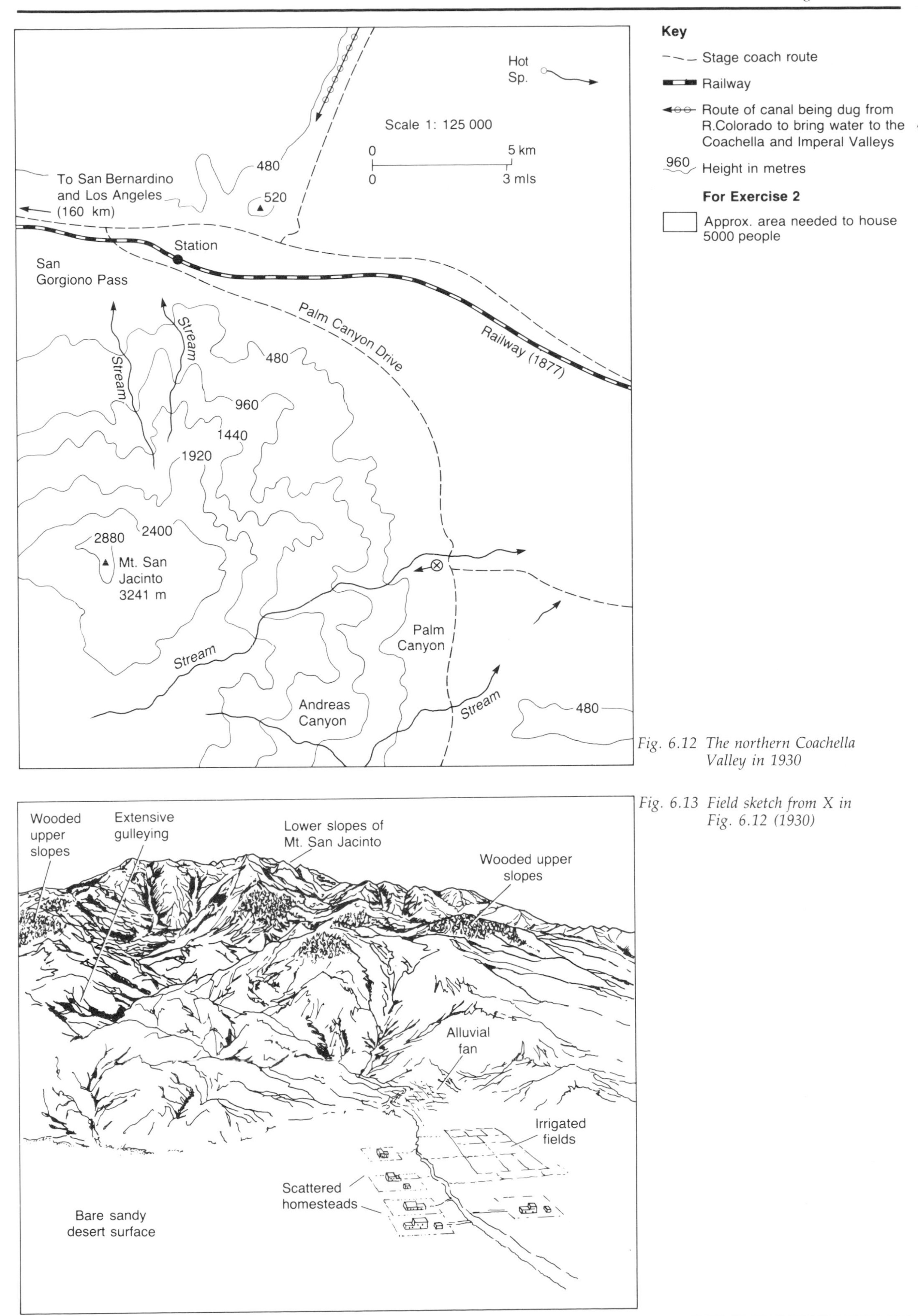

Fig. 6.12 The northern Coachella Valley in 1930

Fig. 6.13 Field sketch from X in Fig. 6.12 (1930)

In fact, the area has been settled and the town of Palm Springs has grown up. It is not a planned new town, but has many of the characteristics of one. The land on which Palm Springs was built was the subject of a treaty between the State of California and the Indians. The land was divided into a grid and the squares numbered. The even numbers belong to the Agua Caliente Indians and remain relatively undeveloped. Palm Springs covers about 25 km² to the north of point X marked on Fig. 6.12.

Fig. 6.14 A letter to the State Governor of California

Fearnley, Pole and Dunstable – Solicitors,
Los Angeles, California

34, First Avenue,Los Angeles.
12th March 1930

Dear Governor,

I write to you to seek a meeting to discuss the case for a settlement in the land rights issue on behalf of my client Chief Rising Water of the Agua Caliente Indians. You will recall that I have already contacted you on this issue (12.2.30) pointing out the urgency of the matter.

From the beginning of time, the Agua Caliente have lived in this area, making their living entirely peacefully by collecting berries and fruit from the vegetation that grows on the lower slopes of the Coast Ranges and the Sierra Nevada, and around the desert springs. They are capable of living off the land in places that we would not consider possible. They have a rich and colorful culture which they fervently wish to maintain but which is under pressure from the newcomers. They intend to resist any changes to their culture and way of life.

The right of the Indians to the land is indisputable. They have become worried by the number of farmers moving into the area following the arrival of the Southern Pacific Railroad in 1877, but they have so far remained calm. It cannot be guaranteed that they will continue to do so if large tracts of their land are taken by new settlers.

It cannot make you proud to know that Chief Rising Water, the elder of the Agua Caliente near the new farming settlement in the valley, actually shed tears as he described to me how his tribe's best collecting area has recently been destroyed by the development of new fields for a vegetable farm, for which permission was neither asked nor granted. The tribe is very bitter and may yet turn to violence in the defence of their rights.

The Agua Caliente, as I have indicated, are a gentle people. They can see the benefits of the new agriculture and they wish to share in its progress. It would be a pity if what could be a mutually beneficial development should turn to violence. Any development of the area for agricultural or urban purposes must be made in the spirit of cooperation and justice. To that end, if the government is not reasonable, we will start legal proceedings to ensure that my clients' wishes, aspirations and rights are not trampled. I wish to appeal to your sense of justice, Governor. An early meeting to discuss these matters would be appreciated, both by my client and by the settlers in the Coachella Valley whose peace could ultimately be threatened.

Yours sincerely,

Gordon G. Fearnley

Gordon G. Fearnley

Inner City Redevelopment

Case-study: Montreal

Table 6.2 gives the average total income for a sample of census tract for Montreal, Canada. The average total income for the metropolitan area was Canadian $8005 in 1971. Fig. 6.15 is a map of the census areas.

The inner city area can be defined as being the area of low average incomes. For Montreal this is the area with incomes below Canadian $5100. In some cities of the developed world it is a continuous belt surrounding the central business district (CBD), but in Montreal it is concentrated to the south-east of the CBD adjacent to the St Lawrence River. The photograph (Fig. 6.16) shows high-rise development of the CBD. Census tract 41 is adjacent

Table 6.2 Average total income in Canadian $ of family head (ATI) by census tract (CT) for the City of Montreal, 1971

CT	ATI	CT	ATI	CT	ATI	CT	ATI	CT	ATI
1	6843	53	4491	105	9295	157	4962	209	5278
2	6738	54	3805	106	8348	158	5264	210	5613
3	6972	55	14179	107	6433	159	5237	211	5398
4	7045	56	2585	108	13 692	160	4588	212	5817
5	6970	57	4361	109	9883	161	3870	213	6035
6	7086	58	3470	110	7369	162	4985	214	5266
7	7059	59	4254	111	8096	163	4375	215	5945
8	7299	60	2490	112	6478	164	4502	216	5039
9	6997	61	2964	113	8938	165	4266	217	5337
10	6501	62	9289	114	17 021	166	5085	218	4823
11	7407	63	6255	115	11 957	167	4613	219	4745
12	6390	64	8148	116	8073	168	4495	220	4749
13	6866	65	8092	117	6491	169	4608	221	5390
14	5884	66	5696	118	6760	170	4573	222	5296
15	5843	67	4141	119	5656	171	4652	223	5621
16	5879	68	3547	120	8918	172	5434	224	4965
17	6466	69	3888	121	13 919	173	5775	225	7459
18	5805	70	4063	122	5714	174	5568	226	6372
19	5361	71	–	123	7171	175	5509	227	5893
20	5332	72	5253	124	6848	176	5374	228	5967
21	6038	73	4711	125	8539	177	5141	229	–
22	5363	74	4908	126	11 754	178	5301	230	6090
23	5635	75	4647	127	8513	179	5829	231	5861
24	5576	76	4797	128	22 089	180	5460	232	5947
25	5184	77	4373	129	19 509	181	5998	233	5293
26	5733	78	4151	130	5647	182	6095	234	5556
27	5190	79	4728	131	5750	183	5515	235	5937
28	5484	80	4451	132	5326	184	6089	236	6213
29	5465	81	5401	133	4872	185	6488	237	5937
30	5911	82	5271	134	4704	186	6091	238	6580
31	6385	83	5337	135	4326	187	6582	239	6140
32	5993	84	4845	136	4619	188	6416	240	5909
33	5484	85	5789	137	4785	189	–	241	5325
34	4733	86	6563	138	4422	190	7930	242	5873
35	5276	87	5606	139	4184	191	8595	243	5628
36	5382	88	6331	140	4878	192	6231	244	5905
37	4702	89	6565	141	5117	193	8030	245	5906
38	4926	90	6701	142	4298	194	6696	246	5544
39	4836	91	–	143	5294	195	9151	247	5652
40	–	92	6406	144	4988	196	9031	248	5620
41	4680	93	5815	145	–	197	6974	249	6552
42	6600	94	5600	146	5568	198	7356	250	6143
43	4545	95	5713	147	5607	199	6421	251	6181
44	4382	96	5724	148	6629	200	6367	252	5904
45	3948	97	6689	149	5464	201	6674	253	6152
46	4119	98	9718	150	5836	202	6599	254	6962
47	4180	99	9098	151	5667	203	6263	255	6213
48	4180	100	8722	152	5650	204	6211	256	6095
49	4594	101	7312	153	6274	205	6503	257	6386
50	4687	102	9830	154	4614	206	6002	258	6153
51	3941	103	10 690	155	5468	207	5964	259	6081
52	3878	104	12 411	156	5461	208	5817	260	6912
261	6396	268	7769	275	6844	282	8219	289	–
262	6186	269	7100	276	6888	283	10 763	290	6607
263	5851	270	7648	277	7333	284	11 904	291	5611
264	9268	271	7344	278	7041	285	8817		
265	6861	272	7995	279	9755	286	9633		
266	6327	273	6685	280	8538	287	9227		
267	6414	274	6341	281	8866	288	14 403		

NB City of Montreal Average $6634
MMA Average $8005

(Source: 1971 Census of Canada, Census tract bulletin, Montreal Series B Table 3 – Income distributions) to this area. Many of the houses in census tract 41 were constructed quickly and cheaply and do not provide adequate living conditions, particularly in a climate which has an annual range of temperature of 30.5°C (−8.9°C in January to 21.6°C in July) and has snow cover for over six months of the year.

Population is declining in the inner city while the population of Montreal continues to grow.

14 Making a tracing of the census areas and draw a choropleth to show the average total incomes in Montreal. Use the following sub-divisions:

Below Canadian $5100

$5101 – $6700

$6701 – $10 000

Above $10 001

15 Compare the distribution with the map on p. 195.

16 Try to explain the similarities and differences.

Fig. 6.15 Map of census area of Montreal

Fig. 6.16 View over the CBD of Montreal

Table 6.3 Age–sex structure of four census areas in Montreal

Age group	MMA M	MMA F	41 M	41 F	45 M	45 F	46 M	46 F	47 M	47 F
0–4	3.9	3.8	3.5	3.1	3.1	3.1	3.3	3.4	3.1	4.0
5–14	10.1	9.7	10.6	10.3	9.6	9.3	7.9	8.7	12.8	11.9
15–24	9.0	9.2	9.9	8.9	8.9	6.7	7.8	8.1	10.2	10.6
25–34	7.6	7.6	5.3	5.0	6.6	5.6	6.4	5.6	5.3	6.2
35–44	6.5	6.5	5.9	5.9	6.8	5.4	5.9	5.3	6.2	5.7
45–54	5.3	5.7	6.4	6.8	5.6	6.8	6.2	6.2	5.3	5.7
55–64	3.8	4.2	4.4	5.3	6.8	6.0	6.7	7.1	4.8	4.0
65 above	2.8	4.1	3.2	5.0	4.4	5.4	6.5	6.5	2.2	2.6
Total	100		100		100		100		100	

(Source: Census of Canada, Census tract bulletins 1966, 1976)

Table 6.4 Declining population of four census areas in Montreal

Date	41	45	46	47
1961	3462	3647	4589	2157
1966	2747	3187	3813	1416
1971	2260	2585	3045	1135
1976	1647	1537	2002	915

(Source: Census of Canada, Census tract bulletins 1966, 1976)

Table 6.5 Census material for Montreal

	MMA	Census areas 41	45	46	47
Total population	2 743 230	2295	2615	3015	1170
Ethnic group:					
Asian	36 500	15	–	–	–
British Isles	438 500	90	155	100	30
French	1 762 690	2140	2405	2860	1070
German	38 440	–	5	–	5
Hungarian	11 480	–	–	–	–
Italian	160 605	40	5	5	10
Netherlands	9 040	–	5	–	–
Polish	20 410	–	5	–	–
Russian	3 605	–	–	–	–
Scandinavian	6 355	–	–	–	–
Ukrainian	18 050	5	5	–	–
Population 5 years and over	2 532 100	2140	2465	2795	1070
Highest level of schooling:					
Less than Grade 9	1 253 385	1510	1860	2045	780
Grades 9–10 – No other training	400 955	395	295	425	185
With other training	89 145	35	40	50	15
Grade 11 – No other training	204 255	55	95	85	25
With other training	89 880	20	25	30	15
Grades 12–13 – No other training	140 655	60	60	80	15
With other training	95 005	20	30	10	–
Some university – No other training	100 210	30	30	40	20
With other training	39 200	5	10	10	5
University degree – No other training	95 640	5	10	10	5
With other training	23 775	5	–	5	–
Industry division:					
Both sexes	1 079 785	805	830	715	305
Divisions 1, 2, 3, 4	6 005	5	5	–	–
Division 5	276 650	230	185	170	60
" 6	50 930	20	30	30	30
" 7	102 805	110	75	75	30
" 8	161 550	115	85	90	30
" 9	61 505	25	20	20	5
" 10	264 205	175	165	125	75
" 11	56 005	25	40	30	10
Occupation major group:					
Males	699 305	525	640	505	200
Group 11	55 995	–	5	10	–
" 27	17 275	–	5	–	–
" 31	12 465	–	–	10	–
Groups 21, 23, 25, 33	52 270	15	10	20	–
Group 41	81 565	60	75	40	15
" 51	81 685	40	35	40	15
" 61	67 445	85	120	35	30
Groups 71, 73, 75, 77	6 900	–	–	–	5
Group 81/82	26 310	10	30	25	5
Groups 83, 85	96 685	95	70	40	20
Group 87	49 815	35	20	35	15
" 91	40 280	40	50	40	30
Other	50 385	90	45	80	20

(For *Key to Table 6.5* see top of p. 206.)

	Census areas				
	MMA	**41**	**45**	**46**	**47**
Occupied dwellings	805 770	665	870	1090	285
Owner-occupied	284 080	30	40	45	5
Tenant-occupied	521 690	630	830	1040	275
Single detached	190 780	15	40	40	5
Single attached	206 235	495	665	835	230
Apartment (flat)	407 780	150	175	205	40
Rooms per dwelling (average)	4.9	4.5	4.4	3.9	5.1
Persons per room (average)	0.67	0.74	0.66	0.70	0.77
Flush toilet (exclusive use)	793 795	655	815	1040	275
Bath or shower (exclusive use)	778 180	555	615	650	230
Dwellings with:					
Automatic clothes dryer	245 485	45	60	65	35
Colour television	110 560	40	40	40	30
Vacation home (owned)	68 320	210	55	65	80
One automobile	453 810	210	305	355	80
2 or more automobiles	88 305	10	25	25	5
Oil as principal heating fuel	618 740	375	420	660	180
Gas as principal heating fuel	113 745	235	340	345	85
Electricity as principal heating fuel	66 335	35	55	70	–
Period of construction:					
Before 1946	250 750	620	770	1025	255
After 1960	261 295	5	30	65	5
Occupied dwellings	805 440	660	830	1120	265
% dwellings rented	84.5	94.7	95.4	95.4	96.5
Tenant-occupied dwellings:					
Average cash rent	99	63	62	54	59
Length of occupancy:					
Less than one year	157 240	145	175	195	65
1–2 years	160 070	145	140	200	45
3–5 years	132 855	55	155	165	55
6–10 years	152 850	130	125	215	50
More than 10 years	202 415	185	235	340	55
Birthplace and immigration:					
Born in Canada	2 337 555	2240	2540	2960	1130
Born outside Canada	405 680	55	80	50	45
Immigrated after 1945	330 335	20	50	30	35
Migration:					
Non-migrants.	1 945 865	1995	2210	2590	915
Same dwelling	1 216 750	1020	1085	1420	415
Different dwelling	729 115	975	1125	1175	495
Migrants	584 415	150	250	210	155
From an MA	314 920	45	55	60	45
Same MA	269 990	40	30	50	45
Different MA – same province	17 685	–	15	10	–
Different province	27 240	–	5	5	–
From a non-MA	118 685	50	80	90	95
Same province	101 885	45	70	80	65
Different province	16 800	5	10	10	30
From outside Canada	115 345	20	40	20	5
Place of residence in 1966 not stated	35 470	30	75	45	5

Key to Table 6.5

Division 1 Agriculture
Division 2 Forestry
Division 3 Fishing and trapping
Division 4 Mines (including Milling), quarries and oil wells
Division 5 Manufacturing industries
Division 6 Construction industry
Division 7 Transportation, communications and other utilities
Division 8 Trade
Division 9 Finance, insurance and real estate
Division 10 Community, business and personal service industries
Division 11 Public administration and defence

Group 11 Managerial, administrative and related occupations
Group 27 Teaching and related occupations
Group 21 Occupations in natural sciences, engineering and mathematics
Group 23 Occupations in social sciences and related fields
Group 25 Occupations in religion
Group 31 Occupations in medicine and health
Group 33 Artistic, literary, recreational and related occupations
Group 41 Clerical and related occupations
Group 51 Sales occupations
Group 61 Service occupations
Group 71 Farming, horticultural and animal husbandry occupations
Group 73 Fishing, hunting, trapping and related occupations
Group 75 Forestry and logging occupations
Group 77 Mining and quarrying, including oil and gas field occupations
Groups 81 and 82 Process occupations
Group 83 Machining and related occupations
Group 85 Product fabricating, assembling and repairing occupations
Group 87 Construction trades occupations
Group 91 Transport equipment operating occupations

Other (for males) Materials handling and related occupations, not elsewhere classified, other crafts and equipment operating occupations and occupations not elsewhere classified

17 Tables 6.3 and 6.4 give information on the total population for four of the census tracts – 41, 45, 46 and 47 and the age–sex structures of these areas.

(i) Draw a line graph to show the population change from 1961 to 1976.

(ii) Draw the age–sex pyramid for Montreal and by working in groups divide up the remaining pyramids, i.e. for areas 41, 45, 46, 47. Indicate on the pyramid for Montreal the cohorts in which there is a surplus or deficit in population in comparison to the population of the census tract.

18 What will be the socio-economic characteristics of these areas? Write a brief description of what you would expect using the following headings:

Standard of living

Tenure and length of occupancy

Number of migrants

Employment based upon manual work

19 Now use the census material (Table 6.5) to test these ideas (hypotheses) on the characteristics of the inner city. You should present your findings complete with appropriate graphs and diagrams.

20 Is there any link between the average income of the area, the population structure and the physical state of the inner city area? Try to construct a model to show how the factors interact and give as much detail as necessary.

21 Make a list of factors you could look for in preparing for fieldwork to look at the state of the urban fabric. How would you collect your data?

Inner City Deprivation

Areas of the inner city are sometimes known as the twilight zones. They may be rather shabby in appearance with deterioration of the urban fabric and an inefficient transport infrastructure. Fig. 6.17 gives a version of the sort of processes that can go on in the twilight zone. These areas may attract poorer, or marginalised groups to occupy property, either to buy or to rent. Migrants may seek accommodation in the inner city on arrival in a country, but may find the discriminating practices maintain the cycle of deprivation, keeping them rooted in the inner city.

It may be difficult for the inner city area to develop as new industrial and infrastructure investment is lured away to 'greenfield' sites in the suburbs. This in turn may mean that small subsidiary industries will not develop to service the larger firms and high rates of unemployment may result.

In extreme cases areas may be dominated by one particular racial group and a ghetto may be formed. The services will change to serve the new community and buildings may change use to cater for the culture of the newcomers. This may lead to the area becoming enriched and rejuvenated. It can also lead to the further deterioration of the environment if this is accompanied by high rates of unemployment. School leavers with little to look forward to may be easy prey for drug pushers and high rates of criminal activity may follow.

22 Use the diagram, Fig. 6.17, to show how the environment of the inner city may deteriorate. Try to give more detail to some of the labels in the diagram

23 Twilight areas may suffer unfairly from having a bad image. How would you set about changing the negative image of the Inner City? Produce a 'rough' for a handout to be distributed to local entrepreneurs pointing out the positive aspects of the area (including its location and potential for new business). Some of the group should produce a handout from the Local Planning Authority and some from a private Development Consortium.

Case-study: Glasgow

Glasgow is a city which has a number of urban development schemes operating. It is a typical example of a city which flourished during the industrial revolution when labour intensive industries developed rapidly and has struggled to cope with the changes of the post-industrial era.

Glasgow industries developed around the river Clyde and featured ship building and marine engineering. A textile industry based on imported cotton and tobacco industries employed many people. With the decline in the industries the inner part of Glasgow developed many features of the model of urban deprivation shown in Fig. 6.17. To counter this the Glasgow Eastern Area Renewal Project (GEAR) was established in 1976. The GEAR Project is an attempt at comprehensive social, economic and environmental regeneration. It also aims to create the conditions for the development of a balanced and thriving community.

GEAR has an area of 1600 ha which is 8% of the area of the Glasgow district and 18% of this was vacant or derelict land (Fig. 6.18). The population of the area in 1976, at the start of the project, was about 45 000. This had shrunk dramatically in the post-war period as a result of clearance. In 1951 there were 145 000 people living in the area; in 1961, 115 000; and in 1971, 82 000 people. The residual population still retained strong links with many old established families in the area; 62% of households had at least one member who had lived in the area for over 40 years. Other indicators of deprivation of the area are shown in Table 6.6.

At the same time the GEAR area was an important industrial area with 42 000 jobs of which 55% were in manufacturing, 37% in servicing and 6% in construction.

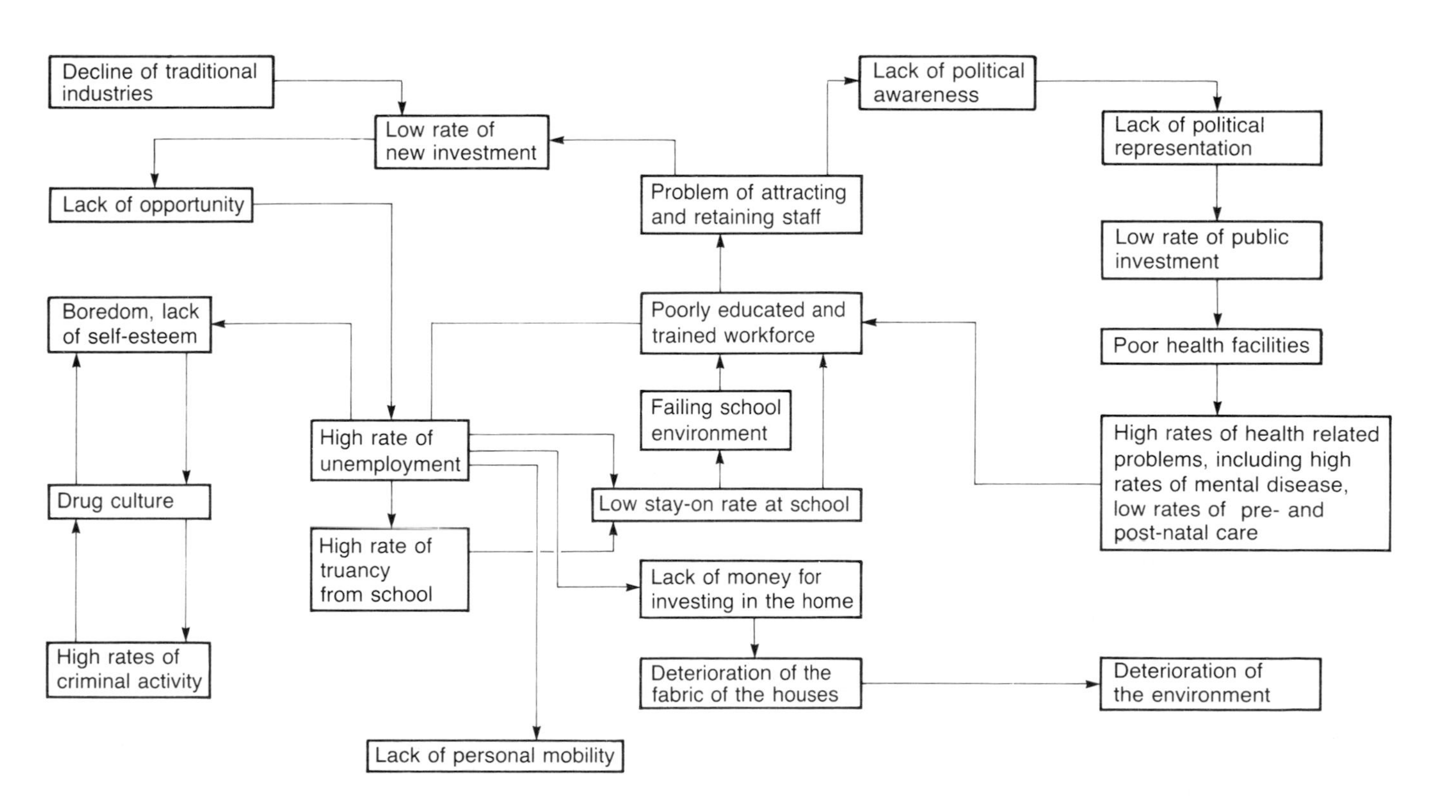

Fig. 6.17 A cycle of urban deprivation

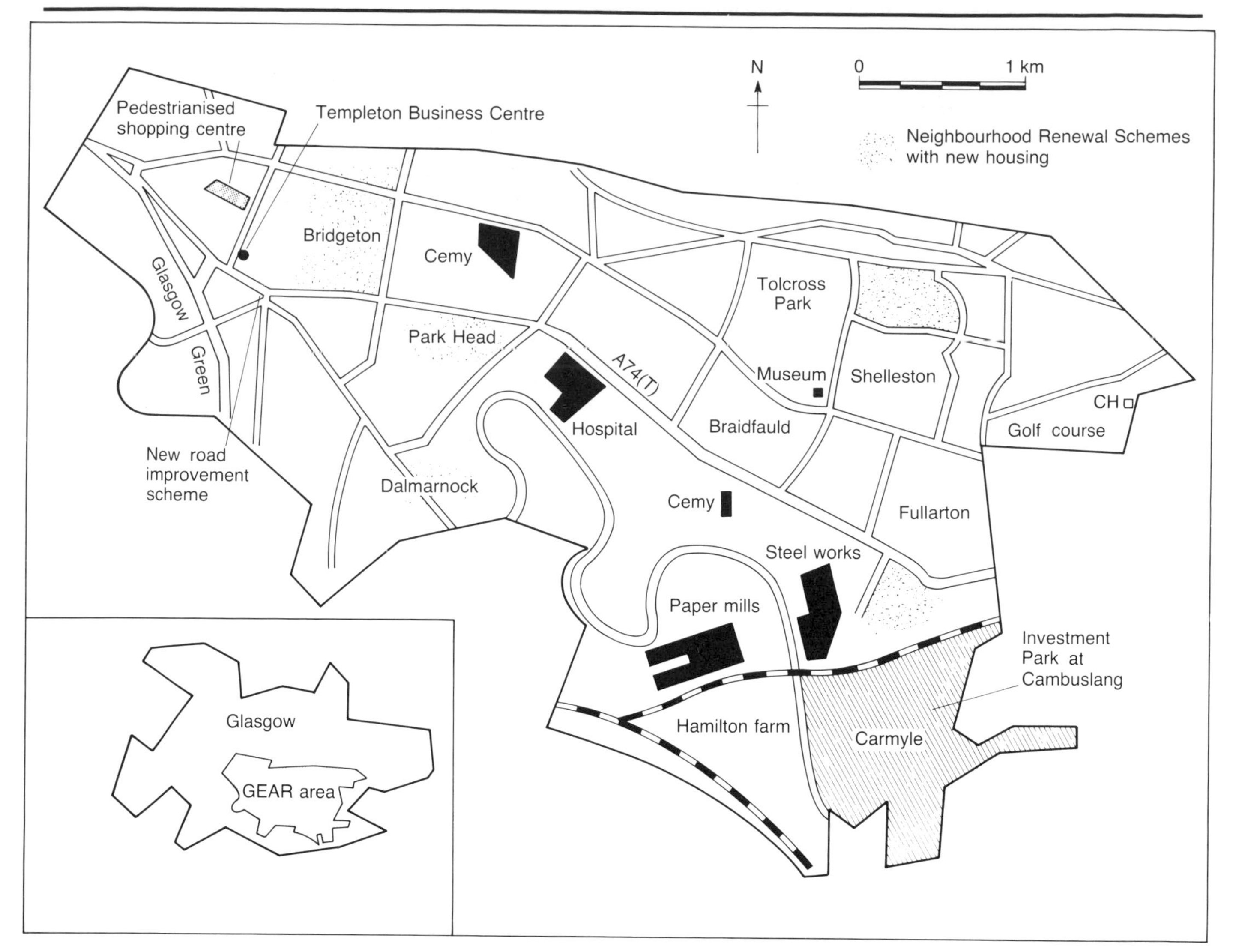

Fig. 6.18 The Glasgow Eastern Area Renewal Project (GEAR)

The GEAR *Strategy and Programme* identified six 'Basic Objectives' for the project:

(i) to increase residents competitiveness in securing employment

(ii) to arrest economic decline and realize the potential of GEAR as a major employment centre

(iii) to overcome the social disadvantages experienced by residents

(iv) to improve and maintain the environment

(v) to stem population decline and engender a better balanced age and social structure

(vi) to foster residents' commitment and confidence.

To assist in the creation of employment a series of new projects was set in motion. 170 hectares of land was made available to establish an 'Investment Park' at Cambuslang for industrial and commercial development. This site was linked to the main motorway network by a 4 km extension of the M74. Investment has been forthcoming from the European Coal and Steel Community to help develop the area. New service industry in the form of a hotel is planned for the area.

Housing Associations have built many new homes in the area and renovation of structurally sound, older properties has been a feature of modern management of the area. There is considerable effort to redevelop the area as a place where young families may live in new accommodation, while some of the old architecture is being renovated to play its part in this mixed environment.

It has been recognized that to produce tangible results from urban regeneration programmes is a very difficult task. It involves massive investment over a long period of time. Some of the components which bring well-being to the community are outside the control of local action, e.g. creation of employment.

Table 6.6 Socio-economic indicators of GEAR area

	GEAR area	Strathclyde region
Households with pensioners	41%	29%
Households including handicapped persons	29%	21%
Households with children	30%	40%
Households with no car	84%	61%
Income < £1750	55%	39%
Male unemployment rate	20%	13%
Female unemployment rate	17%	11%

24 Table 6.7 shows the expenditure in GEAR up to March 1984. Using the sector headings, draw a pie chart to show how the money has been invested.

25 Describe and explain the distribution of investment.

26 Read the article from the *Guardian* (Fig. 6.19) and try to evaluate the success of GEAR.

27 Why has the Glasgow project succeeded while the attempts to regenerate Liverpool appear to have produced less success?

While investment is being directed into the inner city, less is available for investment in other areas. One of the types of area recently identified as being in need is the suburban council estate. Such estates were built in the 1950s and 1960s to house overspill population from the inner city areas.

They are suffering from their own special kind of deprivation in which there is little money circulating in the local economy, little demand for consumer goods and little competition between shops. Unemployment rates are currently over 30% and rents are paid with money from the social security system. With debts rising, services are closing down and going out of business. The houses themselves are in need of repair and suffer from dampness. For some people the level of deprivation is on a par with the inner city. Tables 6.8 and 6.9 give information to compare for such estates – Easterhouse in the suburbs of Glasgow, East Middlesbrough, Kirby and Orchard Park near Hull – with inner city estates.

28 Comment on the indicators and unemployment rates for the four outer estates.

29 Explain what the figures tell you about the problem of deprivation in out-of-town locations.

*Table 6.7 Expenditure in GEAR to March 1984**

Sector	Programme	Participants†	Amount £ millions	% Total Expenditure
Housing	Tenement rehabilitation	GDC/SSHA/HC	49 106	24.30
	New public sector housing	GDC/SSHA/HC	38 203	18.91
	Public sector modernization	GDC/SSHA/HC	32 233	15.95
	Housing built for owner occupation	GDC/SSHA	1136	0.56
Industry	Factory building	SDA	17 782	8.80
	Land assembly site preparation	SDA	9579	4.74
Environment	Environmental improvements	SDA	12 748	6.31
	Anti-pollution measures	SRC/SDA/GDC/SDD/MSC	994	0.49
Infrastructure services	New Infrastructure	SRC/GDC/SDA	8184	4.05
	Local traffic management	SRC	5909	2.92
	Protection services	SRC	4262	2.11
	Local shopping	GDC/SSHA	84	0.04
	Access to employment services	SRC	3280	1.62
Social and community support	Leisure and recreation	SDA/SRC/GDC	7450	3.69
	Health and social services	SRC/GDC/GGHB/VOL ORGS	6471	3.20
	Young people services	SRC/VOL ORGS	2169	1.08
	Community involvement	SDA/HC	1273	0.63
	Promoting welfare benefits	SRC/VOL ORGS	1207	0.60
All		Total	202 070	100

*MSC training programmes excluded; costs not available
†GDC: Glasgow District Council; SRC: Strathclyde Regional Council; SDA: Scottish Development Agency; SDD: Scottish Development Department, SSHA: Scottish Special Housing Association; HC: Housing Corporation; MSC: Manpower Services Commission; VOL ORGS: Voluntary Organizations; GGHB: Greater Glasgow Health Board.

Fig. 6.19 An article from the Guardian.

Consensus reverses urban decay

Peter Hetherington on the inner city transformation that cast politics aside to promote economic revival and aroused envy south of the border where the Minister for Merseyside's ideas were not matched by cash

The Government's apparent determination to press ahead with more Task Force-style projects in the inner cities has been greeted with deep cynicism by some closely involved with the first such initiative on Merseyside.

The Merseyside Task Force was launched as a response to the 1981 Toxteth riots, but never lived up to expectations and has now assumed the role of a Department of the Environment regional office.

Those closely concerned with the initiative—promised so much by the former Environment Secretary, Mr Michael Heseltine, who briefly became the "Minister for Merseyside"—complain that Cabinet concern has not been matched with extra cash.

Indeed, since the Toxteth riots Government spending in a whole range of areas has been reduced—from the Urban Programme (cut again last week) to housing investment cash.

Increasingly, ministers, civil servants, and Task Force officials look to the example set in Glasgow, where a large segment of the city has been transformed in a project now attracting international attention.

Mr James Watson, architect and planner, remembers the awful sight of the inner city slowly dying barely 10 years ago.

"Total decay," he says. "Three hundred acres of derelict sites, an old iron works pouring out filth, terrible housing—you certainly couldn't get a mortgage even if you wanted to buy a house or flat—massive planning blight, and an apparent feeling of hopelessness."

Since 1976, about £470 million has been channelled into the 4,000-acre area, including around £300 million of public money. Now even the population of 45,000 is increasing slightly.

Mr Watson, head of area development with the Scottish Development Agency (SDA) can point to almost 3,000 new private houses—"in total 15,000 have either been built or fully modernised, and 60 per cent of people now have a better home"—plus 250 new factories, workshops and offices, and a large business development centre.

Hundreds of acres have been landscaped, and new parks created, to underline the claim of the Glasgow Eastern Area Renewal Project — Gear, as it is known—that "no other area has witnessed such dramatic changes and improvements."

The English—local councils, MPs, and Government departments alike—are invariably envious of the level of co-ordination brought to the project by the Government-funded SDA, which has no English equivalent.

They are amazed by the political consensus in a Labour city where two local authorities have deferred to a third party, the SDA, in the interests of progress.

Could this, then, be the way forward for English inner cities as the Government considers new initiatives, notably urban renewal agencies for several areas?

Some officials involved with current urban projects believe that an SDA-style approach— co-operating with local councils rather than imposing a solution—provides the only hope of progress, although they concede that it will cost a lot more money.

The SDA, grandly titled an economic regeneration agency, has an annual budget of £90 million and can harness factory building, land renewal, promotional, and industrial investment powers to complement the work of local councils. It is responsible to the Scottish Office

Experience gained in the Gear area has been used to develop projects including Clydebank, Dundee, Greenock, Coatbridge, and the Leith district of Edinburgh.

Dr George Mathewson, the SDA's chief executive, believes that success was possible because of the economic consensus between all parties. "People realised that economics matters more than politics, but that it is not always acceptable to politicians of whatever colour because it means voluntarily giving up some powers and political ideology."

A consensus approach might appear impractical in the political climate of Liverpool. But several prominent Merseysiders believe the Gear example points a way forward while exposing the hollowness of the Government's approach to the English inner city.

Ministers may blame left wing councils, seemingly bent on confrontation, for urban problems. But in Liverpool this obscures the real issue: acute under-funding.

This was recognised over four years ago by Mr Heseltine. But on almost every front government funding has been cut in real terms since the riots: £24 million urban programme cash in 1982–83, for instance, down to £23.3 million in the current year.

Some initiatives emerged — redevelopment of the Anglican cathedral site, a technology park at Wavertree, improvement of peripheral housing estates, renovation of dockland through a development corporation. But Mr Michael Parkinson, director of Liverpool University's Centre for Urban Studies, describes them as "oases in the desert."

The only substantial Liverpool initiative that bears any comparison on scale with Gear is the city council's urban regeneration strategy, where 17 priority areas are being transformed with new and improved housing. But the initiative excludes the private sector.

Mrs Jean McFadden, leader of Glasgow City Council, says her council is determined to be pragmatic. "We may have different political views to the people we co-operate with, but in many cases we put them aside if it is for the good of the city.

Table 6.8 Indicators comparing outer estates and inner areas

	8 inner areas %	4 Non-London inner areas %	4 Outer estates %
Unemployment	15.8	21.8	26.3
Proportion of low-skilled	34.1	37.5	49.0
Overcrowding	8.2	8.1	11.2
Households lacking amenities	8.6	8.1	1.9
New Commonwealth population	17.7	14.4	1.5
Pensioners	15.0	16.0	7.7

(Source: Dept. of Employment, Kirkby Job Centre, Cleveland County Council (Research and Intelligence Dept.))

Table 6.9 Recent unemployment rates in four suburban council estates (January to July 1984)

	Easter-house	East Mid'brough	Kirkby	Orchard Park
Males	39.1	38.9	37.4	n/a
Females	18.5	17.3	16.7	n/a
Total	31.2	30.5	29.5	n/a

(Sources: Dept. of Employment; Kirkby Job Centre, Cleveland County Council (Research & Intelligence Dept.))

Fig. 6.20 Wave Bye Bye to Gran

Easterhouse is one of four large peripheral estates built around Glasgow in the 1950s. It lies six miles north-east of the city centre on high land; 90% of its population lives in the 'barrack-block' – monotonous, three and four tenement blocks. Easterhouse was never considered a success and now has a combination of poor housing, high unemployment and low income. Between 75% and 80% of households have no car, so they are locked in and hampered in attempts to find employment.

Redevelopment of inner city areas by moving out to suburban council estates or redeveloping and rehousing on site can produce new problems in turn. Some of the blame for the failure of these schemes lies with the lack of consultation with the people themselves and in the design of such drab, uninspiring structures. Technical problems of construction have also led to many high-rise solutions being rejected and pulled down.

30 Look at Fig. 6.20 and explain the meaning of the cartoon.

Suburbanization

This is the growth of the suburban area which surrounds the urban area and this growth can be seen as being of two forms. The first is the growth on the edge of the town known as **decentralization**, and the second is the movement away from the urban area to develop the size of smaller towns in the urban hierarchy. This is known as **deconcentration**. These are represented in Fig. 6.21. Areas A and B are additions to the total housing stock of the area.

Fig. 6.21 Model to show suburbanization

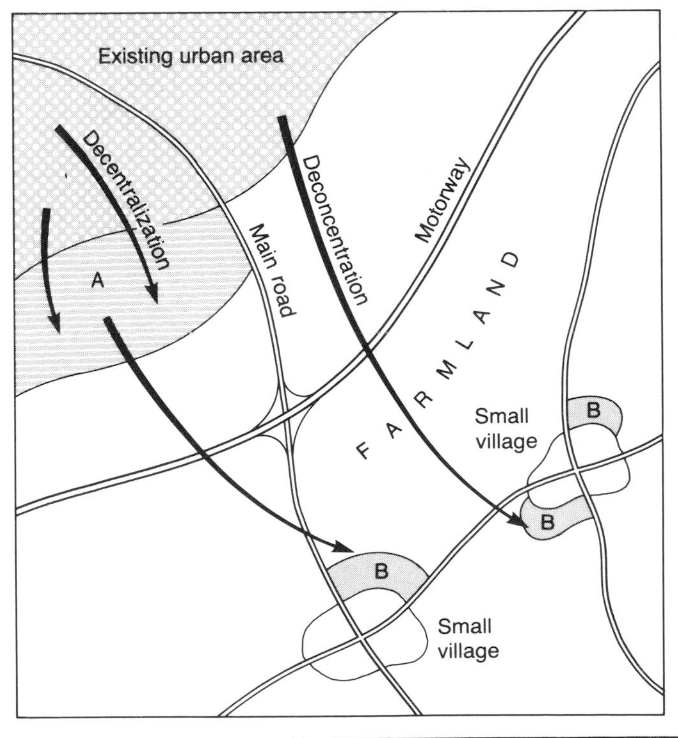

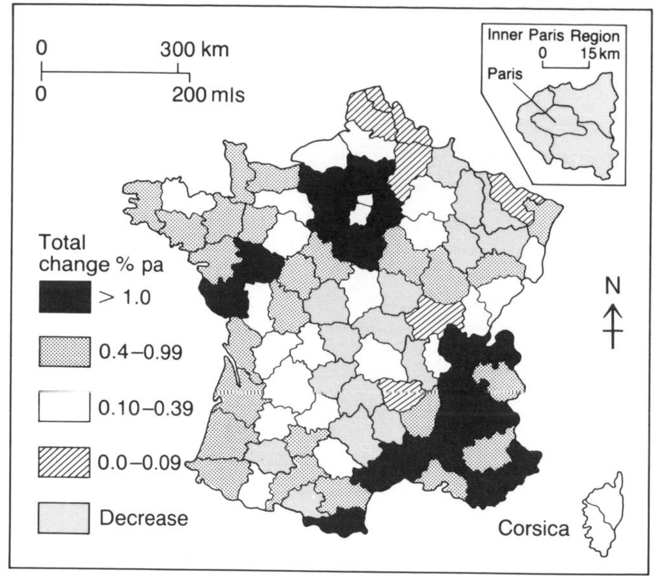

Fig. 6.22 France: population change by département (% p.a.)

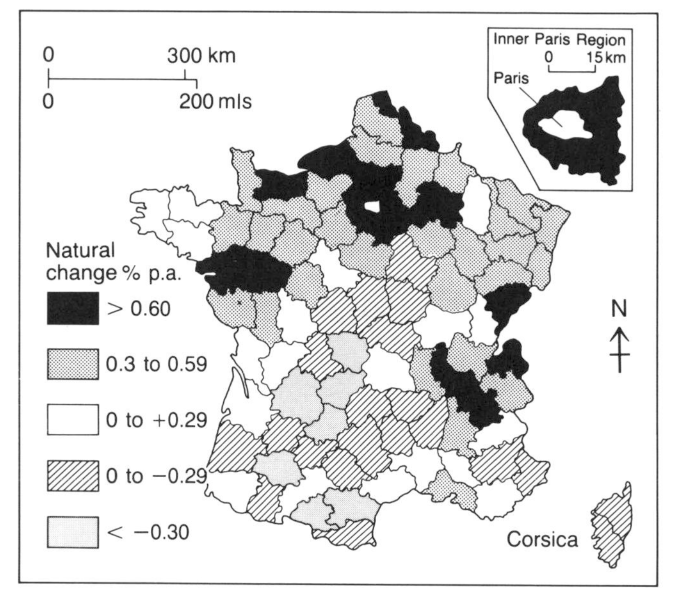

Fig. 6.23 France: natural change (excess of births over deaths), by département (% p.a.)

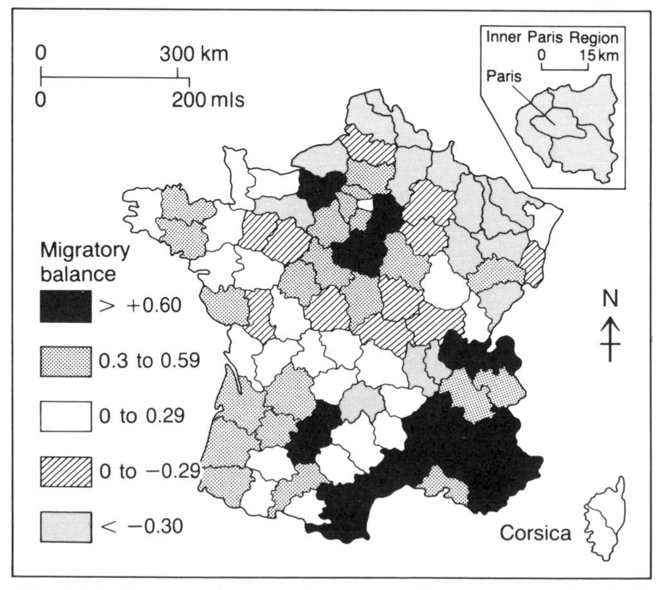

Fig. 6.24 France: migratory balance by département, 1975–1982 (% p.a.)

A has been added to the existing town while areas B in the two villages are additions away from the main urban areas. These processes can be seen to operate on a small, local scale, i.e. around a town, or on a larger regional scale in which additions are found away from the traditional industrial areas. The Lower Earley area, mentioned in Theme 1, p. 35–7, is an example of decentralized suburbanization on the existing urban area of Reading.

Case-study: Paris area

France is a country which has shown considerable suburbanization trends, in common with those shown in UK, USA and New Zealand.

31 Figs. 6.22, 6.23 and 6.24 show the pattern of demographic change in France. Make a tracing of the population change in France (Fig. 6.25) and place the tracing over the other two maps. Try to distinguish the following areas:

(i) areas with high increase and high natural increase

(ii) areas with high increase and high migration increase

(iii) areas with decrease and low natural increase

(iv) areas with decrease and low migration increase.

Devise a key for these and annotate your tracing. Add to your tracing areas where you think deconcentrated growth and decentralized growth are occurring.

There has been considerable decentralized growth around Paris. Fig. 6.26 shows the migration in the Île de France region and Table 6.10 gives information on natural increase and migration.

Table 6.10 Paris: annual % population change

No. on Map (Fig. 6.26)	District	1975–82 Total change	Natural change	Migratory balance
1	Paris	−0.8	0.3	−1.1
3	Hauts de Seine	−0.6	0.6	−1.2
4	Val de Marne	−0.8	0.7	−1.0
2	Seine de St Denis	−0.1	0.8	−0.9
5	Seine et Marne	2.8	0.7	1.7
6	Val d'Oise	1.3	0.9	0.5
7	Yvelines	1.4	1.0	0.4
8	Essonne	1.0	0.9	0.1

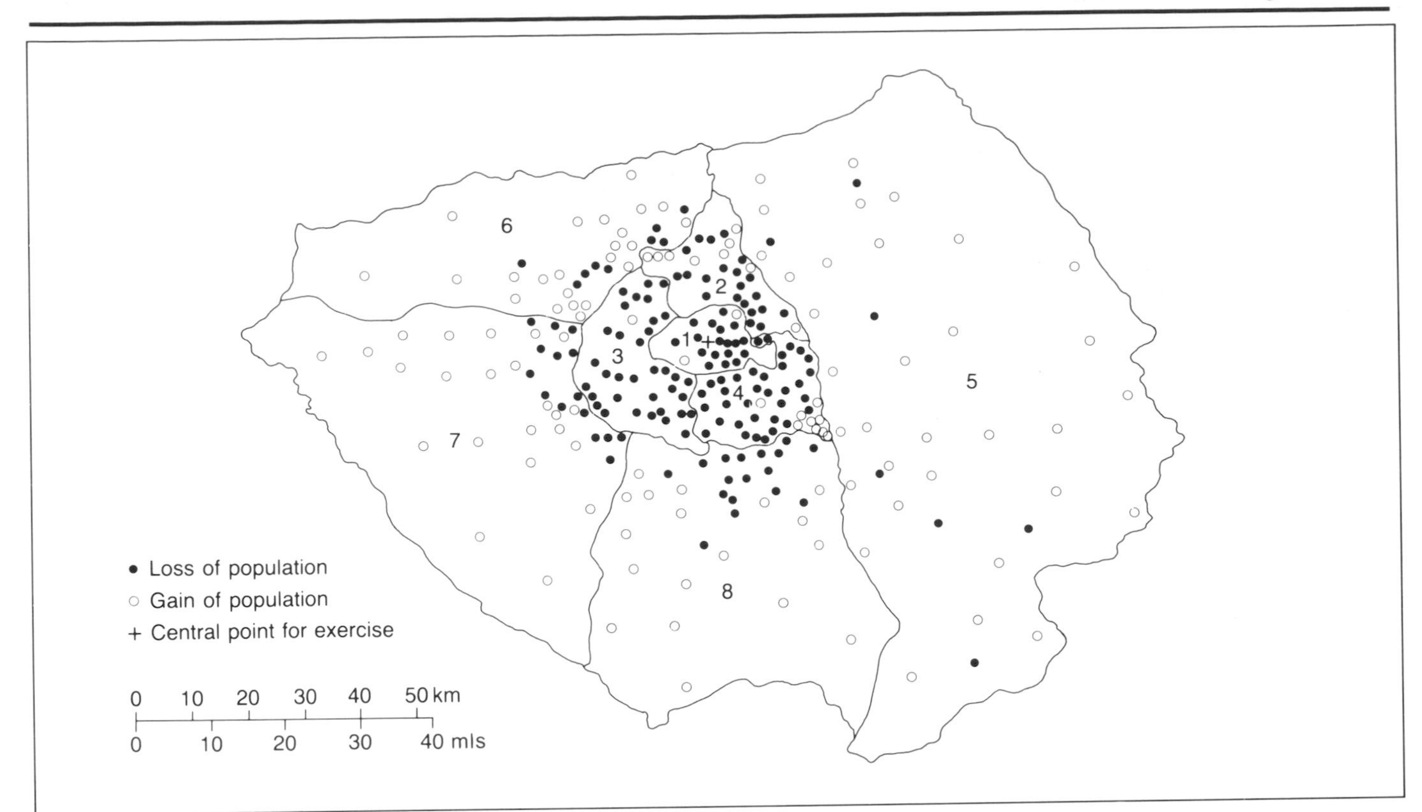

Fig. 6.25 Migration in the inner and outer parts of the Paris region

32 Test the hypothesis that decentralization is taking place around Paris by drawing circles at 10, 20, 30, 40 and 50 km from the centre of Paris (the centre is marked by a cross). In each ring, count the number of settlements registering a loss and those showing a gain. Plot your results on a suitable graph to show the changes as you move from the centre of Paris.

33 Describe your results.

34 On a tracing overlay of the Paris area draw bar graphs to represent the information in Table 6.10 of the respective districts.

The effects of suburbanization can be very significant. With decentralization the land closest to the town, known as the **urban fringe**, can take on a very untidy character. Farmers may be unwilling to spend much time and effort on land which may be re-zoned for housing and the result may be a mixture of land-uses of which 'horseyculture' may be prominent. Land may also be acquired for speculative purposes and development may be many years in coming.

How can these problems be minimized? The introduction of areas of limited or no growth around towns has been used in many countries. The Green Belt has been established around London for many years. However, suburban development sometimes leap-frogs the restricted area and develops beyond.

Decision-making Exercise R: New Development in S.E. England

Background

The development of the South-East region over the last thirty years has been based on the Abercrombie Plan in which the suburbanization of London was stopped by the Green Belt and new towns provided growth poles. These growth poles were to attract industry moving out of London and allow the redevelopment of the capital. However, service industries and office development have grown much faster than anticipated and the population of the South-East region has continued to grow.

This has resulted in pressures along the margins of the Green Belt as land is purchased in anticipation of a relaxation of building restrictions. Increasing densities of population have occurred in the areas enclosed by the Green Belt and leap-frogging of the belt has led to spontaneous development in towns such as Guildford. The new towns have not been big enough to be truly independent

R

of London and do not provide a sufficient range of industries for their populations. An unforeseen change is the increased volume of commuting and transportation produced by improvements to road and rail infrastructures. This has led to increased commuter hinterlands.

Your task
You have to identify an area for development. This development will be a new urban area capable of being independent of London. To do this you will use the sieving technique outlined in Techniques, p. 104.

1. Make a tracing of the boundary of the South-East region and the coastline. Mark on the Greater London boundary (Fig. 6.26).

2. Sieve out the following areas using this sequence:

(i) The Green Belt (Fig. 6.26)

(ii) Areas of significant environmental resources (Fig. 6.27)

(iii) Agricultural areas (Fig. 6.28)

(iv) Recreational pressure areas (Fig. 6.29)

Shade these areas as you remove them from the areas suitable for development.

3. Identify the major areas for potential development (omit the very small areas).

4. Now check to see if these areas are near existing new towns, expanded towns or other large towns. If they are, omit them. (Fig. 6.26.)

Fig. 6.26 New town and infrastructure of S.E. region

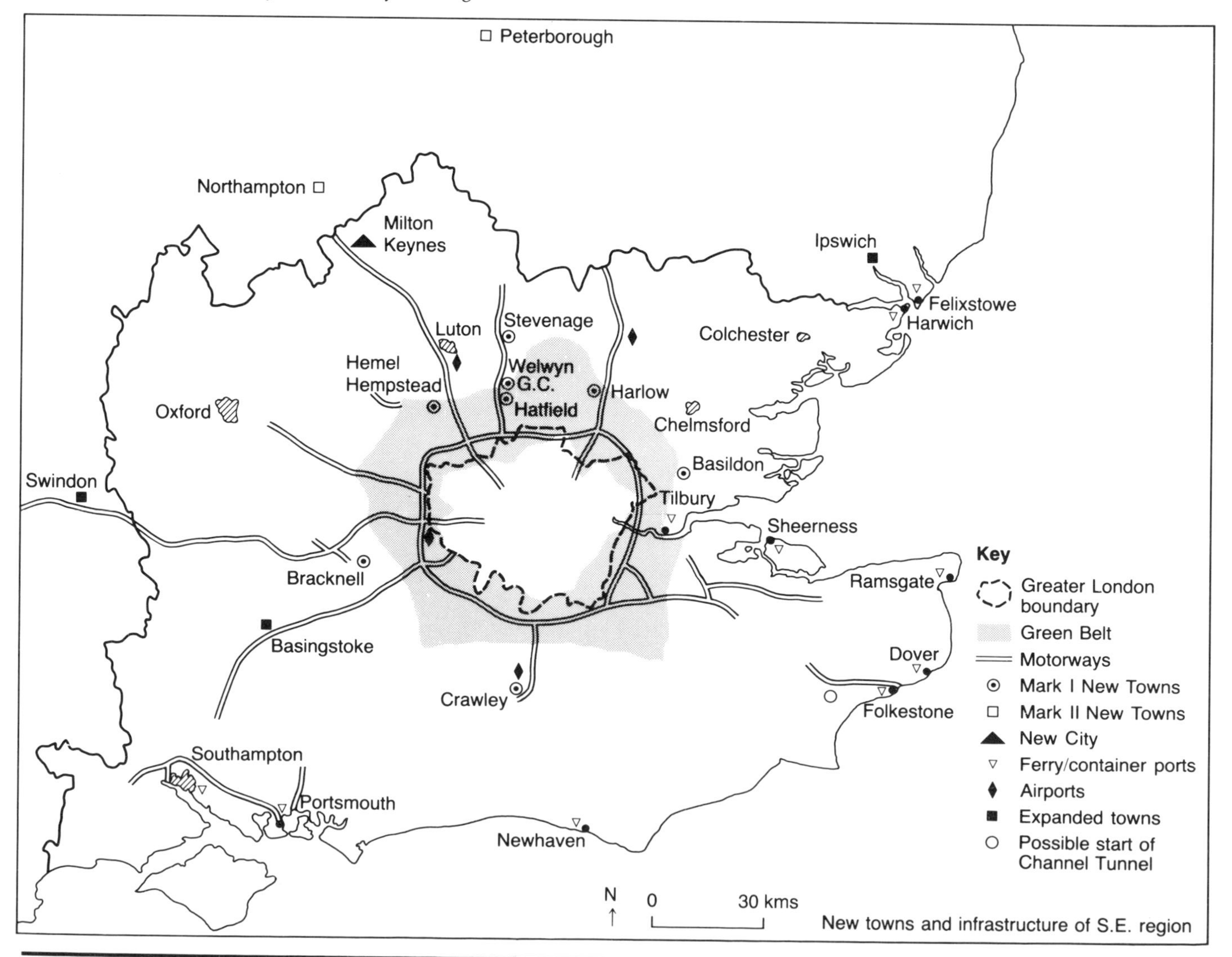

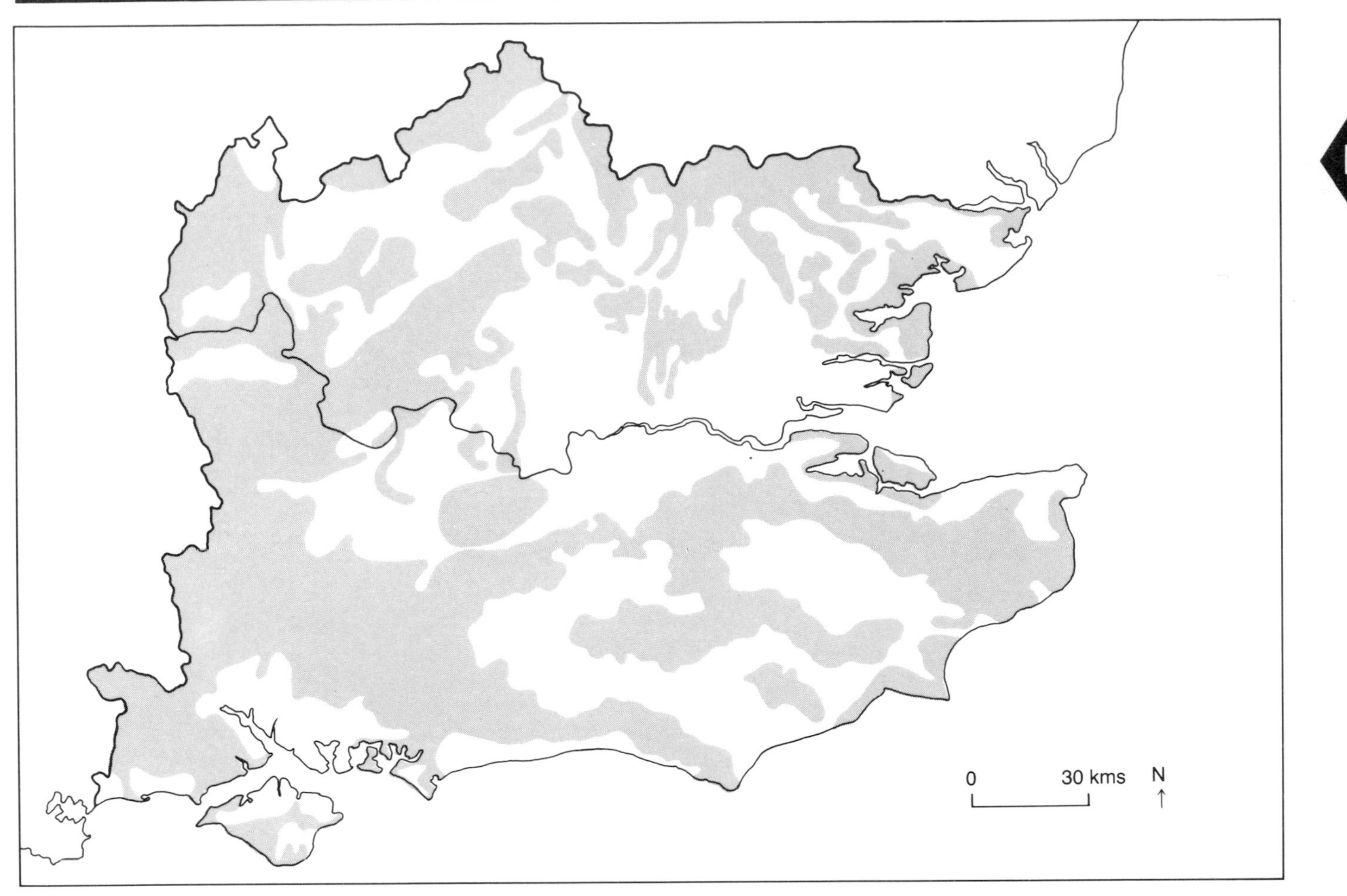

Fig. 6.27 Areas of significant environmental resources

Fig. 6.28 Areas of special significance for agriculture

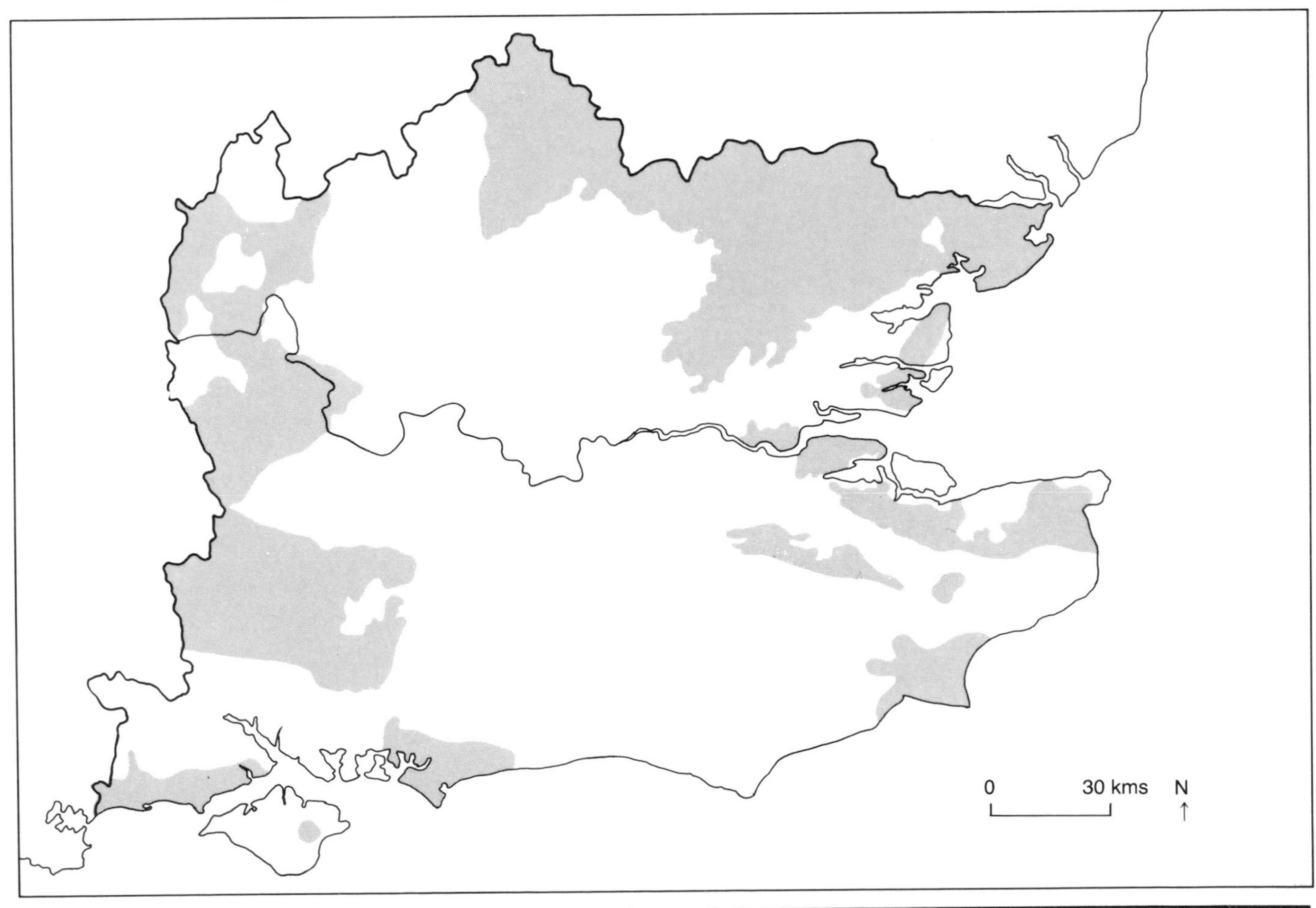

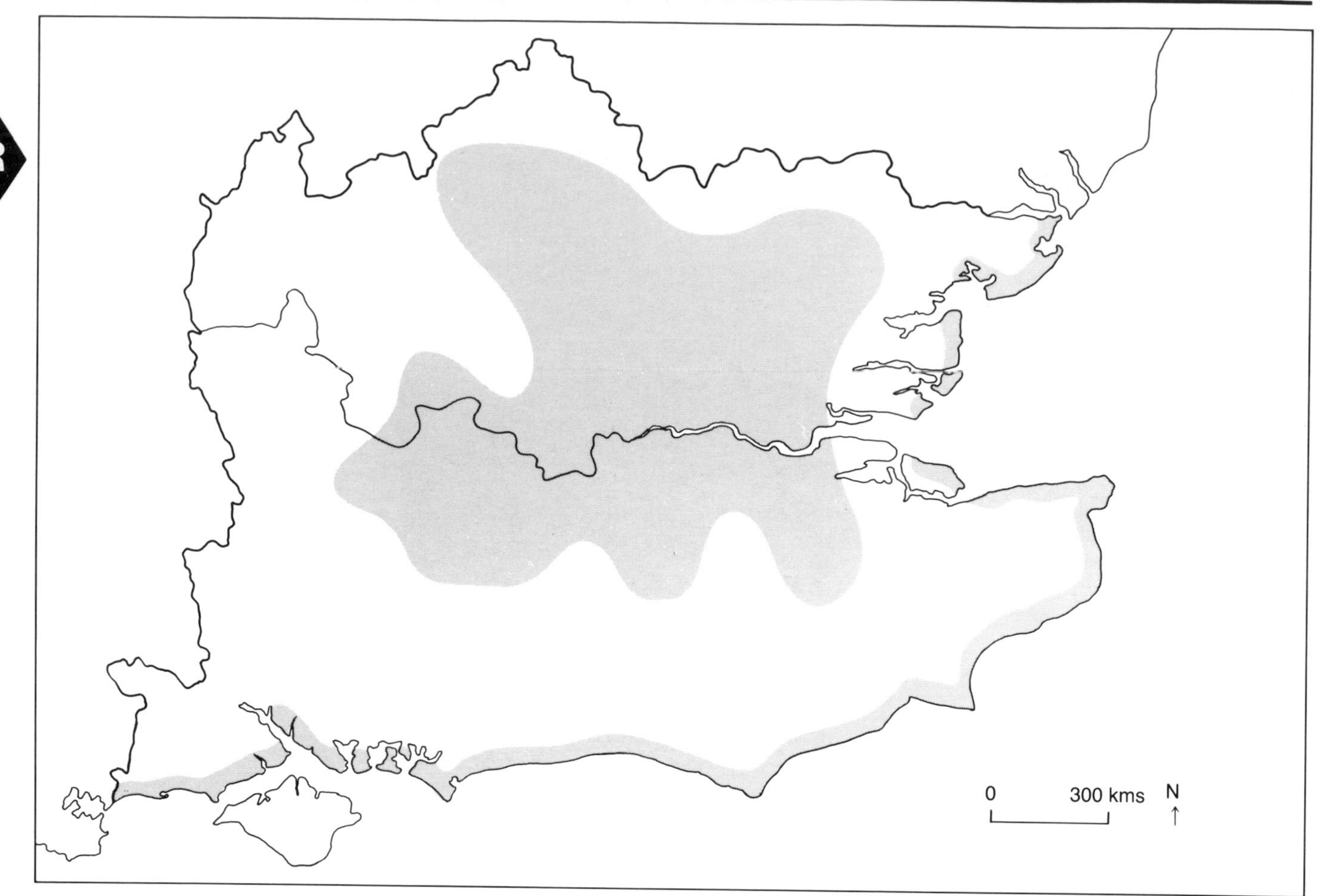

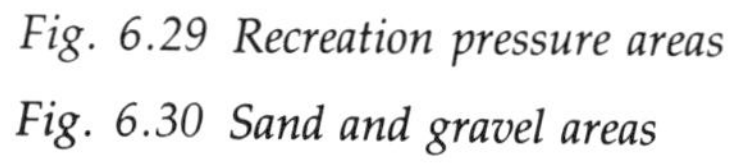

Fig. 6.29 Recreation pressure areas

Fig. 6.30 Sand and gravel areas

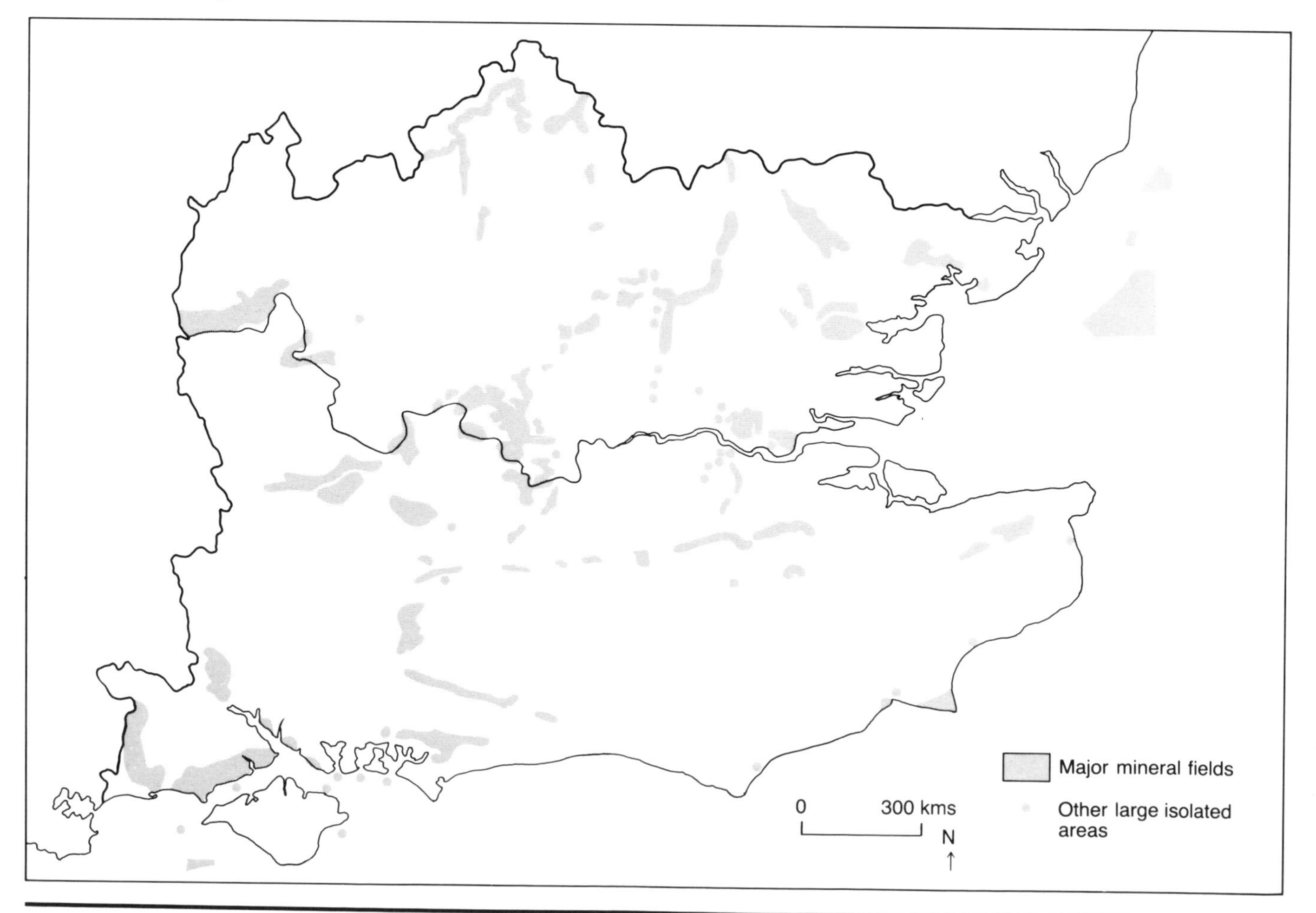

R

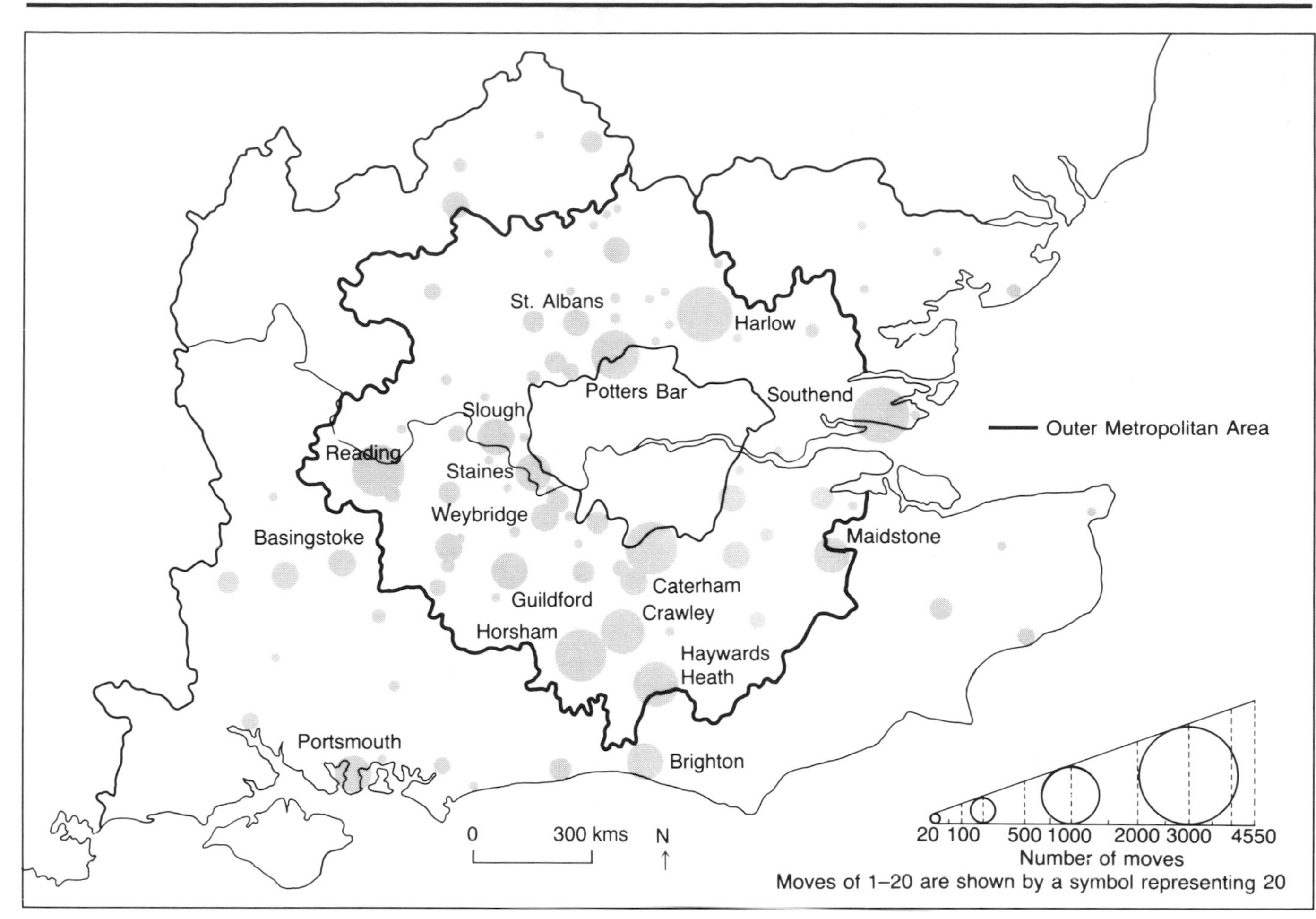

Fig. 6.31 Destination of office moves in the South-East, excluding Greater London

Fig. 6.32 Movements of industry in the South-East, excluding Greater London

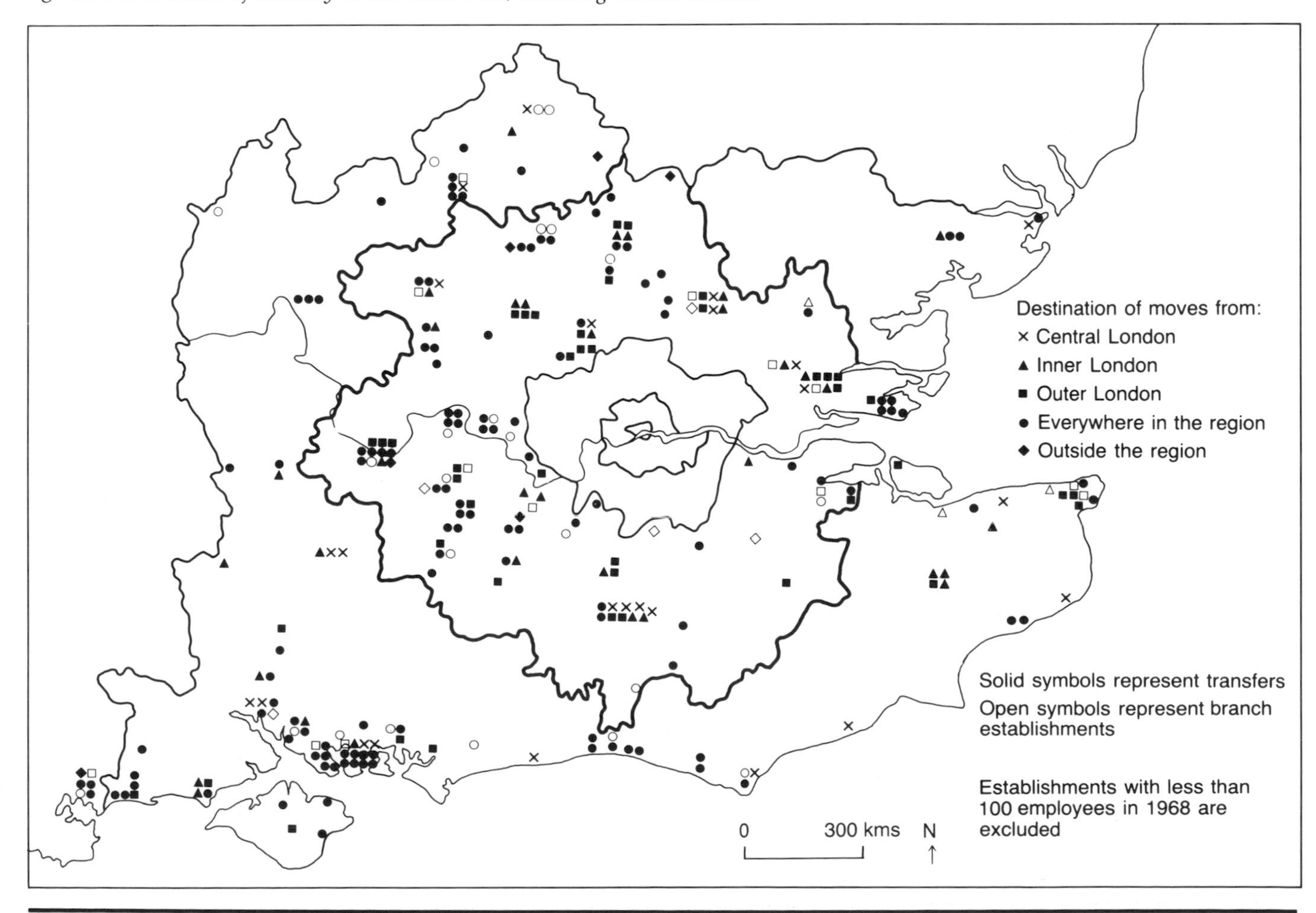

London Docklands

The regeneration of London Docklands is the most important inner city development of the 1980s in Europe. It offers unique opportunities for residents, investors and businesses:
For residents, a wide choice of attractive, well-located homes; for investors and businesses, sites with high investment potential and schemes in which to set up new or re-located enterprises.

The major advantage of Docklands is its closeness to the centre of London. Its unique waterscape environment covers a total area of 20 square kilometers (8 square miles) in which lie the magnificent docks and wharves of the once bustling Port of London.

There are four main areas within Docklands: Wapping, Isle of Dogs, the Royal Docks and Surrey Docks, plus the Enterprise Zone. Each is being planned to exploit the various existing local themes, while leaving considerable flexibility in the development of individual sites, whether for modern industrial, commercial or residential use.

An Enterprise Zone has been established within London Docklands, in which there are special advantages. These are designed to cut red tape and provide major financial and tax incentives during the ten years to 1992. Within Docklands, the Government is currently investing over £200 million to create a modern infrastructure of utilities, transportation and telecommunications that will benefit both businesses and residents. This will integrate the existing infrastructure with the most advanced communication systems, including the Docklands Light Railway (in operation in 1987) and satellite earth stations. London City Airport (a short take off and landing airport) will be operational in 1987, providing quick routes to major UK and European cities, while travel through the Capital will become more pleasant with the advent of new riverbus services.

The Exceptional Place

Numerous environmental improvements, together with new recreational and leisure facilities, many utilising the unique waterscape features, are already contributing to the transformation.

The sheer scale of current investment activity in this new and visionary metropolitan environment of tomorrow, located in the heart of a major capital city, underwrites its potential now as one of the most attractive living, development and business opportunities in the world today.

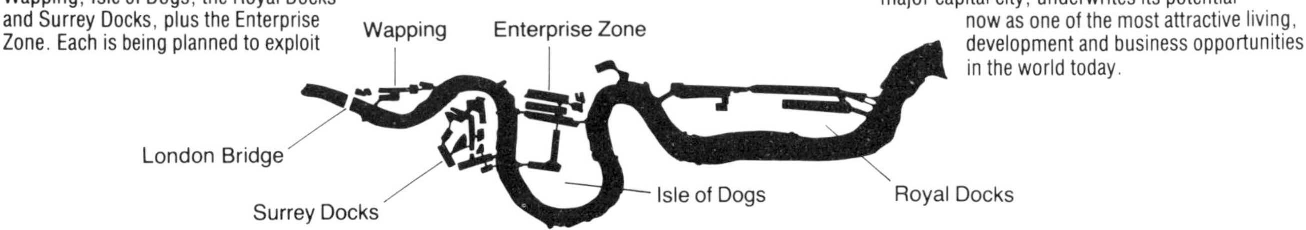

Fig. 6.33 Information produced by the London Dockland Development Board

Fig. 6.34 Newport resources

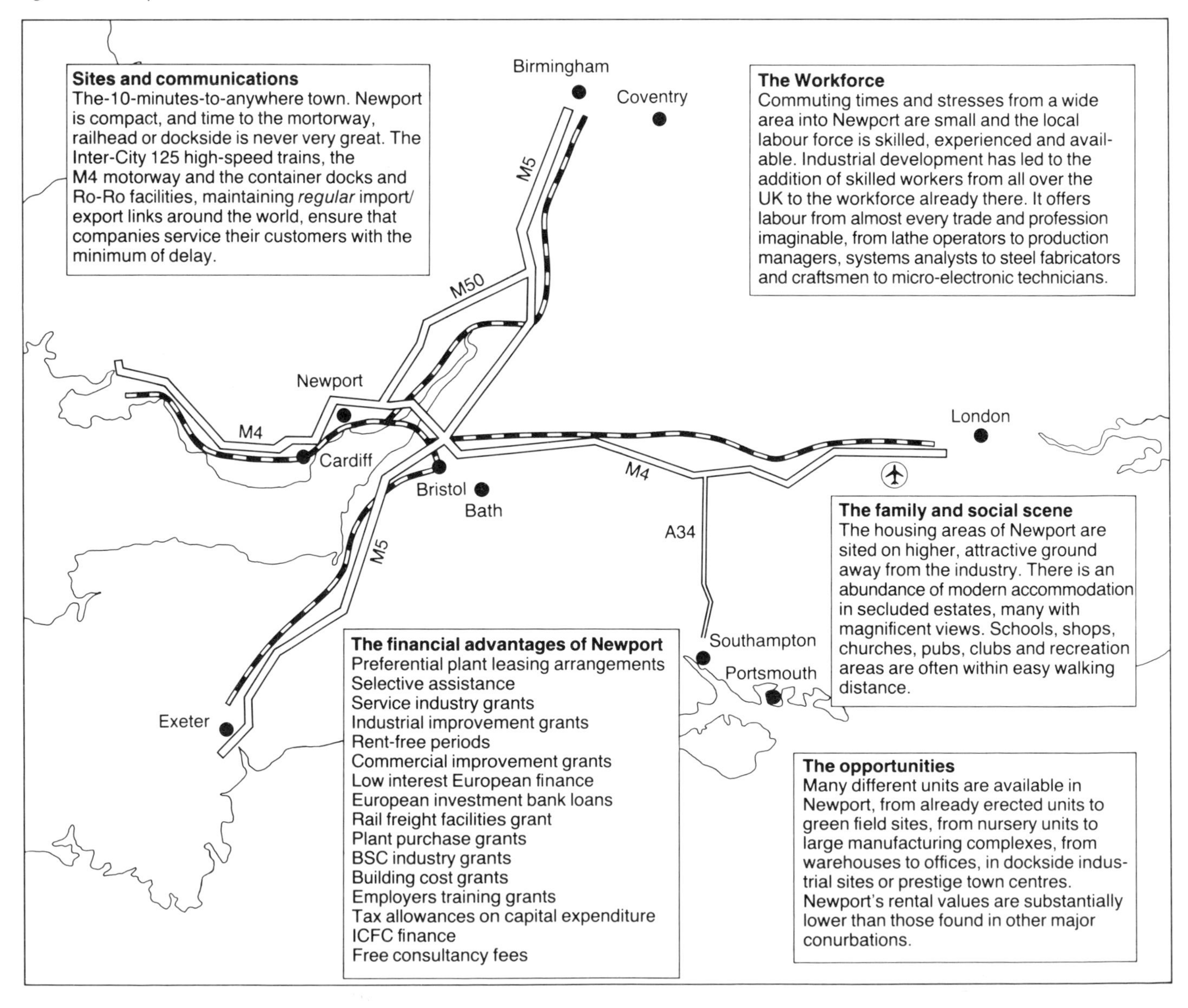

R

5. Now check your areas for nearness to sand and gravel for building materials and existing infrastructure. Omit them if they are far away from these resources. You should aim to be near these, but if not then make a note of the improvements you would have to make. (Figs. 6.26 and 6.30).

6. While your development needs to be independent of other areas, to encourage industry you will need to consider the linkage with other firms and offices. Place your tracing over the map showing office and industrial moves (Figs. 6.31 and 6.32) and make your final judgement on this basis.

Having made your decision, write a brief report indicating your reasons for selecting the site. Use your atlas to give more detail.

Now write a letter to the local conservation group in your chosen area to explain why the site was chosen and explain the necessity for such development.

This development would add to the development of an already prosperous area. As shown in Theme 4, other parts of the country are doing their best to attract industry. Figs. 6.33 and 6.34 show information produced by the London Docklands Development Board and by the development agency for Newport, South Wales. Complete this decision-making exercise by considering the views of representatives of these two areas. Work in pairs, one taking the role of the representative of Newport and the other taking the place of the LDDB. Write paragraphs outlining the objections voiced by these bodies to your proposals, and compare the arguments you have used on behalf of those organizations. Try to consider local as well as national arguments.

A further type of decentralised development has recently emerged in the form of 'private new towns'. Small settlements on new or 'green field' sites have been proposed by consortia of builders in the area around London. The developers call them 'new country towns'. Examples of these are Foxley Wood in North Hampshire, and Tillingham Hall, between Basildon and Upminster in Essex. They would be small, private new towns if built. The planning applications from the builders have produced successful protests from local action groups. The local action groups point out that areas of the countryside will be built on and the advantages of living in the countryside will be lost. The developers argue that it is better to concentrate new development on one centre rather than continue to make a series of small additions to existing villages. They contest that the best solution is one large settlement with all the amenities, rather than small additions to a number of villages with piece-meal provision of new facilities.

At Tillingham Hall the developers, Consortium Developments Ltd, consisted of nine well known national building companies. They argued that there were six advantages to the principle of building on new sites. These included reduction of pressure on the existing settlements to expand, more economic provision of amenities and infrastructure, and reduction of pressure on agriculture due to urban growth. The builders also considered that they could pay more attention to providing an integrated community in which a variety of housing layout and style could be used. This would have the effect, they claim, of increasing the supply of houses and reducing the house price spiral.

The planning applications for Tillingham Hall and Foxley Wood were unsuccessful. Local opposition was particularly strong in both cases. Opposition to local developments be they roads, houses, or a development such as the Channel Tunnel have led to the identification of the NIMBY syndrome. This acronym stands for Not In My Back Yard and is used as a criticism of people who sympathise with the need for local development but do not want it to affect them. However, like the roads in the Introduction, suburban development is going to have to go somewhere.

Shanty Town Development in the Third World

The movement of people from rural areas to seek work in the towns has been discussed in Theme 4. Recent studies have shown that this rural – urban migration is in most cases a much more planned, orderly move into the urban areas than was at one time thought. There is also considerable return migration. Again, this is a deliberate and planned move. An original move may be made by one

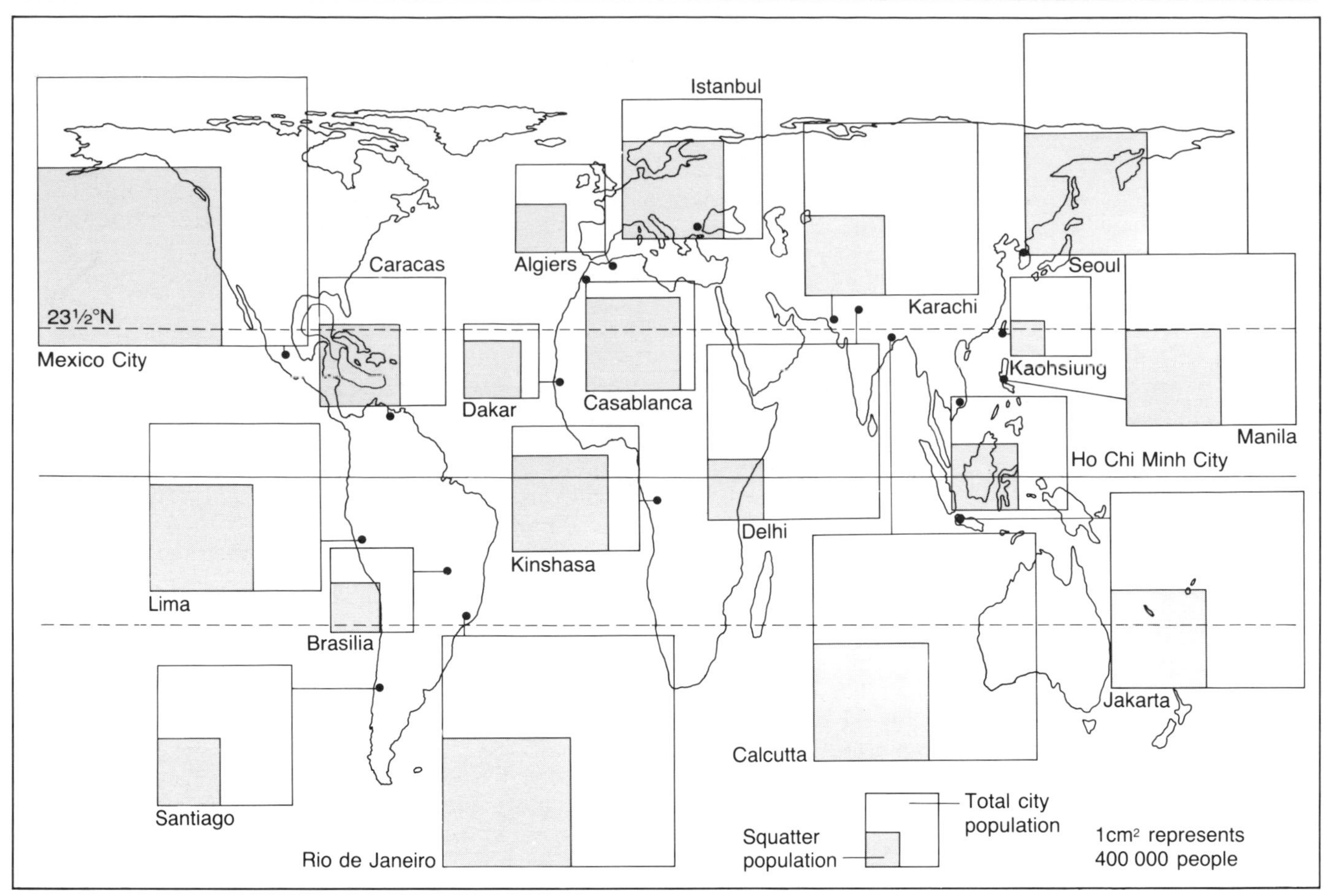

Fig. 6.35 Size of the squatter population in selected world cities

Fig. 6.36 The 'self-help' model

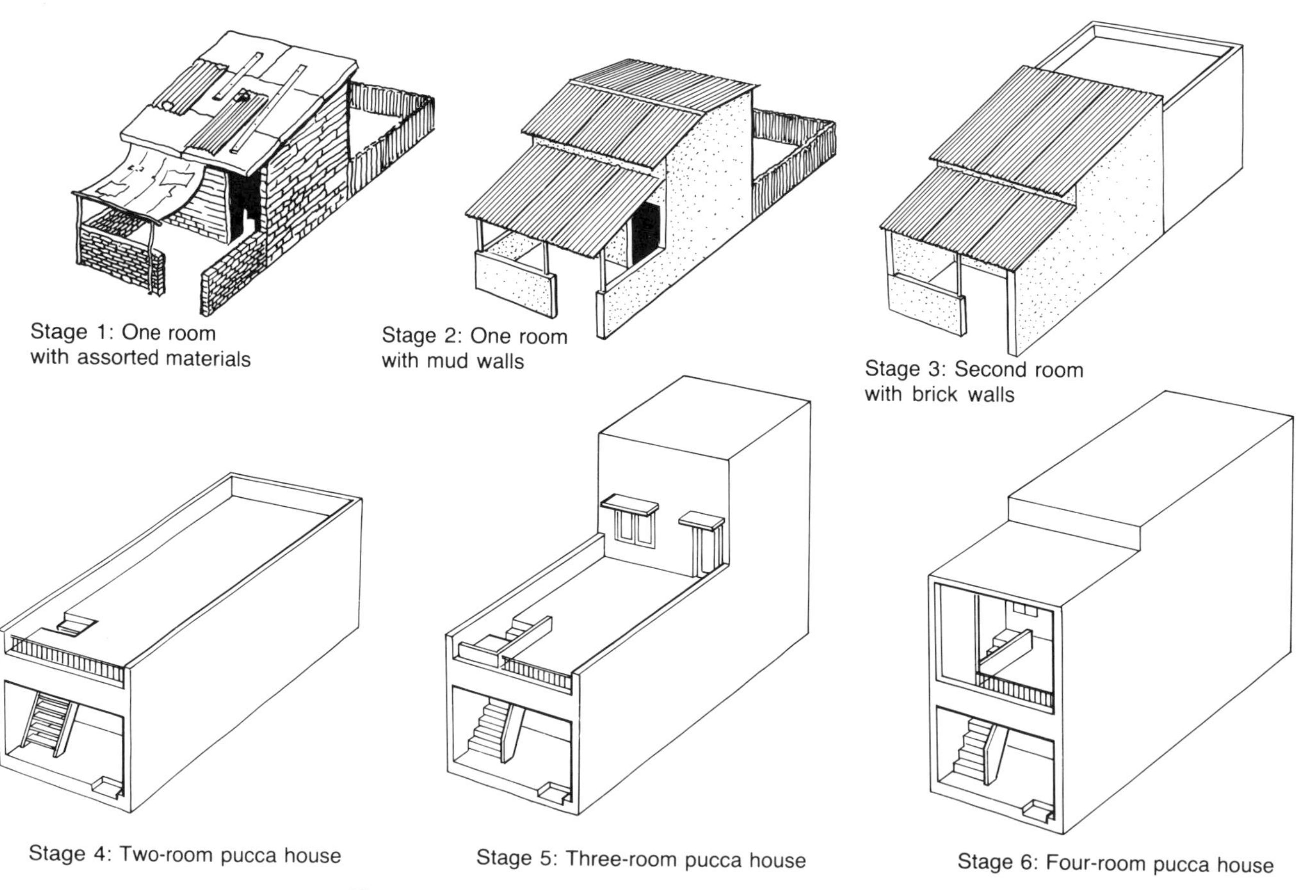

Fig. 6.37 A cartoon from the New Internationalist

member of the family who establishes a base to receive other members of the extended family. The group may grow in this area or it may be a spring-board for further migration into the town, but contact is not lost and the route established may be used by many other relatives.

The basic need on arrival in the urban area, however, will be for shelter, and high rentals and lack of money may force the migrant to build his own shelter or to move in with others in what are known as **shanty towns**. These settlements are known as favelas in Brazil, bustees in India and kampungs in Indonesia, but many of them have common characteristics. The population of the shanty town may be a large proportion of the population of the urban area, as shown in Fig. 6.35.

> **35** Study Fig. 6.35 and try to calculate the percentage of the population living in shanty developments in the selected cities. Make sure you can identify the name of the country in which each of the towns is found.

Shanty developments vary in nature from city to city, and they are often associated with squatter settlements, where people live on land they do not own. Some settlements are of this type with shacks assembled on any available marginal land such as along the railway lines, on the floodplains of rivers or on beaches. However, some settlements are on land owned by a small village engulfed by the growth of the urban area, or on lands owned by the tribes as in the case around some of the African cities.

Shanties will develop where land is available. Some are close to the city centre and others further away. They may grow by swift, highly planned movements as happened in parts of Lima, or they may grow gradually, by incremental growth.

Migrants will be keen to make progress once they have made the move to the urban area. The dwelling may take on many functions, as well as providing shelter. It may be a place where people gather to maintain social and cultural customs. It may be a means of income by providing living space for renting out to newcomers. It may provide space for industries, trades and for storage as a warehouse. Migrants may grow vegetables to sell and it may be a source of prestige and something to take pride in. This last point is illustrated by the fact that many shanty settlements have developed into established suburbs. Fig. 6.39 shows the stages shanty housing can pass through.

> **36** Imagine that you are a migrant to the town. Under what conditions would you improve your dwelling, as shown in Fig. 6.36? Make a list of these conditions and discuss them in your class.

One important drawback to this idea of upgrading through self-help schemes is that many shanty town dwellers do not own the house but rent it. This may be a disincentive to even start upgrading the house, but if the property is improved then it will increase in value. The cartoon (Fig. 6.37) gives some indication of what might happen to the property. If it does not become the site of a hotel, the landlord may increase the rent, evict the tenants and rent it out to more wealthy people. This gentrification process is something which happens in Western cities, as areas become fashionable and land prices rise in response to increased demand.

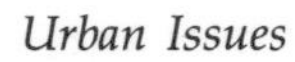

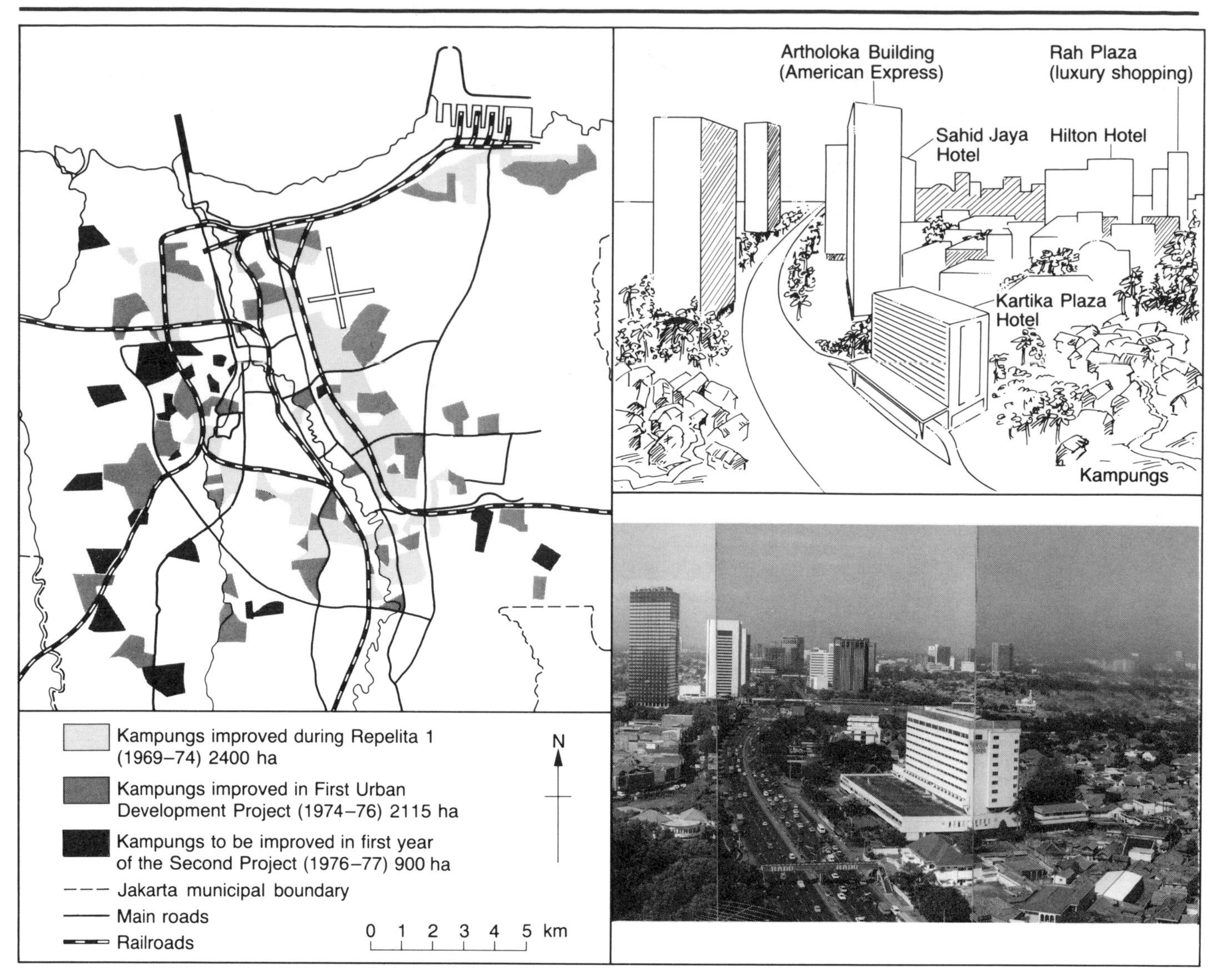

Fig. 6.38 The map shows the location of the kampungs in Jakarta. The line drawing and composite photo show the great contrast in accommodation, with kampungs built in the shadows of ultra-modern hotels

Case-study: kampung improvement in Indonesia

In 1969, the Indonesian government started to implement programmes to upgrade the kampungs. Jakarta, the capital of Indonesia, has a population of 7 191 000 which at current rates of growth will be nearly 17 millions by the year 2000. Unemployment is considered to be between 10% and 15% and poverty is very apparent. The average income in 1976 was $200 per capita for the whole city but at least 65% of the population had incomes below $130 per head, regarded as the absolute poverty line.

The majority of the housing in Jakarta is in the kampung areas (Fig. 6.38). Kampung means compound and was a village outside the town at one time, which has been swallowed up by the outward growth of the town. They are now mixtures of houses, generally of one storey. The kampungs closest to the city centre have high densities of population of between 400 and 500 people per hectare. Plot sizes in the central kampungs are about 45 m^2 while the peripheral kampungs have plot sizes of 160 m^2. Most of the kampung dwellers have legitimate tenure; only those settlements on the land adjacent to the railway and the river are occupied by squatters. Hence there has been considerable upgrading of the properties.

The main problem in the kampungs, however, is the inadequate level of services and infrastructure. Less than one third of the population of Jakarta has piped water with no proper sewerage system. Private or communal latrines with septic tanks, pit privies and the rivers and canals form the sewerage system. Only about one half of the refuse generated by the town is removed. Accessibility is poor for most people, there being only a poorly developed transport system. Flooding is a particular problem in Jakarta as much of the development is on low-lying ground, and the drainage system is unable to cope, particularly when blocked by refuse. The poor sanitation makes health problems acute and Indonesia has one of the highest incidences of cholera in the world.

Other manifestations of poverty include an infant mortality rate of 116–120 per 1000 live births, high rates of malnutrition, with 37% of all children under 7 being classified as malnourished, 17% severely so, and 80% of families have 'grossly inadequate diets' according to the Applied Nutrition Programme. Primary school enrolment averages 70% with some areas being as low as 40%.

Against this background of poverty the Kampung Improvement Programme (KIP) was started in 1969 with the following aims:

The overall aim of the Kampung Improvement Programme was immediately to increase the standard of living of kampung households through implementation of an integrated physical, social and economic programme package which will:

(a) reduce deficits in household needs of essential public services

(b) increase human capacity, incomes and productivity

(c) increase households' and enterprises' control of capital assets

(d) promote social and economic stability and reduce vulnerability within kampungs

(e) promote self-help and self-reliance among kampung people.

To do this, the KIP was essentially concerned with the provision of infrastructure:

Local roads The construction of new access roads, and/or the upgrading of existing ones, to provide access to neighbourhoods, particularly for emergency vehicles
Footpaths New or improved footpaths to provide access to all properties
Drainage New or improved drainage channels to dispose of surface water from roads and footpaths into the main collector canals
Water To provide improved access to adequate supplies of clean water
Sanitation To construct family pit privies where possible, and to construct group toilet, washing and bathing facilities, connected to a septic tank
Solid waste To construct household garbage boxes, and to provide equipment (hand carts and trailers) to enable the community to improve the collection of solid waste and its removal to central collection points
Schools The construction of primary schools (*Sekolar Dasar*) where school facilities are lacking in an area
Clinics The construction of health clinics (*Puskesmas* and *Poskesehatan*) where needed in kampungs.

Fig. 6.39 Stages of the KIP procedure

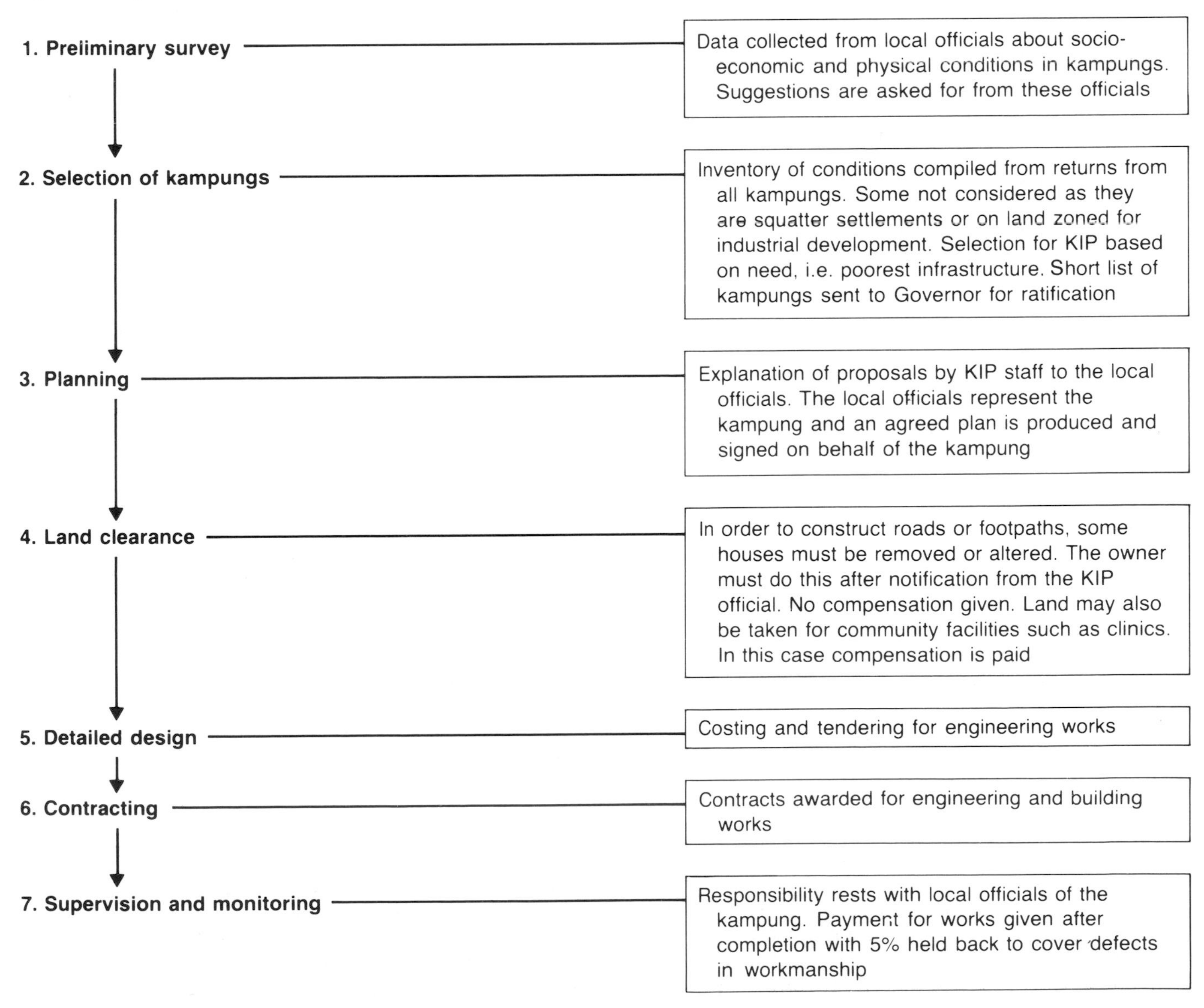

> **37** Consider Fig. 6.42, which gives the procedure for the Kampung Improvement Programme. Are there any drawbacks to the procedure? Will the really poor benefit? Who is the key figure in this programme? Are there any dangers in this arrangement? Discuss these questions with your class.

The role of the community is threefold. It must comply with land clearance plans, maintain the infrastructure and ensure that refuse is collected from all households.

Finances
Most of the services are provided free with some exceptions:

(i) *Water* People are charged for the installation of standpipes and the water is then metered. This money can be recouped by selling the water

(ii) *Communal lavatories and washrooms* There is a levy on the inhabitants of the kampung to pay for cleaning and maintenance

(iii) *Refuse collection* Local levy to pay refuse collectors

(iv) *Property tax* A very low tax payable by all people to the government.

Implementation and evaluation
The KIP was implemented in three stages. Repelita I was the first national development plan which ran from 1969 to 1974 and Urban I and II ran up to 1979 and were financed by the World Bank.

> **38** Table 6.11 shows some of the quantitative achievements of the KIP. Calculate the costs per person and the costs per hectare for the three periods of time. (The currency is Rupiah with US$ 1 = 415 Rp.)
>
> **39** Take the role of the government official and write the speech he or she might give in evaluating the scheme. Try to take a city-wide perspective and check back to the original aims to see if they have been achieved.

The KIP was a relatively cheap development, using only 5% to 10% of the city's budget for 10 to 12 years. If the city has been improved this is a small price to pay for such development. It should be remembered that it did not cater for all the urban poor of Jakarta but it did win the Aga Khan Award for Architecture in 1980. Its contribution has been summed up as follows:

The Kampung Improvement Programme measurably improved the lives of a half million urban squatters living in densely populated communities (kampungs). Sponsored by Jakarta's municipal government, the KIP built much needed access roads and sewage and drainage systems around existing structures. Though extensive, the programme was at the extremely low cost of $60 per capita. The result was that the quality of kampung life was greatly enhanced, while the cultural fabric of these communities was undisturbed. Indeed, once the access and social services were created, kampung residents began making improvements to their own homes, as well as contributing to the economic life of Jakarta at large.

Table 6.11 Quantitative achievements of the KIP

	Jakarta		
	Repelita I (1969–74)	**Urban I (1974–5)**	**Urban II (1976–9)**
Total area (ha)	2385	1975	3316
Total population	1 173 000	765 000	1 097 000
Average population density (persons/ha)	494	387	331
Total cost (Rp.m.)	6484	15 963	29 360
Total cost per capita (Rp.)			
Total cost per hectare (Rp. × '000)			
Quantities			
Roads (km)	179	179	567
Footpaths (km)	270	330	350
Drainage channels (km)	145	79	186
Public taps/wells	78	839	588
Communal lavatories and washrooms	71	135	71
Pit privies	–	–	14 045
Garbage carts	207	265	516
Health posts and centres	21	25	43
Schools		54	96

Industrial Development
As well as providing housing, the shanty development provides an important industrial base. Since the early 1970s, the industry of the shanty dwellers has been recognized as playing an important role in the economy of the country. Table 6.12 gives information from Indonesia of the sort of employment taken by the shanty dwellers. Many of

> **40** Study Table 6.12 and attempt a classification of the jobs shown.
>
> **41** Describe how they could be beneficial to the economy of the country.

Table 6.12 Occupational clustering of migrants and commuters from fourteen West Javan villages working in urban areas, 1973

Village	Number of migrants and commuters	in most common occupation	(%)	in second most common occupation	(%)	in two most common occupations (%)
I	74	Groundnut hawker	65	Government army	15	80
II	55	Cooked food/cigarette hawker	35	Day labourer	22	57
III	91	Cooked food hawker	43	Jewellery hawker	21	64
IV	70	Becak driver	57	Day labourer	16	73
V	82	Becak driver	41	Factory worker	34	75
VI	100	Labourer	35	Hospital worker	13	48
VII	87	Kerosene hawker	32	Household domestic	15	47
VIII	77	Airline/hotel worker	32	Household domestic	10	42
IX	87	Miscellaneous goods hawker	60	Government/army	12	72
X	88	Driver	27	Government/army	26	53
XI	87	Becak driver	38	Construction worker	20	58
XII	92	Carpenter	49	Government/army	28	77
XIII	99	Barber	31	Bamboo worker	20	51
XIV	104	Bread hawker	42	Driver	32	74

(Source: Hugo (1977:64). Reproduced by permission of the Editor, *Bulletin of Indonesian Economic Studies*).

these people would aim to be employed by the government or to join the army due to the stable nature of these institutions. However, most have to fall back on other ways of making money.

In fact, analysis of employment in the so-called informal sector has shown high rates of growth which can act as multipliers developing other parts of the national economy. They are labour-intensive, use indigenous resources and are extremely ingenious in recycling materials, as well as providing useful services.

Decision-making Exercise S: Developing Port Moresby, Papua New Guinea

S

Fig. 6.40 shows the extent of Port Moresby, the fast growing capital of Papua New Guinea. You will notice that, while considerable areas have been improved, there are still some unimproved settlements and urban villages.

Your task is to take the area shown in Fig. 6.41 and prepare a plan for its development. Fig. 6.42 gives you some possibilities. You may use them all or you may decide to be selective in your use of them. You should complete your work by producing:

(a) a master plan showing developments with annotation to make your proposal clear,
(b) flow chart indicating your procedures in making the scheme work. For example, at what stage would you consult the people living in the areas, if at all?

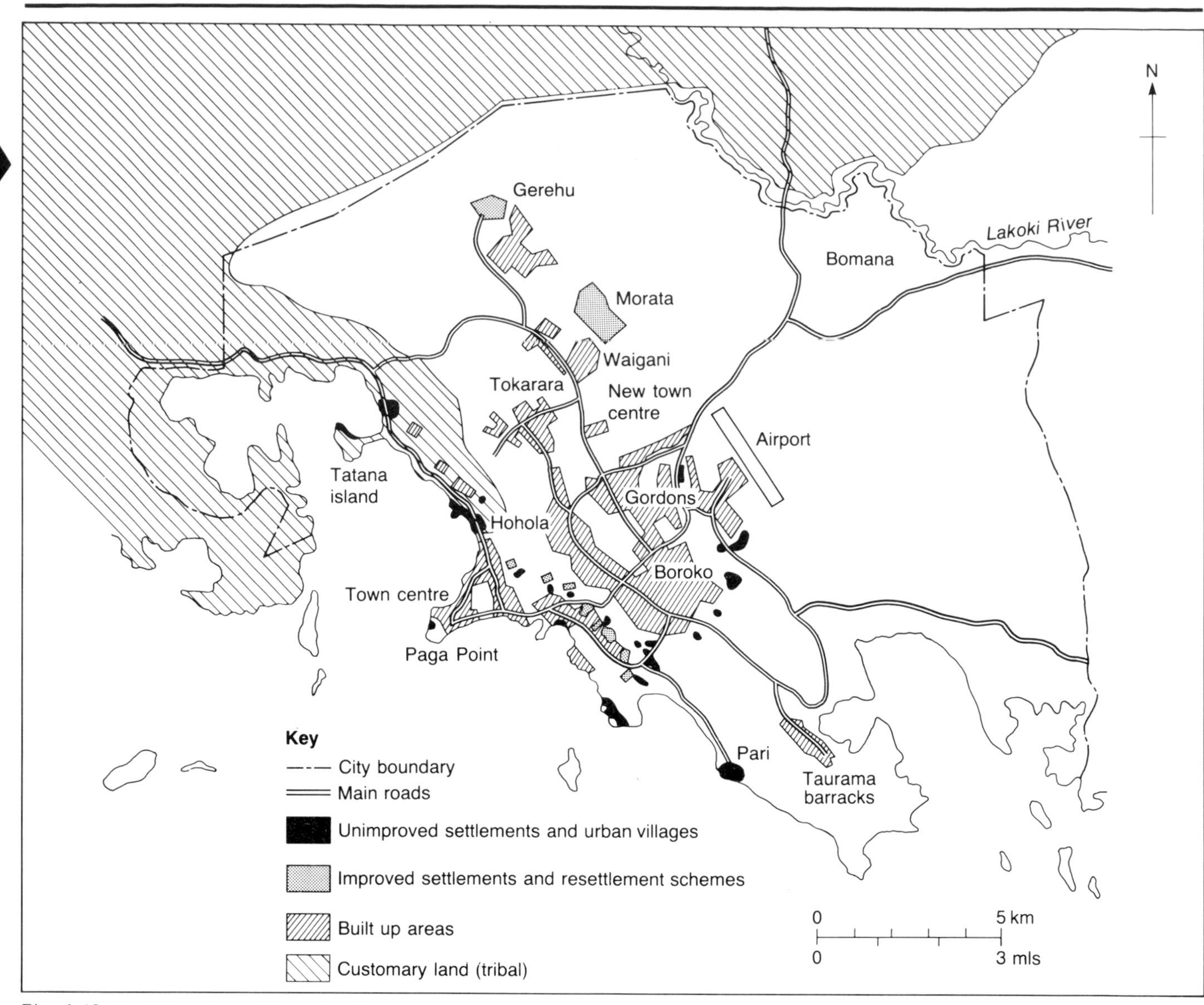

Fig. 6.40 Port Moresby: settlement patterns

Fig. 6.41 Port Moresby: conditions prior to improvement

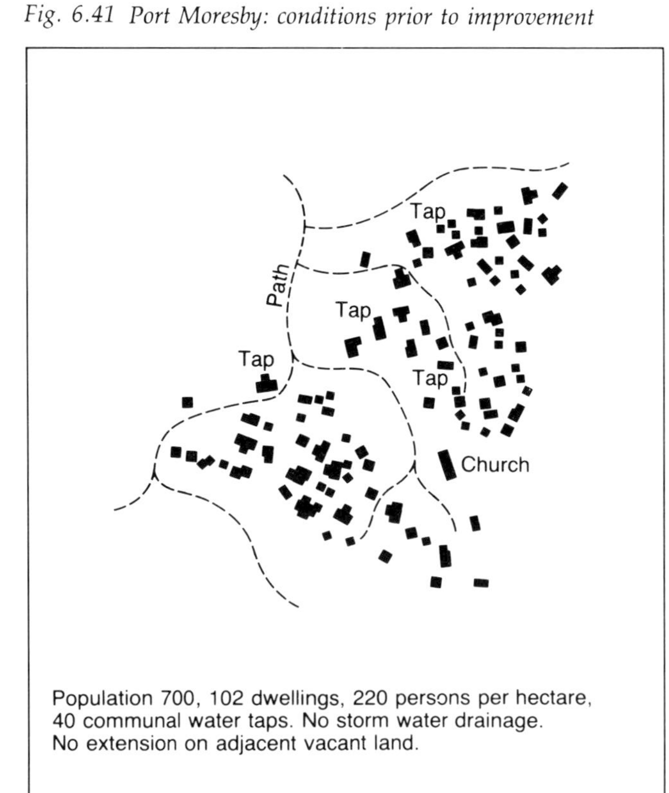

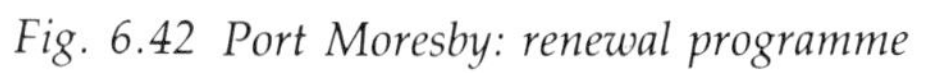

Fig. 6.42 Port Moresby: renewal programme

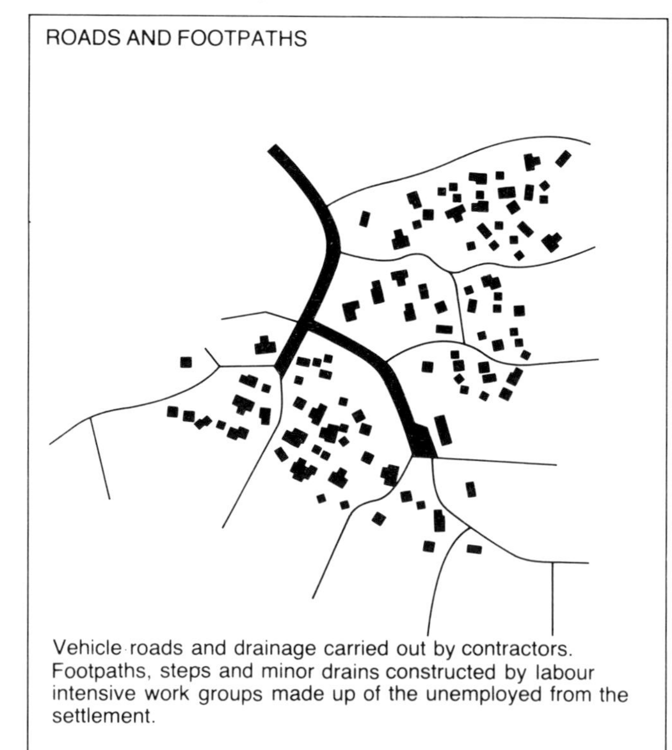

Fig. 6.43 Port Moresby: renewal programme (cont)

WATER SUPPLY

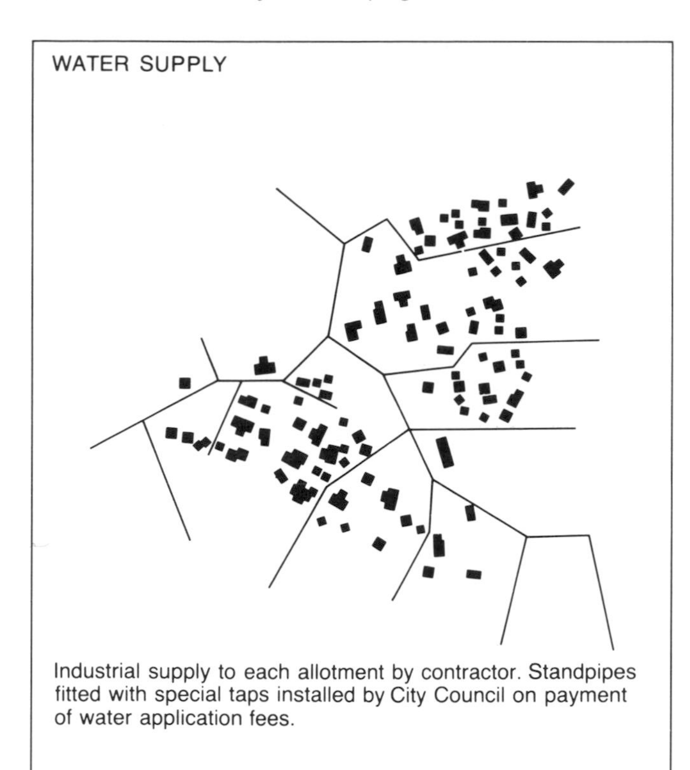

Industrial supply to each allotment by contractor. Standpipes fitted with special taps installed by City Council on payment of water application fees.

SANITATION

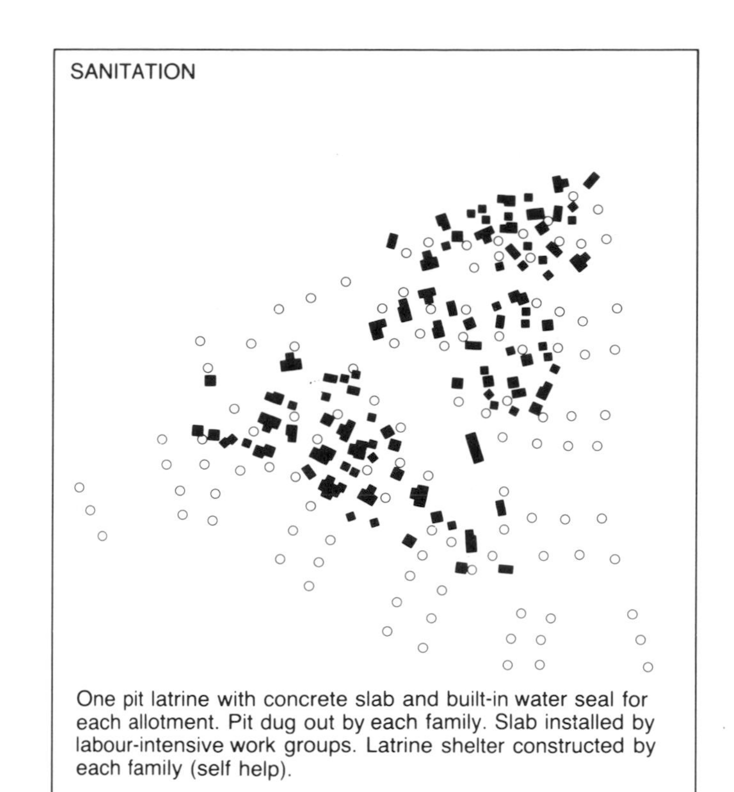

One pit latrine with concrete slab and built-in water seal for each allotment. Pit dug out by each family. Slab installed by labour-intensive work groups. Latrine shelter constructed by each family (self help).

SUBDIVISION

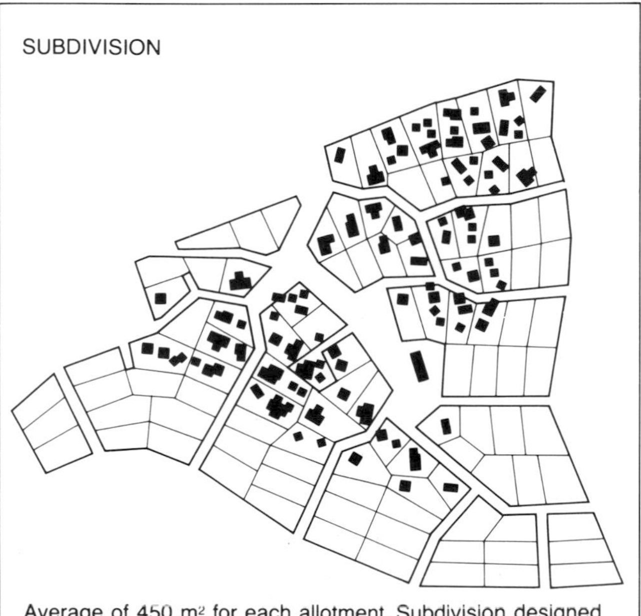

Average of 450 m^2 for each allotment. Subdivision designed around the location of existing dwellings. One permanent dwelling only allowed on each allotment. The surplus must relocate on vacant allotments provided on adjacent land. Each allotment registered with Lands Dept. Security of tenure through sub-lease from Housing Commission.

RELOCATION AND CONSTRUCTION

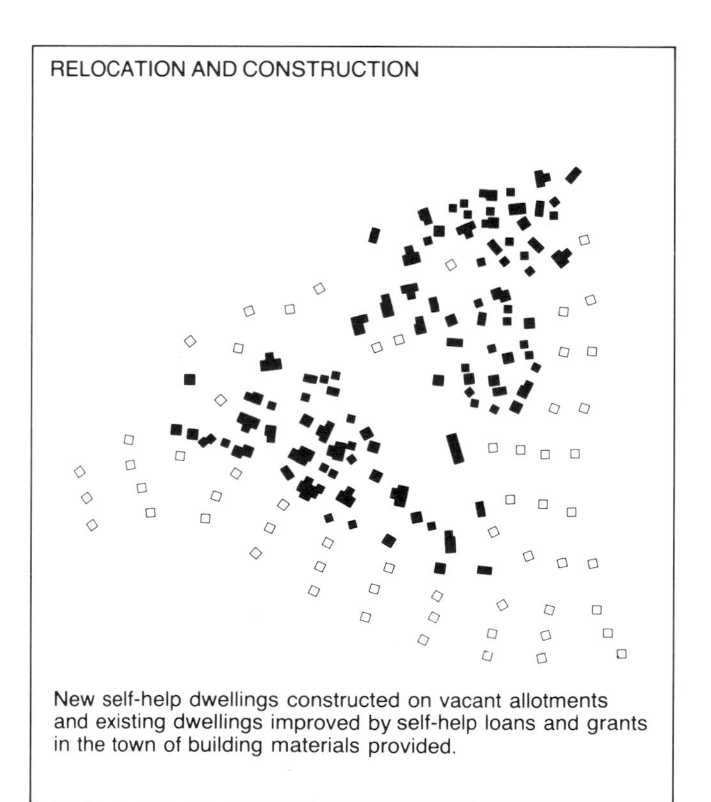

New self-help dwellings constructed on vacant allotments and existing dwellings improved by self-help loans and grants in the town of building materials provided.

Mapping techniques

Two mapping techniques which are very useful to illustrate trends are discussed below.

Proportional symbols

A map should provide a clear picture of what it is trying to show. One way in which this can be achieved is to use **three dimensional symbols**. The map is treated as though you were viewing it at an angle. **Proportional symbols** are then located on the map to illustrate the data.

In this case, we will use circular columns. The top and bottom of such figures are circles, but viewed at an angle, a circle becomes an ellipse. If you wish to show a figure of 200 in this way, first draw an ellipse where you wish the total to be shown. Then choose your scale. Let us suppose that the scale chosen is 1 cm = 100. Draw a line 2 cm vertically upwards from one side of the ellipse, and the same on the other side. Now draw another ellipse at the top of the column. This is easiest using a template, such as can be bought for computing or science. Now shade in the column, missing out the part of the basal ellipse which would be hidden if the figure were indeed solid. A world map of urban population by areas in 1980 has been drawn by this method (Fig. T6.1), using the data in Table 6.1, p. 194.

Fig. T6.1 Urban population in 1990

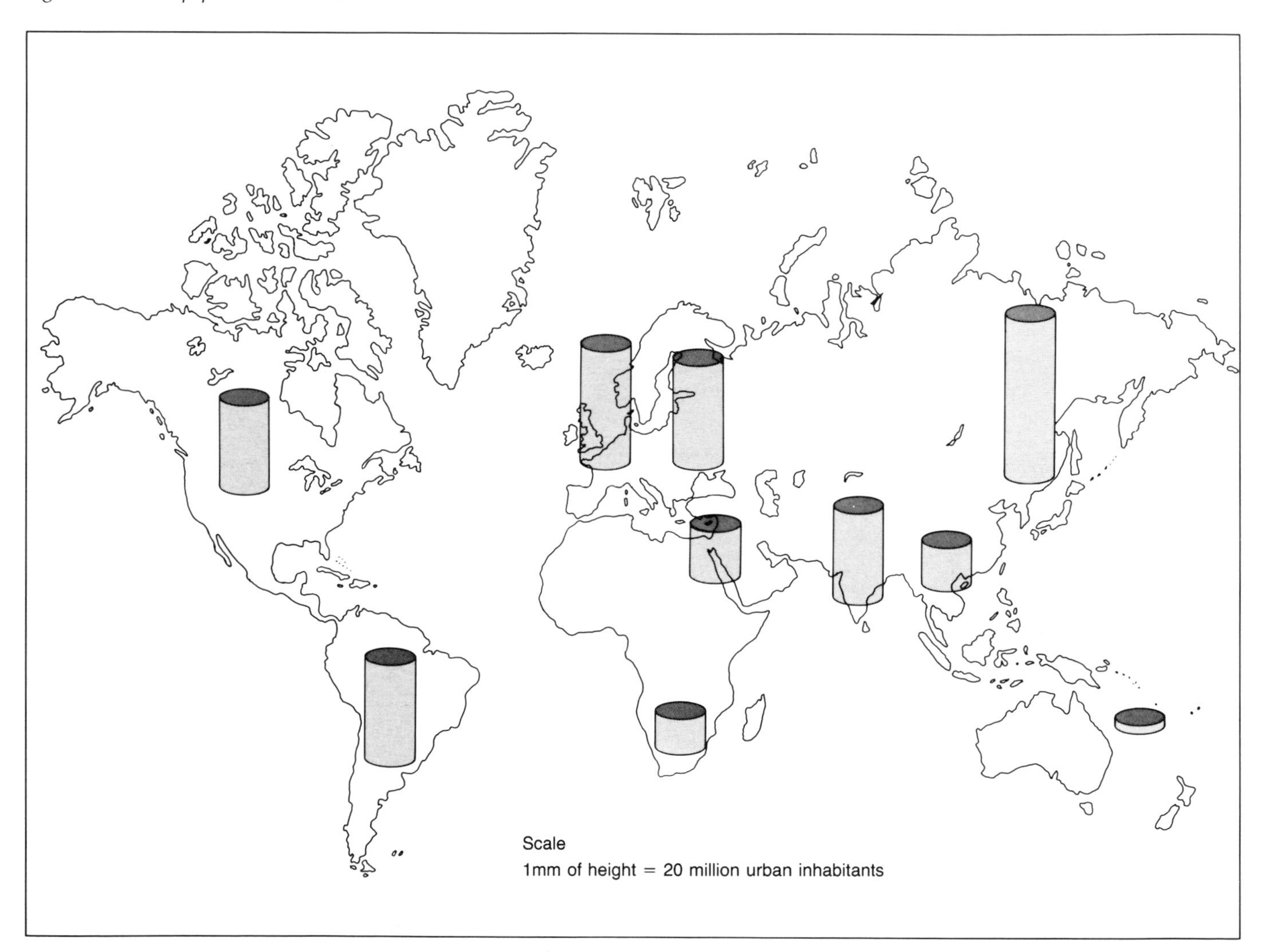

Topological mapping

The idea of a **topological map** is to represent the approximate shape of an area, while drawing that area in proportion to some variable such as population size, production of oil, GDP per capita etc.

In the example shown in Fig. T6.2 the world has been represented according to the size of urban population in that country in 1980. It was drawn on to graph paper using a scale of one small square to 1 million people living in urban areas. The squares were shaded on the graph paper until a shape was achieved which looked like North America, although it was too difficult to represent the northern islands in this fashion. The other areas were shaded in until a satisfactory resemblance to the world map was achieved.

1 Which of these two methods of showing the world's urban population do you think is more effective?

Fig. T6.2 A topological map of the world to show size of urban population

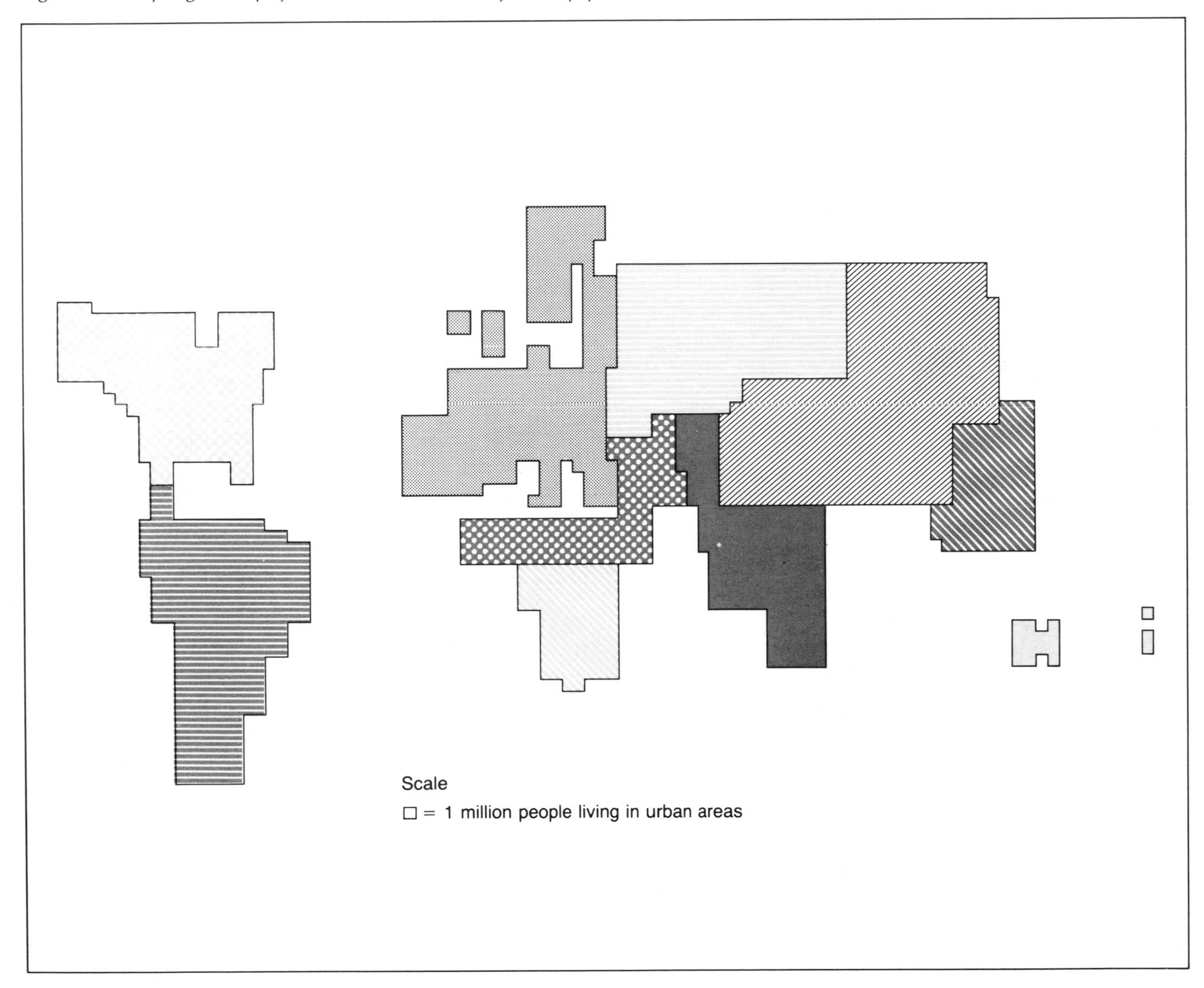

Index